Emerging Applications of Carbon Nanotubes and Graphene

This book comprehensively reviews recent and emerging applications of carbon nanotubes and graphene materials in a wide range of sectors. Detailed applications include structural materials, ballistic materials, energy storage and conversion, batteries, supercapacitors, smart sensors, environmental protection, nanoelectronics, optoelectronic and photovoltaics, thermoelectric, and conducting wires. It further covers human and structural health monitoring, and thermal management applications.

Key selling features:

- Exclusively takes an application-oriented approach to cover emerging areas in carbon nanotubes and graphene
- Covers fundamental and applied knowledge related to carbon nanomaterials
- Includes advanced applications like human and structural health monitoring, smart sensors, ballistic protection and so forth
- Discusses novel applications such as thermoelectrics along with environmental protection related application
- Explores aspects of energy storage, generation and conversion including batteries, supercapacitors, and photovoltaics

This book is aimed at graduate students and researchers in electrical, nanomaterials, chemistry, and other related areas.

Emerging Materials and Technologies

Series Editor: Boris I. Kharissov

The *Emerging Materials and Technologies* series is devoted to highlighting publications centered on emerging advanced materials and novel technologies. Attention is paid to those newly discovered or applied materials with potential to solve pressing societal problems and improve quality of life, corresponding to environmental protection, medicine, communications, energy, transportation, advanced manufacturing, and related areas.

The series takes into account that, under present strong demands for energy, material, and cost savings, as well as heavy contamination problems and worldwide pandemic conditions, the area of emerging materials and related scalable technologies is a highly interdisciplinary field, with the need for researchers, professionals, and academics across the spectrum of engineering and technological disciplines. The main objective of this book series is to attract more attention to these materials and technologies and invite conversation among the international R&D community.

4D Imaging to 4D Printing
Biomedical Applications
Edited by Rupinder Singh

Emerging Nanomaterials for Catalysis and Sensor Applications
Edited by Anitha Varghese and Gurumurthy Hegde

Advanced Materials for a Sustainable Environment
Development Strategies and Applications
Edited by Naveen Kumar and Peter Ramashadi Makgwane

Nanomaterials from Renewable Resources for Emerging Applications
Edited by Sandeep S. Ahankari, Amar K. Mohanty, and Manjusri Misra

Multifunctional Polymeric Foams
Advancements and Innovative Approaches
Edited by Soney C George and Resmi B. P.

Nanotechnology Platforms for Antiviral Challenges
Fundamentals, Applications and Advances
Edited by Soney C George and Ann Rose Abraham

Carbon-Based Conductive Polymer Composites
Processing, Properties, and Applications in Flexible Strain Sensors
Dong Xiang

Nanocarbons
Preparation, Assessments, and Applications
Ashwini P. Alegaonkar and Prashant S. Alegaonkar

Emerging Applications of Carbon Nanotubes and Graphene
Edited by Bhanu Pratap Singh and Kiran M. Subhedar

For more information about this series, please visit:
www.routledge.com/Emerging-Materials-and-Technologies/book-series/CRCEMT

Emerging Applications of Carbon Nanotubes and Graphene

Edited by

Bhanu Pratap Singh and Kiran M. Subhedar

CRC Press
Taylor & Francis Group
Boca Raton London New York

CRC Press is an imprint of the
Taylor & Francis Group, an informa business

Designed cover image: © Shutterstock

First edition published 2023
by CRC Press
6000 Broken Sound Parkway NW, Suite 300, Boca Raton, FL 33487-2742

and by CRC Press
4 Park Square, Milton Park, Abingdon, Oxon, OX14 4RN

CRC Press is an imprint of Taylor & Francis Group, LLC

ISBN: 9781032140155 (hbk)
ISBN: 9781032140186 (pbk)
ISBN: 9781003231943 (ebk)

DOI: 10.1201/9781003231943

Typeset in Times
by Newgen Publishing UK

Contents

Preface

Carbon has been known to humans since ancient times. Despite this, the science of carbon is still intriguing to the scientific community. In more recent times, its marvelous allotropes in the form of fullerenes, carbon nanotubes and graphene have been discovered. These carbon nanomaterials have a unique place in nanoscience owing to their exceptional thermal, electrical, optical, chemical, and mechanical properties and have found applications in diverse areas. Several applications in daily life related to carbon materials include, electric motors: carbon brushes, current collector, overhead sliding contacts; water purifier, filters, absorbent; catalyst support; roads: coal tar as binder; and lead pencil, etc.

Until recently these applications used carbon in bulk form. Advancement in electron microscopes gave the scientist a tool to observe objects as small as molecules and atoms and this led to the discovery of many carbon materials in nano forms e.g., fullerenes, carbon nanotubes, graphene, carbon filaments, graphite whiskers, carbon nano horns, carbon onions, peapods, etc. The technique also proved very useful in understanding their structure and morphology. In the class of nanomaterials, carbon is the only element in the periodic table that has isomers from o-dimensions (0-D) to 3-dimensions (3-D). It is interesting that sp^2 hybridization which forms a planar structure in two-dimensional graphite also forms a planar local structure in the closed polyhedra (0-D) of the fullerene family and the cylinders(1-D) called carbon nanotubes and single atom thick 2-D graphene sheet. Graphene and carbon nanotubes' analogous sister allotropes have been explored extensively in recent times and scientists have realized their unique electronic structure with their honeycomb crystal structure revealing special physical and chemical properties, and related applications in a wide range of fields have evolved.

Carbon nanotubes and graphene are two of the fascinating materials in the carbon family of material science of engineering, specifically in the area of nanoscience and nanotechnology. A lot of fundamental studies on synthesis and properties of these materials have been carried out to date, which have introduced it into several high-end futuristic commercial applications. These materials can become an integral part in most of the advanced future technologies. Therefore, it is highly necessary to discuss emerging applications in a single dedicated book. The present book covers the comprehensive review of most recent and emerging applications of these materials in a wide range of sectors such as energy, structural, electronic, optical, optoelectronic devices, smart sensors, conductive composites, aerospace conducting wires, human and structural health monitoring, and thermal management applications. It aims to provide an opportunity for researchers from different backgrounds and disciplines, such as chemists, material scientists, and physicists, in a wide spectrum of applications based on carbon nanotube and graphene. Each chapter introduces a particular application of carbon nanotube and graphene.

The book will acquaint the readers with the collective information about the wide range of applications of carbon nanotubes and graphene in emergent areas pertaining to the modern-day technological breakthroughs.

Bhanu Pratap Singh
Kiran M. Subhedar

Preface

Editor Biographies

Dr. Bhanu Pratap Singh has received MTech (Chemical Engineering) degree from the Indian Institute of Technology, Kanpur, India and PhD from the Indian Institute of Technology, Delhi, India on "Studies on Carbon Nanotubes based Epoxy Composites." After completing the MTech degree, he joined CSIR-National Physical Laboratory, New Delhi as Scientist B, in 2004. Presently, he is Senior Principal Scientist and Deputy Head in the Advanced Carbon Products Section of Advanced Materials and Devices Division, and Professor in Academy of Scientific and Industrial Research (AcSIR). Dr. Singh has more than eighteen years of research experience on the development of advanced carbon products. His research interest is on large-scale synthesis of carbon nanostructures and their advanced nanocomposites for structural applications, anode materials for Li-ion batteries, efficient EMI shielding materials and ballistic materials. He has carried out exhaustive studies on the mechanical, electrical and thermal properties of MWCNTs and graphene-based polymer composites. Presently he is focusing on the establishment of continuous process for the preparation of oriented CNT fiber/yarn and sheets for personal armour applications. Dr. Singh teaches students on the masters and PhD programmes in AcSIR. More than thirty students have completed their dissertation work for their MTech/MSc/BTech degrees under his supervision/co-supervision. He has also supervised three PhD students and currently eight PhD students are working under him.

Dr. Singh has published 138 research articles in international reputed journals. These articles are also highly cited (total citations > 7800 with h-index of 48, i10 index of 99 as per google scholar data). He has also co-authored one book entitled *Carbon Nanomaterials: Synthesis, Structure, Properties and Applications* (CRC Press) and contributed ten chapters in edited books on this work besides three patents (granted) on the process/products developed. He has also successfully completed several national/international projects as PI/Co-PI. The research carried out by him is having practical impact for societal, environmental, and strategic sectors. He is the recipient of the NRDC National Innovation Award 2018, IEI Young Engineer Award 2017, CSIR-Young Scientist Award 2015 in Engineering Sciences, and HEAM Young Scientist Award 2014. He is the Academic Editor of the *Journal of Nanomaterials*. He is an executive member of Indian Carbon Society, a life member of the Materials Research Society of India, and a life member of both the Metrology Society of India and the Society for Polymer Science, India. He was a senate member of the Academy of Innovative and Scientific Research (AcSIR) from 2017 to 2019.

Dr. Kiran M. Subhedar joined as Scientist at CSIR-NPL in 2012 and is currently working in the Advanced Carbon Products Section of Advanced Materials and Devices Metrology Division, CSIR-National Physical Laboratory (CSIR-NPL). He completed his postdoctoral research at the Israel Institute of Technology-Technion, Israel from 2009 to 2012, where he worked on graphene-based devices. He has also worked as visiting fellow at TIFR Mumbai from 2007 to 2009. He received his PhD degree in the area of superconductivity from Shivaji University, Kolhapur, India in 2007. His research interests are synthesis, characterization, and application of novel carbon materials and its related product development. His current research activities involve graphene growth by the CVD technique, and studies the involved growth mechanism of CVD-grown graphene. Other research activities include activated carbon, synthesis of carbon nanotubes yarn from vertically aligned carbon nanotubes grown by the CVD technique for different applications such as strong and lightweight body armor material.

Contributors

Ajay K. Verma, Research Scholar, Advanced Carbon Products and Metrology, CSIR-National Physical Laboratory, New Delhi-110012, India.

Bhanu Pratap Singh, Senior Principal Scientist, Advanced Carbon Products and Metrology, CSIR-National Physical Laboratory, New Delhi-110012, India.

Bhasker Gahtori, Principal Scientist, Advanced Carbon Products and Metrology, CSIR-National Physical Laboratory, New Delhi-110012, India.

Bharat B. Kale, Scientist-G, Centre for Materials for Electronics Technology, Panchavati, Opp. Dr. Homi Bhabha Road, Pashan, Pune 411008, India.

Deepshikha Gupta, Associate Professor Amity Institute of Applied Sciences, Amity University, Sector-125, Noida 201313, India.

Gaurav Kumar, Assistant Professor Department of Chemistry, Shri Varshney College, Aligarh 202001, India.

Guruvandra Singh, Research Scholar, CSIR-National Physical Laboratory, New Delhi-110012, India.

Harsharaj S. Jadhav, Research Scientist, Centre for Materials for Electronics Technology, Panchavati, Opp. Dr. Homi Bhabha Road, Pashan, Pune 411008, India.

Indu Elizabeth, Scientist, CSIR-National Aerospace Laboratories, HAL Airport Road, Bangalore-560017, India.

Jeevan Jyoti, Postdoctoral Researcher, Centre of Advanced Study in Physics, Panjab University, Chandigarh 160014, India.

Jinsub Park, Professor, Department of Electronic Engineering, Hanyang University, Seoul 04763, South Korea.

Kiran M. Subhedar, Senior Scientist, Advanced Carbon Products and Metrology, CSIR-National Physical Laboratory, New Delhi-110012, India.

Kriti Tyagi, Scientist, Advanced Carbon Products and Metrology, CSIR-National Physical Laboratory, New Delhi-110012, India.

Mamta Rani, Research Scholar, Advanced Carbon Products and Metrology, CSIR-National Physical Laboratory, New Delhi-110012, India.

Manish Pal Chowdhury, Assistant Professor, Department of Physics, IIEST, Shibpur, India.

Manoj Sehrawat, Research Scholar, Advanced Carbon Products and Metrology, CSIR-National Physical Laboratory, New Delhi-110012, India.

Mukesh Jewariya, Senior Scientist, CSIR-National Physical Laboratory, New Delhi-110012, India.

Navneet Kumar, Postdoctoral Researcher, Department of Electronic Engineering, Hanyang University, Seoul 04763, South Korea.

Pallvi Dariyal, Research Scholar, Advanced Carbon Products and Metrology, CSIR-National Physical Laboratory, New Delhi-110012, India.

Pankaj Kumar, Principal Scientist, CSIR-National Physical Laboratory, Dr K. S. Krishnan Marg, New Delhi-110012, India.

Ramchandra S. Kalubarme, Scientist, Centre for Materials for Electronics Technology, Panchavati, Opp. Dr. Homi Bhabha Road, Pashan, Pune 411008, India.

Ranjit S. Kate, Research Scholar, Centre for Materials for Electronics Technology, Panchavati, Opp. Dr. Homi Bhabha Road, Pashan, Pune 411008, India.

Reena Goyal, Post Doctorate Fellow, Graduates School of Science and Engineering, Regional Environment System, Superconducting Material Laboratory, Shibaura Institute of Technology, Toyosu, Koto-Ku, Tokyo 135–8548, Japan.

Rohit Verma, Assistant Professor, Amity Institute of Applied Sciences, Amity University, Sector-125, Noida 201313, India.

Sanjay R. Dhakate, Chief Scientist, Advanced Carbon Products and Metrology, CSIR-National Physical Laboratory, New Delhi-110012, India.

Santwana Pati, Scientist, CSIR-National Physical Laboratory, Dr K. S. Krishnan Marg, New Delhi-110012, India.

Subhash Nimanpure, Pool Scientist, CSIR-National Physical Laboratory, New Delhi-110012, India.

Sukanta Das, PhD – Engineer, Space Walker, Tokyo, Japan.

Surya Kant Tripathi, Professor, Centre of Advanced Study in Physics, Panjab University, Chandigarh 160014, India.

Sushant Sharma, Post Doctorate Fellow, School of Chemical Engineering, University of Ulsan, Daehak-ro 93, Nam-gu, Ulsan 44610, Republic of Korea.

Suyog A. Raut, Research Associate, Centre for Materials for Electronics Technology, Panchavati, Opp. Dr. Homi Bhabha Road, Pashan, Pune 411008, India.

Swati Verma, Postdoctoral Researcher, Department of Civil and Environmental Engineering, Hanyang University, Seoul 04763, South Korea.

Tejendra K Gupta, Assistant Professor, Amity Institute of Applied Sciences, Amity University, Sector-125, Noida 201313, India.

Vipin Kumar, PhD – R&D Associate Staff Member, Manufacturing Science Division, Oak Ridge National Laboratory, Knoxville, TN 37932, USA.

Xianhui Zhao, PhD – R&D Associate Staff Member, Environmental Sciences Division, Oak Ridge National Laboratory, 1 Bethel Valley Road, Oak Ridge, Tennessee 37830, United States.

Yu Zhou, PhD – Project Researcher, Aoki & Yokozeki Laboratory, Department of Aeronautics and Astronautics, The University of Tokyo.

1 Introduction to Carbon Nanotubes and Graphene and their Emerging Applications

Kiran M. Subhedar and Bhanu Pratap Singh

CONTENTS

DOI: 10.1201/9781003231943-1

1.1 INTRODUCTION

Due to their fascinating structures and several exceptional properties, carbon nanotubes (CNTs) and graphene have been star materials for many years. Graphene and CNTs both include alternate single and double bonds in their hexagonal structures. They exhibit certain exceptional physical properties, including high mechanical properties such as tensile strength, Young's modulus, percentage elongation, high electrical and mechanical properties and high specific surface area.

S. Iijima of the NEC laboratory in Japan made the discovery of CNTs in 1991 while researching the soot left over from the manufacture of fullerenes [1, 2]. Since then, CNTs have been the subject of extensive investigation by scientists all around the world. They have a huge aspect ratio because of their long length (up to several microns) and small diameter (a few nanometers). As a result, it is anticipated that these materials will also have intriguing mechanical and electrical properties. CNTs can be compared to lengthy graphene sheets that have been wrapped. Primarily CNTs are of two types: i) single-walled carbon nanotubes (SWCNTs) and ii) multi-walled carbon nanotubes (MWCNTs).

SWCNTs are thought to be prepared by rolling a single layer of graphite (also known as graphene) into a seamless cylinder. Similar to this, a MWCNT is an assembly of SWCNT cylinders that are coaxially stacked one inside the other. The distance between layers of graphite and concentric layers of tubes are almost same. Van der Waals bonding keeps these concentrically arranged nanotubes together. MWCNTs are made of a variety of wall types, structures, and properties, as well as extra features like tips, internal closures in the center of the tube, which creates a so-called "bamboo" type of structure, and Y-junctions of MWCNTs. The length to diameter ratio of nanotubes is typically around 1000 [3]. Thus, they can be regarded as structures that are almost one-dimensional. Their characteristics change depending on the structure.

1.2 STRUCTURE AND SYNTHESIS OF CNTs

1.2.1 Carbon Nanotube Structure

As shown in Figure 1.1 [4], a SWCNT can be produced by rolling a sheet of graphene into a cylinder along a (m, n) lattice vector in the material.

The diameter and chirality of nanotubes, which are their essential properties, are determined by the m and n indices. SWCNTs come in two different varieties: semiconductors and metal-like conductors. As they develop, nanotubes are capped at both ends by hemispherical caps that are created when hexagons in the graphite sheet are replaced with pentagons, causing the structure to curve [5].

A SWCNT is made up of two distinct areas, each with unique physical and chemical characteristics. The sidewall of the tube is the first, and the end cap is the second. Due to the graphite sheet being rolled up in different orientations, the sidewall may once again have multiple chiralities (Figure 1.1). The end cap structure resembles or was inspired by a smaller fullerene, such as C_{60} [6].

1.2.2 Synthesis of CNTs

1.2.2.1 Arc-Discharge Technique

The three primary production techniques for CNTs are chemical vapor deposition (CVD) laser ablation, and d.c. arc discharge. The most popular and possibly simplest technique for production of CNTs is the d.c. arc discharge, which was principally used for production of fullerenes (C_{60}). This technique involves placing two carbon rods, spaced roughly 1 mm apart, in a chamber that is typically filled with inert gas at low pressure between 50 and 700 mbar.

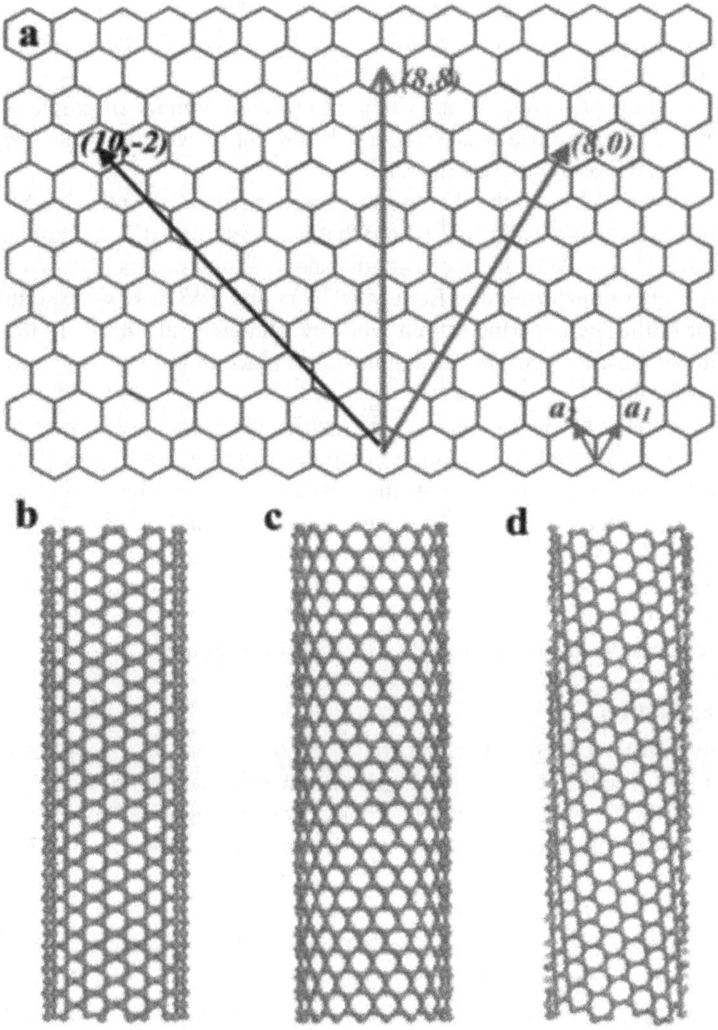

FIGURE 1.1 (a) Schematic honeycomb structure of a graphene sheet. SWCNTs can be formed by folding the sheet along lattice vectors. The two basis vectors a_1 and a_2 are shown. Folding of the (8,8), (8,0), and (10,-2) vectors leads to (b) armchair, (c) zigzag and (d) chiral tubes, respectively. Reprinted with permission from [4]. Copyright {2002} American Chemical Society.

A high temperature (~4000 K) discharge between the two electrodes is produced by a direct current of 50 to 100 A driven by around 20 V. One of the carbon rods (the anode) is vaporised by the discharge, and the other rod develops a tiny rod-shaped deposit (cathode). Ebbesen and Ajayan reported on the large-scale synthesis of MWCNTs using a modified version of the conventional arc-discharge approach [7]. In a helium environment, an 18 V dc potential was applied between two thin graphite rods. The number of nanotubes produced was 75% compared to the graphitic material used as rods.

According to the TEM study, the material was nanotubes with two or more concentric shells of carbon. The nanotubes were several micrometers long and ranged in diameter from 2 to 20 nm. Typically, pentagons served as the tube tips' caps. If SWCNTs are preferred, then metal catalysts such as Fe, Co, Ni, Y, or Mo or their combinations must be added to the anode. According to experimental findings, the diameter distribution is influenced by the catalyst's composition, the growth temperature, and a number of other growth factors.

MWCNTs will be the primary product if both electrodes are made of graphite. No catalyst is used in this process; hence no heavy acidic purification step is required. This suggests it is possible to synthesize MWCNTs with few defects.

The majority of research uses the d.c. arc-discharge method to synthesize SWCNTs, which involves putting the catalyst powder into a hole that has been drilled in a graphite electrode that serves as the anode. Under ideal chamber circumstances, arcing occurs between the anode and a cathode made entirely of pure graphite. Both MWCNTs and SWCNTs were simultaneously and selectively produced in one experiment in a work by Mathur et al. [8, 9]. In the experiment, a catalyst/graphite composite electrode was employed in place of putting the catalyst powder into a hole in the graphite rod.

In accordance with the proper ratios, natural graphite, coke, catalyst powder, and binder pitch were fully mixed in a ball mill before being compressed into green blocks. The catalyst was a combination of Ni and Co powders. In an inert atmosphere, the green blocks were heated to 1200°C to generate carbonized blocks with various coke, natural graphite powder, Ni, and Co compositions. These blocks were served as the anodes in the arcing process, and the cathode was a block of high-density graphite. During the arcing process (dc voltage 20–25 V, current 100–120 A, and 600 torr helium), a stepper motor was used to maintain a consistent gap of 1–2 mm between the electrodes. In this instance, SWCNTs yield was doubled.

1.2.2.2 Laser Ablation

In the laser ablation method, a graphite target containing tiny amounts of metal catalyst is vaporised by a pulsed laser [10]. The target is kept in a high temperature furnace with an inert environment that is heated to about 1200°C. As the vaporised carbon condenses, the nanotubes are deposited on the reactor's colder surface. This method produces a yield of nanotubes that is around 70% [11].

1.2.2.3 Chemical Vapor Deposition

A simple way to synthesize CNTs using the CVD approach is through the pyrolysis of organometallic precursors like metallocenes (for example, ferrocene) in a furnace. Various hydrocarbon, catalyst, and inert gas combinations have been used in the past by numerous researchers to produce CNTs using the CVD process [12]. In one of the studies by Sharma et al., [13] toluene (1:8) was injected into the quartz tube. MWCNTs were produced in a three-zone furnace at 760°C

FIGURE 1.2 (a) CVD set-up at CSIR-NPL, and (b) MWCNTs produced in a single batch [13].

as illustrated in Figure 1.2(a), which produced about 60 gm of MWCNTs in each batch of 9 hours and flowing at a rate of 20 ml/h precursor in an inert medium.

The growth rate in CVD can be controlled by a number of factors, as opposed to other ways of CNT synthesis which includes temperature, feed gasses, precursor and inert gas flow rates, type and amount of catalyst. The length, orientation, and diameter of the synthesized CNTs can be somewhat controlled by altering the growth parameters. The techniques used in the gas phase have been found to yield CNTs with fewer impurities and to be best suited for mass manufacturing. The ability to scale up the synthesis of CNTs for the processing of composites is therefore highest when using gas phase techniques like CVD [11].

1.3 STRUCTURE AND SYNTHESIS OF GRAPHENE

1.3.1 Graphene and Its Structure

Graphene is a single layer of carbon atoms in which the carbon atoms arranged on a honeycomb lattice and forms hexagonal structure which resemble a chicken wire. The carbon with its peculiar property undergoes sp^2 hybridization in case of planer structure of graphene which possesses both the strong covalent bonding as well as weakly coupled pie electrons. The strong covalent bonding leads to its superior mechanical properties and very high melting temperature. Graphene is 100 times stronger than steel. On other hand the weakly coupled pie electrons leads to excellent electronic properties such as very high electric mobility. These excellent electronic properties of graphene are consequence of its unique electronic structure. In its energy dispersion there exist points known as K and K′ originating from existence of the non-equivalent carbon atoms, where the dispersion relation is conical, and with the most important characteristic as the energy is linearly proportional to the wave number at Fermi energy [14]. These exceptional mechanical, electrical and optical properties of graphene make it a wonder material.

1.3.2 Graphene Synthesis

To prepare the graphene there are several different routes available depending upon the required dimensions, scalability, defect and disorder level, kind of application and also its cost. Based on these factors the graphene is categorized into two groups viz graphene on substrate and graphene in the form of powder or suspension, and accordingly the following routes are generally used to synthesize the graphene.

1.3.2.1 Micro-Mechanical Exfoliation

Micro-mechanical exfoliation was the first method employed by Geim and Novoselov to isolate single-layer graphene for the first time and for which they got the Nobel prize. In this method the piece of graphite is fixed to scotch tape and after repeated peeling of the layers of the graphite it gets thinner and thinner, and finally, when stamped on to a substrate like SiO_2/Si, at some places there can be single-layer of graphene fixed to the substrate. As in the graphite, graphene layers are stacked and bound with each other by weak Van der Waal force and can be easily peeled off even from the monolayer. This process of micro-mechanical exfoliation is quite simple and cost effective and most importantly it produces graphene with very high quality. However, the method is limited with low yield and only useful for small scale laboratory usage, especially for fundamental research.

1.3.2.2 Epitaxial Growth of Graphene on Silicon Carbide Surface

Epitaxial growth of graphene on single crystalline silicon carbide (SiC) surface is one of the most interesting methods for synthesis of graphene and has been the subject of interest to many researchers in the early research work of graphene. In this method of graphene synthesis, the single crystalline SiC substrate is usually heated to very high temperature, over 1650°C, under argon atmosphere. Because of the very low vapor pressure of carbon as compared to silicon, the silicon gets sublimated from surface layers of SiC and the remaining carbon gets crystallized onto the SiC substrate as graphene. Bommel et al. [15] were first to demonstrate graphite formation on both 6H-SiC(0001) and (0001) surfaces. The crystal structure, crystal plane orientation and crystal plane stacking information of SiC was reported in 2008 by Hass, de Heer [16] and Conrad, and de Heer's group reported 1–3 monoatomic layer fabrication on silicon terminated (0001) face of single crystal 6H-SiC substrate and 1–2 layer fabricated on (0001) face of 6H-SiC wafer using thermal decomposition method [17]. Usually, smaller grain size is found to grow with this method, however Emtsev et al. reported wafer-size graphene layers, atmospheric pressure graphene synthesized on 6H-SiC (0001) surface which have larger domain size [18].

1.3.2.3 Synthesis with Chemical Route

The synthesis of graphene oxide (GO) and reduced graphene oxide route involving the oxidation of graphite to produce graphene oxide to separate layers within the graphite has been known for several years. With this again there are different methods of GO synthesis available like (i) the Brodie method, (ii) the Staudenmaier method, and (iii) the Hummers and Offeman method. Basically, these methods use strong acids and oxidants. Different reaction conditions such as temperature, type of oxidants, decide the degree of oxidation. In 1958, Hummers reported [19] the synthesis of GO by mixing graphite with sodium nitrite, sulphuric acid and potassium permanganate and it becomes the popular Hummer's method. In this process, graphite was mixed with sodium nitrite and reacted with sulphuric acid. Subsequently after one hour of stirring, gradually and slowly adding potassium permanganate to the solution usually raises the temperature and needs to be controlled by maintaining the temperature to 10–20°C. The further reaction needs to be stopped by the addition of hydrogen peroxide which leads to formation of graphene oxide after washing water. Synthesized GO was reduced using hydrazine hydrate and annealed to get reduced graphene oxide.

1.3.2.4 Chemical Vapor Deposition

For many applications like in electronic and optoelectronic devices, both the quality and size or area of the graphene is very much important. Though the mechanical exfoliation and epitaxial growth methods are producing the graphene with high quality, the yield and size of the graphene is very limited. On the other hand, in the chemical route, graphene synthesis with good yield can be produced however the quality in terms of the presence of defects, structural disorders, and impurities especially in the form of attached functional group, grain boundaries may become a matter of concern. So, CVD growth has emerged as an alternative method and further it has also matured significantly over the period of time. The CVD can produce graphene with large area and with reasonably very high quality. Moreover, growth of even graphene single crystal can be possible. Typically in CVD method the source hydrocarbon in the form of gas or liquid is decomposed in the presence of a metal catalyst in the form of substrate at temperature around 1000°C and so produced carbon atoms get crystallized into graphene layer onto surface of the catalyst [20–22]. The CVD growth proceeds through most important stages like nucleation and growth. The CVD process involves a vast number of process parameters. Depending on the type of catalyst, its crystallographic properties and solubility of carbon in it there can be different kinds of growth mechanism which leads to graphene with different quality. The solubility of carbon in copper is very much negligible and hence growth can be restricted to surface only and

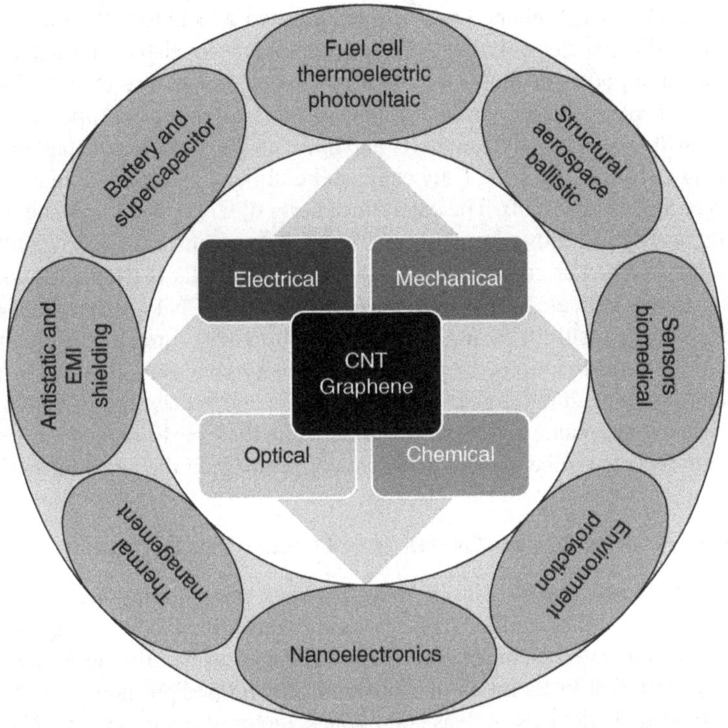

FIGURE 1.3 Applications of carbon nanotubes and graphene.

self-limited single-layer growth can be possible on copper substrate. In contrast, the solubility for nickel is considerably higher and may result in multilayer growth. Further, the CVD growth of graphene can be possible in very high vacuum to even atmospheric pressure also.

1.4 APPLICATIONS OF CNTs AND GRAPHENE

Because of the outstanding properties of CNTs and graphene, they have important application in many areas including composites [23–26], energy generation and storage [27–33], electrocatalysts [34], ballistic [13, 35–38], electromagnetic interference (EMI) shielding[39–49], sensors and detectors [50–52], optoelectronics [27, 53, 54], thermoelectric[55], water purification[56], biomedical [57–59] and nanoelectronics applications[60–67] etc., as shown in Figure 1.3.

1.4.1 Carbon Nanotubes and Graphene in Solar Cells and Fuel Cells

CNTs and graphene are intriguing carbon nanomaterials because of their large surface area, high electrical conductivity, visual transparency, thermal and chemical stability, and mechanical flexibility. As a result, they could be useful in energy conversion systems such as organic solar cells, dye-sensitized solar cells (DSSCs), perovskite solar cells, and fuel cells. They were investigated as transparent electrodes for organic and perovskite solar cells, as well as electron and hole collection buffer materials. These applications have the potential in reduction of cost and/or enhance the stability of solar cells, as well as enabling the production of flexible solar cells. They have a great potential in replacing the Pt as the catalytic counter electrode in DSSCs due to their catalytic and conductive capabilities. The cost of DSSCs could be significantly reduced as a result of this. CNTs and graphene have also been tested as an auxiliary binder for TiO_2 paste and a gelator for DSSC gel electrolyte. CNTs and graphene have also been intensively researched as fuel cell catalyst supports or metal-free catalysts. They can reduce catalyst loading by reducing nanostructured catalyst aggregation during catalyst preparation and operation. These carbon nanomaterials can also boost catalytic activity and make catalysts more resistant to poisoning from intermediates and products like CO. They can also be utilized as catalysts in chemical reactions such as oxygen reduction reactions. The catalytic activity of CNTs and graphene is determined by defects, doping, and shape. However, their practical applications for energy conversion are still facing challenges. The power conversion efficiency of solar cells is frequently lower when they are employed as transparent electrodes than when ITO is used. CNTs and graphene are expensive than carbon black. As a result, their use as a catalyst support is limited. Although they are catalytic for specific chemical reactions, they have lesser catalytic activity than Pt. Nonetheless, with the development of new technologies in the manufacture and processing of CNTs and graphene, the price of CNTs and graphene has been lowering, while their performance has improved. CNTs and graphene are expected to be used in practical applications in the near future [27].

1.4.2 Carbon Nanotubes and Graphene in Lithium-Ion Batteries

A battery is the most energy-efficient alternative among the existing energy storage technologies because of its portability and lack of environmental impact [68]. Batteries are reasonably priced and have a high rate of conversion. They can be used for a wide range of applications from big storage batteries in factories and hospitals for immediate power in the incident of a power outage, to portable devices such as cell phones, laptop, camcorders etc. Electric traction is becoming more common in environmentally sensitive locations in the transportation sector, and major automakers are working on electric and hybrid electric vehicles, which need very large capacity batteries.

The primary focus of current research is on developing Li-ion batteries (LIBs) with increased capacity, which can be employed in high-tech applications like the storage of renewable energy and electric vehicles. The materials for the cathode and anode of LIBs are the subject of active study in all areas. Conventional LIBs use $LiCoO_2$, a transition metal oxide as the cathode and graphite as the anodes. Graphite has a low theoretical capacity of 372 mAh g^{-1}, it is imperative to switch out graphite for other high-capacity materials in order to increase the energy density of LIBs [69].

As a result, significant progress in developing alternative anode materials with higher specific capacity is required to satisfy industrialized production. Graphene [70] is another carbon substance that has received a lot of attention recently as an anode for LIB. Graphene is a two-dimensional sheet of sp^2 hybridized single-layer carbon. It has incredible properties such as high electrical conductivity, huge surface area, higher charge mobility, and so on, making it an excellent contender for LIB anode [70, 71]. Electrochemical experiments show that graphene has a larger gravimetric capacity than graphite because it has more sites for Li storage, such as edges and imperfections. Various researchers have projected capacities ranging from 790 to 1000 mAh g^{-1} [72, 73].

CNTs are being widely researched as an anode for LIBs. Distinct structure of CNTs, combined with their excellent electrical conductivity, high aspect ratio and strength, makes them ideal for the manufacture of self-supporting conducting electrodes for LIBs. Based on LiC_2 stoichiometry, SWCNT has a maximum theoretical reversible capacity of 1116 mAh g^{-1} [74–78].

Despite the fact that CNTs and graphene have a much higher theoretical capacity than graphite, these materials have significant irreversibility and thus a low life cycle. A further issue with such anodes is the low voltage plateau at the time of discharge of battery. Unlike graphitic anodes, CNT anodes often exhibit large voltage variations as the cell discharges. As a result, they may be challenging to utilize in most devices that require a consistent voltage source [78].

Because of their restricted gravimetric capabilities, intercalation anodes cannot meet the power need for high energy gadgets, electric vehicles, and other sophisticated applications such as renewable energy storage. As a result, specific capacity is a crucial consideration for improved anode materials. Materials with very high specific capacities include Si, Ge, Sn, SnO_2, SnO, and others. They undergo an alloying/de alloying process with Li. Si, for example, has a theoretical specific capacity of 4200 mAh g^{-1} [79].

The main issue with these materials is their poor cyclability, which is due to large volume expansion during the alloying/dealloying cycle. Two ways have primarily been used to address the issue of volume expansion. One method is to reduce the size of anode material particles and use nanosized materials with various morphologies such as nanowires, nanotubes, and so on [80, 81]. Another extensively used strategy is the use of conductive matrix material, which can buffer volume expansion and so increase cycle life. CNTs and graphene are commonly utilized as a buffering material with different metal/metal oxide [29, 30, 32, 33, 82].

CNTs and graphene, which have a variety of appealing features, are being studied extensively for improving the performance of LIBs. These are recognized as not being suitable active lithium storage materials, but rather as regulators: they serve to regulate the lithium storage behavior of a specific electroactive material and increase the storage capacity [83].

1.4.3 Carbon Nanotube and Graphene for Supercapacitors

Supercapacitors (SCs) offer excellent electrochemical performance in terms of power density, quick charging and discharging capabilities, and a long cycle life, are considered as potential energy storage systems for variety of applications.

The best candidates are new materials like CNTs and graphene, with sp^2 hybridized structure are free of surface-dangling connections. The high exohedral surface area of CNTs and graphene, in addition to their chemical stability is useful for creating a significant electrolyte-electrode contact and accomplishing energy storage based on double-layer capacitance [84]. Additionally, the covalent sp^2 connections between individual carbon atoms have a unique physical structure that promotes high electrical conductivity, which significantly lowers the system's resistance. The exohedral surface configuration and abundant mesopores of these carbon nanomaterials have significantly improved the electrochemical performance specifically in terms of power density and energy density. As a result, the capacitance performance of these materials under high voltage is encouraging. However, manufacturing of electrode materials for high-performance supercapacitor device is still hampered by the high liquid intake and low density of such nanomaterials.

This necessitates a thorough comprehension of pore structure and its stability under compression, as well as the sustainable development of preparation technology, in order to produce electrodes that can balance surface area, bulk density, pore structure, and structural stability in liquids. Finally, because the current electrode activated carbon is so inexpensive, the high cost is another concern regarding commercialization of supercapacitor based on these carbon nanomaterials [84].

1.4.4 Carbon Nanotubes and Graphene for Aircraft and Space Vehicles

Lightweight, visible, greater speed and maneuverability will be more stringent requirements for next-generation aircraft, unmanned aerial vehicles, and missiles. These needs, on the other hand,

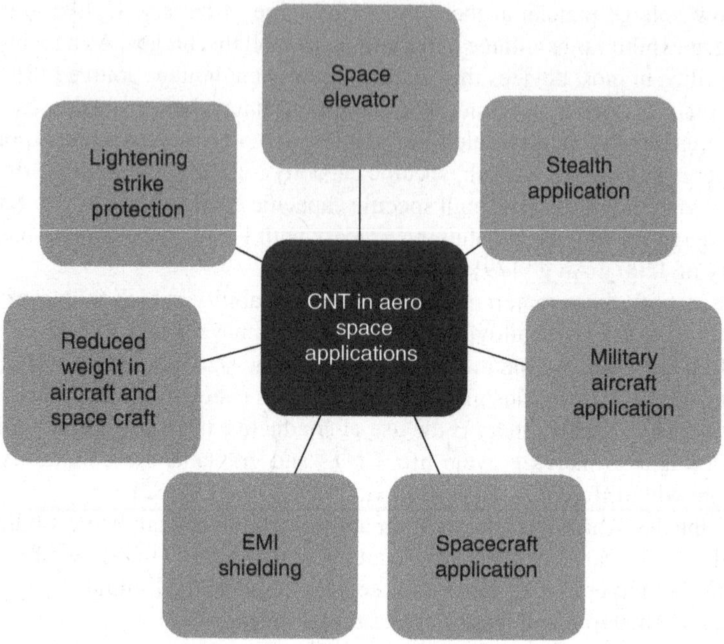

FIGURE 1.4 CNTs in aerospace applications.

necessitate the development of sophisticated materials and systems capable of incorporating these features. CNTs are good applicants to address these needs because they can be used in a variety of applications. Because of the unique properties of CNTs and graphene, such as improved mechanical, electrical, and thermal properties, the potential implementations of these novel materials, particularly in aerospace applications, are expected to have a substantial impact on future aircraft and space vehicles. All kinds of aircrafts and aerial vehicles such as commercial, military, unmanned, satellites, and space launch vehicles are all expected to use CNTs in the near future [85]. This includes their use in airframe, lightweight conducting wires, aircraft icing, propulsion, lightning strike protection, EMI shielding, sensors for structural health monitoring. Applications of CNTs in aerospace are shown in Figure 1.4.

1.4.5 Carbon Nanotubes and Graphene in Space Elevator

The use of CNTs in a proposed space elevator concept is one of the technologies that have sparked a lot of interest in their use in space applications [85]. Tsiolkovsky [86] first conceived this concept in 1895, and Artsutanov [87] and Pearson[88] were the first to implement it technically. Isaacs and his colleagues [89] suggested the notion of a skyhook in 1966. The space elevator's originators [90] extended the concept to lunar applications based on the technical depiction of the space elevator. Because of CNTs' exceptional properties, multiple scientific papers on the space elevator idea with CNTs have been published [91–99]. The concept of a space elevator has piqued the curiosity of experts as well as the younger generation in space travel. The space elevator has long been a renowned notion for space travel, thanks to introducing of this concept in science-fiction by Clarke [100] and Sheffield [101] with increasing breakthroughs in the development of new materials. There are currently several of the scientific papers on the space elevator [102–105], its stability [106, 107], dynamics [108, 109], radiation shielding [110], the effect of defects [61], design [91], different configuration situations [111], and policy about space research [112].

The space elevator is a link between the Earth's surface and a Geostationary Orbit at an altitude of roughly 36,000 kilometers. The use of the space elevator is motivated by the high expense of placing tiny payloads into Low Earth Orbit. Other inspiring aspects include the complex nature of present space operations, their accompanying risks, and the fact that bringing equipment back from orbit to Earth is difficult without using a reusable launch vehicle [85].

The advantages of using a space elevator after it has been built are the low cost and quickness with which goods can be launched [113]. The cross-sectional area on the ground must be 2–10 times larger than at Geostationary Orbit [88] in order for it to support itself on the ground. There are many space elevator designs, according to Bolonkin [114]. Aside from the cable, the majority of these features include a base station, a big nature counterweight in Geostationary Orbit, and climbers. The base station is referred to as a mobile base station if it is installed on a large ocean-going vessel. Stationary base stations are also located at high altitudes on the planet. The payload is launched into orbit using climbers driven by power beaming. Another cable lowered from Geostationary Orbit to Earth is also used to achieve orbital height. The space elevator's technological problems, such as susceptibility to lightning strikes, probable meteorite damage, Low Earth Orbit cable collisions, atomic oxygen damage, and microscale defects [91, 94], have prompted academics to reconsider their space tether alternatives. A self-supporting core structure with pneumatically inflated portions that enable active control for structure support and balancing against external disturbances is one such method [115]. Given that CNTs make up the physical cable material on which magnetic levitating vehicles can travel for freight and crew placement in space, their involvement in realizing the space elevator arrangement could be significant. The space elevator is depicted schematically in Figure 1.5.

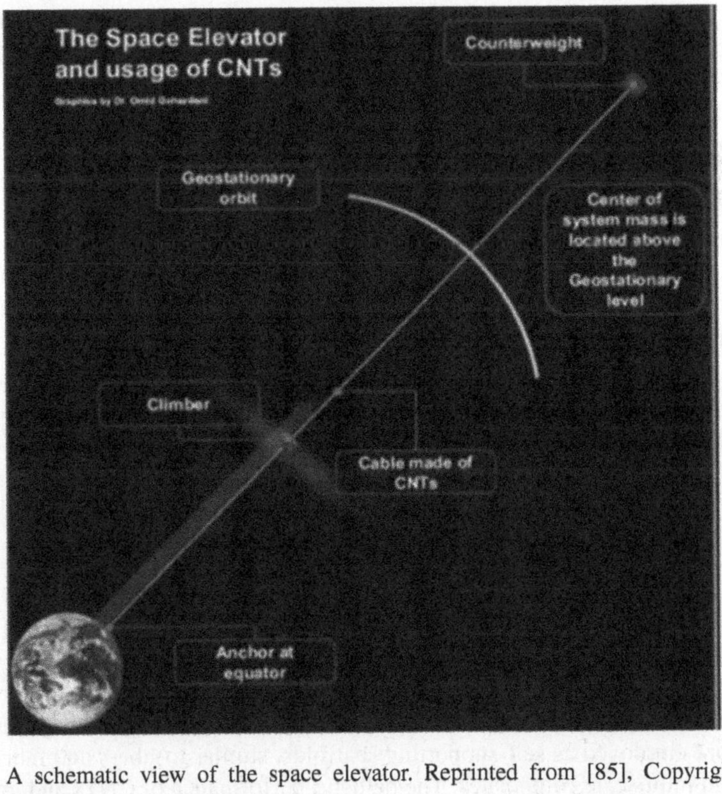

FIGURE 1.5 A schematic view of the space elevator. Reprinted from [85], Copyright (2014), with permission from Elsevier.

1.4.6 Carbon Nanotubes and Graphene in Biomedical Applications

Owing to the unique characteristics of CNTs and graphene, they have rapidly become the potential candidates for nano-bio researchers to explore their usage in biomedical applications. Ultra high surface area, versatile surface functionalization and good biocompatibility make them an ideal component for flexible biomedical devices. There have been remarkable efforts to investigate the biomedical utilization of CNTs and graphene in applications such as biosensing, drug delivery, bioimaging, tissue engineering and cancer therapy. The recent progress in development of CNT and graphene-based biomedical devices with an emphasis on fabrication and biomimetic behavior of these carbon materials in bridging the gap between nature and synthetically designed materials have been done by various researchers. Further, possible challenges and future perspectives in this rapidly developing domain are also owing to the current opinions within the scientific community on the toxicity and limitations in translating these carbon materials into advanced clinical tools.

1.4.7 Carbon Nanotubes and Graphene for Antistatic and Electromagnetic Interference Shielding Applications

Electromagnetic interference (EMI) is a serious hazard since electronics and sensors are developing so quickly. It may prevent equipment from functioning normally or even result in its full failure. To insulate appliances from the negative effects of electromagnetic (EM) noise, a shielding system must be developed. Traditional metal-based EMI shielding materials are less desirable than carbon-based EMI shielding materials due to their heavier weight, susceptibility to corrosion, and processing difficulties [116, 117].

In polymer composites, carbon black is frequently employed as a conducting filler among the numerous carbon fillers (such as graphite, carbon black, or carbon fibers) [117, 118]. One significant disadvantage of using carbon black as a filler is that it takes a lot of carbon black, up to 30 to 40 percent, to achieve the best conductivity, which compromises the mechanical properties of the polymer [119]. Since their discovery, carbon nanomaterials like CNTs and graphene have been shown to be effective at producing conductive composites because they offer conductivity at low loading.

Due to their high aspect ratio, high conductivity, and mechanical strength, CNTs and graphene make excellent building blocks for conductive composites used to create high-performance EMI shielding materials [44, 120–123]. Studies on the potential for EMI shielding applications of CNTs and graphene-based composites have been conducted in a big way [47–49].

1.4.8 Carbon Nanotube and Graphene for Ballistic Protection

The development of ballistic protection systems with higher protection levels and lighter weight receives significant attention due to the rising risks such as warfare, military conflicts, counterinsurgency, and terrorism in border and inner-city areas. In addition to their many technological uses, carbon nanomaterials (CNMs) including CNTs and graphene are now being examined for use in personal protection. Numerous experimental and computational studies on these CNMs' behavior under impact conditions have shown how well they are mechanically stable, capable of absorbing energy, and able to withstand multiple impacts.

They are strong candidates to replace fibers like aramid, ultra high molecular weight polyethylene, and other polymer-based fibers because they are also lightweight, tough, and stiff. These nanoparticles are employed as self-supporting scaffolds, similar to fibers and fabrics, as well as reinforcement in multiscale composites. The ballistic-performance of CNTs and graphene-based nanomaterials, scaffolds, and their reinforced composites have addressed this issue. CNTs and graphene are in significant demand for high-performance lightweight personal-armor applications [36–38, 124, 125].

1.4.9 Carbon Nanotube and Graphene for Smart Sensors

A significant increase in interest in the field of flexible electronics has been seen in recent years. In order to detect human body movement, flexible textile sensors, a crucial component of wearable electronics are typically attached on wearable devices. For tracking human mobility, strain sensors with high sensitivity, stretchability, durability, and quick response/recovery are crucial. However, due to the brittleness of the sensing materials, typical strain sensors based on metal foils or semiconductor films have limited flexibility and stretchability.

Therefore, these sensors are not appropriate for situations that call for both high sensitivity and a wide stretchy range. Researchers have worked extremely hard to create high-performance strain sensors by using many types of conductive materials as the sensing elements and found CNTs [126–128] and graphene [129], and reduced graphene oxide [130, 131] based sensors as a very high potential material in the field of flexible smart sensors [132].

1.4.10 Carbon Nanotube and Graphene for Thermoelectric Applications

Organic thermoelectric materials take advantage of benefits like low thermal conductivity, low cost, environmental friendliness, easy to process, low weight, mechanical flexibility, and roll-to-roll production, which are helpful for the production of portable and wearable electronics devices within built power source. The low figure of merit of polymeric TE materials is a drawback, mostly because of their typical low electrical conductivity. To improve the thermoelectric performance of organic thermoelectric materials, many attempts have been done [133].

Although power factor values in the range of 103 $\mu W\ m^{-1}\text{-}K^{-2}$are frequently observed for inorganic thermoelectric materials, it is very difficult to have power factor (PF) over 1000 $\mu W\ m^{-1}\text{-}K^{-2}$ for organic thermoelectric materials [134]. It is due to that organic thermoelectric materials have low electrical conductivity, which results in their low carrier mobility, which is typically less than 1 $cm^2\ V^{-1}\text{-}s^{-1}$, is the cause of their low power factor [135]. Organic materials are frequently intensively doped in order to boost electrical conductivity, which will also significantly reduce the Seebeck coefficient (S).

Due to their exceptional electrical transport capabilities, thermoelectric systems based on CNTs and graphene have been suggested as an effective alternative to achieve self-powered devices [136].

However, it has been noted that several organic hybrid materials exhibit high PF of above 1000 $\mu W\ m^{-1}\text{-}K^{-2}$ [92–94]. Due to the superior electrical characteristics of CNTs, those high power factor organic hybrid materials are typically built with them.

A recent method for achieving high PFs of 1800 $\mu W\ m^{-1}\text{-}K^{-2}$for p-type and 1000 $\mu W\ m^{-1}\text{-}K^{-2}$ for n-type in flexible MWCNT films is reported by Sun et al. in their study [137]. By utilizing the anisotropic electrical conductivity and isotropic Seebeck coefficient properties of 1D CNTs as well as the subsequent doping and cold-pressing to enhance the electrical conductivity of MWCNT films, the high power factor is obtained. To demonstrate the materials' potential to convert heat into energy, a thermoelectric generator with an assembly structure is created. This TEG, which is made of CNTs, has the greatest areal output power of 27 W/m^2. Similarly, organic flexible thermoelectric devices have been developed with graphene as an important material [136, 138–141]. A multifunctional superelastic thermoelectric sponge based on graphene was disclosed by Zhang et al. [136] for wearable electronics and thermal control. The sponge exhibits a high compressive strain of 98% and a Seebeck coefficient of 49.2 μV/K. The sponge exhibits remarkable mechanical and thermoelectric stability after 10,000 cyclic compressions at 30% strain. A wearable sponge array thermoelectric device was created to use heat energy from the human body to power medical devices for detecting physiological signs. Additionally, a 4x4 array thermoelectric device installed on the surface of a CPU that is normally operating may produce a stable voltage and lower the CPU temperature by 8 K, offering a workable method for concurrent power generation and thermal control.

The thermoelectric transport characteristics of centimeter-sized monolayer CVD graphene that was electrostatically regulated by a high-capacity ionic gel were described in a different work by Schrade et al. [142]. For the conduction of electrons and holes, respectively, the power factor was 7 and 5.4 mW m^{-1} K^{-2}, respectively.

1.4.11 Carbon Nanotube and Graphene in Environmental Remediation

In recent years, the scientific community has grown increasingly concerned about environmental challenges like air and water pollution. Water pollution is one of them, and because of its terrible and disastrous impacts, it has received greater attention globally. Severe water pollution is brought on by the direct discharge of industrial waste from sectors including paper, printing, leather, textile, dye synthesis, cosmetics, and electroplating that contains various pollutants such heavy metals, crude oils, pigments, synthetic dyes, and organic solvents [56, 143].

The research community has investigated a broad variety of adsorbent materials, particularly hybrids and composites of new materials such as metal/metal oxide nanoparticles, CNTs [56], and graphene [144].

The development and application of carbon nanomaterials, specifically MWCNTs, SWCNTs, graphene, and graphene oxide-based nanomaterials, as adsorbents for wastewater treatment and purification has demonstrated significant potential [145], particularly for the treatment and purification of industrial- and pharmaceutical-laden wastes.

Environmental pollution is a serious problem, but there are few effective remediation methods. More and more, CNTs and graphene are being employed to purify water and air. It has been discovered that CNTs and graphene are effective at removing inorganic and organic pollutants from the environment [146].

1.4.12 Carbon Nanotubes and Graphene in Thermal Management

The creation of the newest electrical and optoelectronic devices requires the use of effective advanced thermal management materials (TMMs). Reliability, flexibility, and the lifespan of electronic devices are the main problems that thermal management (TM) must deal with. The TMMs not only display great thermal conductivity and power density, but also exhibit flexibility in their structure, are light in weight, and have strong mechanical properties.

At the nanoscale, CNTs and graphene are important components of TM. For TM at the nanoscale, the exceptional thermal characteristics of CNTs and graphene have been thoroughly examined. Due to their exceptional mechanical flexibility and thermal conductivity, CNTs and graphene are used in TMMs. Heat transport characteristics of CNTs and graphene are influenced by their size, shape, domain size, and coupling strength. For TMMs, it is thought to be crucial how the mean free path length and phonon-phonon scattering affect the thermal properties.

Electronic devices have gotten faster, smaller, better performers, and more integrated during the past few decades, thanks to the rapid development of microelectronic techniques. TM is a crucial condition for the energy storage system's performance throughout an appropriate temperature range. Due to the limited ionic conductivity, the devices' performance degrades in the low-temperature region. Because of a number of adverse side effects, dependability problems, and safety concerns at high temperatures, the constituent elements tend to deteriorate. Due to the rise in operational frequency and transistor density, heat dissipation and thermal conductivity (TC) are significant problems for modern electronic devices. Due to their superior heat dissipation and TC, CNTs and graphene have attracted tremendous scientific interest across the globe. It includes their applications in batteries, supercapacitors, solar energy, and structural.

1.4.13 Carbon Nanotube and Graphene for Nanoelectronic Applications

The CNTs and graphene with their unique and truly 1D and 2D natures arising from at least one of the dimensions at nano scale, coupled with their featured electronic properties, become special material for nanoelectronics device applications. It is reasonable to expect the devices based on CNTs and graphene with far superior electrical and optical properties may outperform those based on conventional microelectronic materials such as silicon and gallium arsenide. The ever-increasing demand for miniaturized devices along with high processing speed requires the search for alternative materials. In this context, CNTs and graphene are the best potential candidates for these applications. CNTs and graphene have applications in nanoelectronics as interconnects, Schottky diodes and field-effect transistors (FETs). The metallic CNT, with its characteristic long mean free path, high current carrying capacity, and resistance to electromigration, can become the best option as interconnects.

The CNTs also have applications as Schottky diode. The diode specially working in the range of 30 GHz to 3 THz has application as a detector, mixers and frequency multiplier. The CNTs with high switching speed are most suitable for low voltage and high current application. CNTs application as field-effect transistor (FET) also has been demonstrated long back. Usually in a FET architecture consists of metal electrodes as source and drain which contacted to semiconducting channel. The gate as a third electrode isolated from channels controls the conductivity of channel. In case of CNT and graphene-based FET the conventional semiconducting channel is replaced with CNT and graphene [61].

Graphene possesses extraordinary electronic properties such as very high charge carrier mobility, about 100,000 $cm^2 V^{-1} s^{-1}$, apart from its excellent mechanical and optical properties. Its characteristic electronic structures lead to various fascinating quantum phenomena and makes graphene potential material for nanoelectronic devices and terahertz application. Basically, graphene is a semimetal which does not possess an energy band gap. Hence, although it has very high electronic mobility, the FET with graphene channel has a very low on-off ratio, about 10 only, when used as a switch in digital devices. Hence for its effective utilization in FET as a switch, different strategies have been used to induce a band gap in graphene, such as making its nanoribbon [64], application of electric field in bilayer graphene and other modifications. The on-off current ratio of 10^5–10^7 was obtained for the FET based on graphene channel nanoribbon with width <10 nm at room temperature. Further, the band gap in graphene can also be induced by applying an electric field to AB stacked graphene along transverse direction [67]. This kind of FET has application in optoelectronics and tuneable lasers etc. On the other side, the graphene, with its high electronic mobility, makes it suitable for analogue electronics especially high speed RF transistors where high on-off ratio is not needed. IBM demonstrated utilization of graphene for high frequency transistors and the highest cut-off frequency has been boosted from 26 GHz to 350 GHz [65, 66]. This has applications in the field of wireless communications, networking, radar and imaging etc. The graphene being only a single layer, the electronic charge on its surface can be very sensitive its environment hence it has application in graphene-based FET (GFET) which can high sensitivity. Graphene devices with new concept like bilayer pseudo spin field-effect transistor (BiSFET) [63], tunneling transistors with two graphene layers sandwiching either BN or MoS_2 etc have been demonstrated [62].

1.5 CONCLUSIONS

Carbon nanotubes and graphene are two advanced allotropes of carbon which have emerged as fascinating materials. Their unique atomic and electronic structures enable them to acquire the diverse and featured properties in the form of mechanical, electrical, optical and chemical properties. Graphene is 200 times stronger than steel materials while it has a charge carrier mobility one to two order more than conventional semiconductor silicon. Having a very thin single atomic layer it has high transparency to light along with a very high surface area. On other hand, the CNTs possess excellent aspect ratio and hence have high capability to withstand the strain before its failure. Both of these materials possess very high thermal conductivity. With these superior properties of diverse nature, both CNT and graphene become potential and advanced nanomaterials for a wide range of emerging applications such as energy storage, generation and conversion including batteries, supercapacitors, fuel cell, and photovoltaics. Their excellent mechanical properties make them suitable for light weight structural and ballistic material applications. Because of their special optical and electronic properties, graphene and CNTs find their applications in photonics, optoelectronics, nanoelectronics, alternative electrical conductors and sensors. Their other applications include antistatic and electromagnetic interference shielding, thermoelectric, along with applications in biomedical and environmental protection. Overall, CNTs and graphene, with their extraordinary and diverse properties, have great potential for many emerging applications.

REFERENCES

1 Iijima, S., *Helical microtubules of graphitic carbon*. Nature, 1991. **354**(6348): p. 56–58.

2 Novoselov, K.S., et al., *Electric field effect in atomically thin carbon films*. Science, 2004. **306**(5696): p. 666–669.

3 *The Wondrous World of Carbon Nanotubes*. 2003; Available from: http://students.chem.tue.nl/ifp03/Wondrous%20World%20of%20Carbon%20Nanotubes_Final.pdf.

4 Dai, H., *Carbon Nanotubes: Synthesis, Integration, and Properties*. Accounts of Chemical Research, 2002. **35**(12): p. 1035–1044.

5 Mittal, V., *Polymer NanotubeNanocomposites Synthesis, Properties, and Applications*, first edition, 2010: Scrivener Publishing LLC.

6 Harris, P.J.F., *Carbon Nanotubes and Related Structure*, 2004: Cambridge University Press.

7 Ebbesen, T.W. and P.M. Ajayan, *Large-scale synthesis of carbon nanotubes*. Nature, 1992. **358**(6383): p. 220–222.

8 Mathur, R.B., et al., *Co-synthesis, purification and characterization of single- and multi-walled carbon nanotubes using the electric arc method*. Carbon, 2007. **45**(1): p. 132–140.

9 Mathur, R.B., et al., *Process for the simultaneous and selective preparation of single-walled and multi-walled carbon nanotubes*, 2011, Google Patents.

10 Guo, T., et al., *Catalytic growth of single-walled nanotubes by laser vaporization*. Chemical Physics Letters, 1995. **243** (1–2): p. 49–54.

11 Choudhary, V., B. Singh, and R. Mathur, *Carbon nanotubes and their composites*. Syntheses and Applications of Carbon Nanotubes and Their Composites, 2013(9): p. 193–222.

12 Mathur, R.B., B.P. Singh, and S. Pande, *Carbon Nanomaterials Synthesis, Structure, Properties and Applications*, 2017. Taylor & Francis.
13 Sharma, S., et al., *Excellent mechanical properties of long multiwalled carbon nanotube bridged Kevlar fabric*. Carbon, 2018. **137**: p. 104–117.
14 Wallace, P.R., *The band theory of graphite*. Physical Review, 1947. **71**(9): p. 622–634.
15 Van Bommel, A.J., J.E. Crombeen, and A. Van Tooren, *LEED and Auger electron observations of the SiC (0001) surface*. Surface Science, 1975. **48**(2): p. 463–472.
16 Hass, J., W.A. de Heer, and E.H. Conrad, *The growth and morphology of epitaxial multilayer graphene*. Journal of Physics: Condensed Matter, 2008. **20**(32): p. 323202.
17 de Heer, W.A., et al., *Epitaxial graphene*. Solid State Communications, 2007. **143**(1): p. 92–100.
18 Emtsev, K.V., et al., *Towards wafer-size graphene layers by atmospheric pressure graphitization of silicon carbide*. Nature Materials, 2009. **8**(3): p. 203–207.
19 Hummers, W.S. and R.E. Offeman, *Preparation of graphitic oxide*. Journal of the American Chemical Society, 1958. **80**(6): p. 1339–1339.
20 Subhedar, K.M., I. Sharma, and S.R. Dhakate, *Control of layer stacking in CVD graphene under quasi-static condition*. Physical Chemistry Chemical Physics, 2015. **17**(34): p. 22304–22310.
21 Sharma, I., S.R. Dhakate, and K.M. Subhedar, *CVD growth of continuous and spatially uniform single layer graphene across the grain boundary of preferred (111) oriented copper processed by sequential melting-resolidification-recrystallization*. Materials Chemistry Frontiers, 2018. 6(2), p. 1137–1145.
22 Borah, M., et al., *The role of substrate purity and its crystallographic orientation in the defect density of chemical vapor deposition grown monolayer graphene*. RSC Advances, 2015. **5**(85): p. 69110–69118.
23 Sushant Sharma, A.K.A., S.R. Dhakate, Bhanu Pratap Singh, *Mechanical Properties of CNT Network-Reinforced Polymer Composites*, in *Organized Networks of Carbon Nanotubes*, 2020. CRC. p. 75–105.
24 Singh, B.P., S. Teotia, and S.R. Dhakate, *Process for the preparation of carbon fiber-carbon nanotubes reinforced hybrid polymer composites for high strength structural applications*, 2019, US Patent 10,400,074.
25 Kumar, V., et al., *Interleaved MWCNT buckypaper between CFRP laminates to improve through-thickness electrical conductivity and reducing lightning strike damage*. Composite Structures, 2019. **210**: p. 581–589.
26 Sharma, S., et al., *Enhanced Thermomechanical and Electrical Properties of Multiwalled Carbon Nanotube Paper Reinforced Epoxy Laminar Composites*. Composites Part A: Applied Science and Manufacturing, 2017.
27 Ouyang, J., *Applications of carbon nanotubes and graphene for third-generation solar cells and fuel cells*. Nano Materials Science, 2019. **1**(2): p. 77–90.
28 Jyoti, J., et al., *Recent advancement in three dimensional graphene-carbon nanotubes hybrid materials for energy storage and conversion applications*. Journal of Energy Storage, 2022. **50**: p. 104235.
29 Elizabeth, I., B.P. Singh, and S. Gopukumar, *Electrochemical performance of Sb2S3/CNT free-standing flexible anode for Li-ion batteries*. Journal of Materials Science, 2019. **54**(9): p. 7110–7118.
30 Maheshwari, H.P., et al., *Carbon nanotube-metal nanocomposites as flexible, free standing, binder free high performance anode for Li-ion battery*, 2018, US Patent 10,003,075.
31 Pandit, B., et al., *Free-standing flexible MWCNTs bucky paper: Extremely stable and energy efficient supercapacitive electrode*. Electrochimica Acta, 2017. **249**: p. 395–403.
32 Elizabeth, I., et al., *In-situ conversion of multiwalled carbon nanotubes to graphene nanosheets: An increasing capacity anode for li ion batteries*. Electrochimica Acta, 2017. **231**: p. 255–263.
33 Elizabeth, I., et al., *Multifunctional Ni-NiO-CNT composite as high performing free standing anode for Li ion batteries and advanced electro catalyst for oxygen evolution reaction*. Electrochimica Acta, 2017. **230**: p. 98–105.

34 Akula, S., et al., *Simultaneous Co-doping of nitrogen and fluorine into MWCNTs: an in-situ conversion to graphene like sheets and its electro-catalytic activity toward oxygen reduction reaction*. Journal of The Electrochemical Society, 2017. **164**(6): p. F568.

35 Sharma, S., et al., *Improved static and dynamic mechanical properties of multiscale bucky paper interleaved Kevlar fiber composites*. Carbon, 2019. **152**: p. 631–642.

36 Sharma, S., et al., *Synergistic bridging effects of graphene oxide and carbon nanotube on mechanical properties of aramid fiber reinforced polycarbonate composite tape*. Composites Science and Technology, 2020: p. 108370.

37 Sushant Sharma, S.R.D., Bhanu Pratap Singh, *Carbon Nanotubes in Protective Fabrics: A Short Review*. Trends in Textile Engineering & Fashion Technology, 2018. **3**(3): p. 1–3.

38 Gago, I., et al., *Graphene-based nanocomposites with improved mechanical and ballistic protection properties*, in *Advanced Materials for Defense*, 2020. Springer. p. 45–54.

39 Sharma, S., et al., *Design of MWCNT bucky paper reinforced PANI–DBSA–DVB composites with superior electrical and mechanical properties*. Journal of Materials Chemistry C, 2018. **6**(45): p. 12396–12406.

40 Chaudhary, A., et al., *Lightweight and easily foldable MCMB-MWCNTs composite paper with exceptional electromagnetic interference shielding*. ACS Applied Materials & Interfaces, 2016. **8**(16): p. 10600–10608.

41 Verma, M., et al., *Barium ferrite decorated reduced graphene oxide nanocomposite for effective electromagnetic interference shielding*. Physical Chemistry Chemical Physics, 2015. **17**: p. 1610–1618.

42 Singh, B.P., et al., *Microwave shielding properties of Co/Ni decorated single walled carbon nanotubes*. Journal of Materials Chemistry A, 2015.

43 Teotia, S., et al., *Multifunctional, robust, light-weight, free-standing MWCNT/phenolic composite paper as anodes for lithium ion batteries and EMI shielding material*. RSC Advances, 2014. **4**(63): p. 33168–33174.

44 Singh, B., et al., *Effect of length of carbon nanotubes on electromagnetic interference shielding and mechanical properties of their reinforced epoxy composites*. Journal of Nanoparticle Research, 2014. **16**(1): p. 1–11.

45 Gupta, T.K., et al., *MnO2 decorated graphene nanoribbons with superior permittivity and excellent microwave shielding properties*. Journal of Materials Chemistry A, 2014. **2**(12): p. 4256–4263.

46 Singh, B., et al., *Enhanced microwave shielding and mechanical properties of high loading MWCNT–epoxy composites*. Journal of Nanoparticle Research, 2013. **15**(4): p. 1–12.

47 Liu, H., et al., *Lightweight leaf-structured carbon nanotubes/graphene foam and the composites with polydimethylsiloxane for electromagnetic interference shielding*. Carbon, 2022. **191**: p. 183–194.

48 He, X., et al., *Vertically aligned carbon nanotube@ graphene paper/polydimethylsilane composites for electromagnetic interference shielding and flexible joule heating*. ACS Applied Nano Materials, 2022. **5**(5): p. 6365–6375.

49 Wang, X.-Y., et al., *Electromagnetic interference shielding materials: recent progress, structure design, and future perspective*. Journal of Materials Chemistry C, 2022.

50 Kumar, R., B. Singh, and V. Singh, *Exploring the possibility of using MWCNTs sheets as an electrode for flexible room temperature NO2 detection*. Superlattices and Microstructures, 2022: p. 107165.

51 Dariyal, P., et al., *Recent trends in gas sensing via carbon nanomaterials: outlook and challenges*. Nanoscale Advances, 2021. **3**(23): p. 6514–6544.

52 Bhattacharyya, B., et al., *Highly responsive broadband photodetection in topological insulator-carbon nanotubes based heterostructure*. Journal of Alloys and Compounds, 2020. **851**: p. 156759.

53 Shankar, U., et al., *Optically transparent and lightweight nanocomposite substrate of poly (methyl methacrylate-co-acrylonitrile)/MWCNT for optoelectronic applications: an experimental and theoretical insight*. Journal of Materials Science, 2021. **56**(30): p. 17040–17061.

54 Shankar, U., et al., *A facile way to synthesize an intrinsically ultraviolet-C resistant tough semi-conducting polymeric glass for organic optoelectronic device application.* Carbon, 2020. **168**: p. 485–498.
55 Bharti, M., et al., *Free-standing flexible multiwalled carbon nanotubes paper for wearable thermo-electric power generator.* Journal of Power Sources, 2020. **449**: p. 227493.
56 Yadav, S.K., S. Dhakate, and B.P. Singh, *Carbon nanotube incorporated eucalyptus derived activated carbon-based novel adsorbent for efficient removal of methylene blue and eosin yellow dyes.* Bioresource Technology, 2022. **344**: p. 126231.
57 Jyoti, J., et al., *Improved nanomechanical and in-vitro biocompatibility of graphene oxide-carbon nanotube hydroxyapatite hybrid composites by synergistic effect.* Journal of the Mechanical Behavior of Biomedical Materials, 2021. **117**: p. 104376.
58 Singh, K., et al., *Fabrication of amperometric bienzymatic glucose biosensor based on MWCNT tube and polypyrrole multilayered nanocomposite.* Journal of Applied Polymer Science, 2012. **125**(S1): p. E235–E246.
59 Dhand, C., et al., *Preparation of polyaniline/multiwalled carbon nanotube composite by novel electrophoretic route.* Carbon, 2008. **46**(13): p. 1727–1735.
60 Dariyal, P., et al., *A review on conducting carbon nanotube fibers spun via direct spinning technique.* Journal of Materials Science, 2021. **56**: p. 1087–1115.
61 Dmitriev, V., F. Gomes, and C. Nascimento, *Nanoelectronic devices based on carbon nanotubes.* Journal of Aerospace Technology and Management, 2015. **7**: p. 53–62.
62 Britnell, L., et al., *Electron tunneling through ultrathin boron nitride crystalline barriers.* Nano Letters, 2012. **12**(3): p. 1707–1710.
63 Su, J.-J. and A.H. MacDonald, *How to make a bilayer exciton condensate flow.* Nature Physics, 2008. **4**(10): p. 799–802.
64 Stankovich, S., et al., *Electronic confinement and coherence in patternedepitaxial graphene.* Carbon, 2007. **45**.
65 Jenkins, K.A., et al., *Linearity of graphene field-effect transistors.* Applied Physics Letters, 2013. **103**(17): p. 173115.
66 Lin, Y.-M., et al., *Operation of Graphene Transistors at Gigahertz Frequencies.* Nano Letters, 2009. **9**(1): p. 422–426
67 Oostinga, J.B., et al., *Gate-induced insulating state in bilayer graphene devices.* Nature Materials, 2008. **7**(2): p. 151–157.
68 Dunn, B., H. Kamath, and J.-M. Tarascon, *Electrical energy storage for the grid: a battery of choices.* Science, 2011. **334**(6058): p. 928–935.
69 Xu, J., et al., *Spheres of graphene and carbon nanotubes embedding silicon as mechanically resilient anodes for lithium-ion batteries.* Nano letters, 2022. **22**(7): p. 3054–3061.
70 Yoo, E., et al., *Large reversible Li storage of graphene nanosheet families for use in rechargeable lithium ion batteries.* Nano letters, 2008. **8**(8): p. 2277–2282.
71 Wang, G., et al., *Graphene nanosheets for enhanced lithium storage in lithium ion batteries.* Carbon, 2009. **47**(8): p. 2049–2053.
72 Hou, J., et al., *Graphene-based electrochemical energy conversion and storage: fuel cells, supercapacitors and lithium ion batteries.* Physical Chemistry Chemical Physics, 2011. **13**(34): p. 15384–15402.
73 Hwang, H.J., et al., *Multilayer graphynes for lithium ion battery anode.* The Journal of Physical Chemistry C, 2013. **117**(14): p. 6919–6923.
74 Xiong, Z., Y. Yun, and H.-J. Jin, *Applications of carbon nanotubes for lithium ion battery anodes.* Materials, 2013. **6**(3): p. 1138–1158.
75 DiLeo, R.A., et al., *Enhanced capacity and rate capability of carbon nanotube based anodes with titanium contacts for lithium ion batteries.* ACS Nano, 2010. **4**(10): p. 6121–6131.
76 Raffaelle, R.P., et al. *Carbon nanotube anodes for lithium ion batteries.* in *MRS Proceedings.* 2001. Cambridge Univ Press.
77 Ng, S., et al., *Single wall carbon nanotube paper as anode for lithium-ion battery.* Electrochimica Acta, 2005. **51**(1): p. 23–28.

78 de las Casas, C. and W. Li, *A review of application of carbon nanotubes for lithium ion battery anode material.* Journal of Power Sources, 2012. **208**(0): p. 74–85.
79 Zhang, W.-J., *A review of the electrochemical performance of alloy anodes for lithium-ion batteries.* Journal of Power Sources, 2011. **196**(1): p. 13–24.
80 Gu, J., et al., *Template-free preparation of crystalline ge nanowire film electrodes via an electrochemical liquid–liquid–solid process in water at ambient pressure and temperature for energy storage.* Nano letters, 2012. **12**(9): p. 4617–4623.
81 Kasavajjula, U., C. Wang, and A.J. Appleby, *Nano- and bulk-silicon-based insertion anodes for lithium-ion secondary cells.* Journal of Power Sources, 2007. **163**(2): p. 1003–1039.
82 Elizabeth, I., et al., *Development of SnO2/multiwalled carbon nanotube paper as free standing anode for lithium ion batteries (LIB).* Electrochimica Acta, 2015. **176**: p. 735–742.
83 Fang, R., et al., *The regulating role of carbon nanotubes and graphene in lithium-ion and lithium–sulfur batteries.* Advanced Materials, 2019. **31**(9): p. 1800863.
84 Yang, Z., et al., *Carbon nanotube-and graphene-based nanomaterials and applications in high-voltage supercapacitor: A review.* Carbon, 2019. **141**: p. 467–480.
85 Gohardani, O., M.C. Elola, and C. Elizetxea, *Potential and prospective implementation of carbon nanotubes on next generation aircraft and space vehicles: A review of current and expected applications in aerospace sciences.* Progress in Aerospace Sciences, 2014. **70**: p. 42–68.
86 Tsiolkovsky, K., *Grezy o zemle i nebe (i) Na Veste (Speculations between Earth and Sky, and On Vesta; science-fiction works).* Moscow, izd-vo AN SSSR, 1895: p. 35.
87 Artsutanov, Y., *Kosmos na Elektrovoze, Komsomol-skaya Pravda, July 31 (1960); contents described in Lvov, V.* Science, 1967. **158**: p. 946–947.
88 Pearson, J., *The orbital tower: a spacecraft launcher using the Earth's rotational energy.* Acta Astronautica, 1975. **2**(9–10): p. 785–799.
89 Isaacs, J.D., et al., *Satellite elongation into a true "sky-hook".* Science, 1966. **151**(3711): p. 682–683.
90 Pearson, J., *Anchored lunar satellites for cislunar transportation and communication.* Journal of the Astronautical Sciences, 1979. **27**(1): p. 39–62.
91 Edwards, B.C., *Design and deployment of a space elevator.* Acta Astronautica, 2000. **47**(10): p. 735–744.
92 Bolonkin, A., *Non-rocket Earth-Moon transport system.* Advances in Space Research, 2003. **31**(11): p. 2485–2490.
93 Pearson, J., et al. *The lunar space elevator.* In *55th International Astronautical Congress of the International Astronautical Federation, the International Academy of Astronautics, and the International Institute of Space Law.* 2004.
94 Pugno, N.M., *On the strength of the carbon nanotube-based space elevator cable: from nanomechanics to megamechanics.* Journal of Physics: Condensed Matter, 2006. **18**(33): p. S1971.
95 Wörle-Knirsch, J., K. Pulskamp, and H. Krug, *Oops they did it again! Carbon nanotubes hoax scientists in viability assays.* Nano Letters, 2006. **6**(6): p. 1261–1268.
96 Avnet, M.S., *The space elevator in the context of current space exploration policy.* Space Policy, 2006. **22**(2): p. 133–139.
97 Dempsey, J.G., *System and method for space elevator*, 2006, Google Patents.
98 Pugno, N.M., Space elevator: out of order? Nano Today, 2007. **2**(6): p. 44–47.
99 Van Pelt, M., *Space Tethers and Space Elevators*, 2009: Springer.
100 Clarke, A.C., *The Fountains of Paradise*, 1979: Victor Gollancz.
101 Sheffield, C., *Web Between Worlds*, 1988: Random House.
102 Edwards, B.C. and E.A. Westling, *The Space Elevator*, 2003: Bc Edwards.
103 Edwards, B. and P. Raglan, *Leaving the Planet by Space Elevator*, www.leavingtheplanet.com/, 2006.
104 Swan, P.A., P. Swan, and C.W. Swan, *Space Elevator Systems Architecture*, 2007: Lulu. com.
105 Swan, C. and P. Swan, *Space Elevator Survivability Space Debris Mitigation*, 2015: Lulu. com.
106 Perek, L., *Space elevator: stability.* Acta Astronautica, 2008. **62**(8–9): p. 514–520.

107 Pugno, N., et al., *On the stability of the track of the space elevator*. Acta Astronautica, 2009. **64**(5–6): p. 524–537.
108 Woo, P. and A.K. Misra, *Dynamics of a partial space elevator with multiple climbers*. Acta Astronautica, 2010. **67**(7–8): p. 753–763.
109 Takeichi, N., *Geostationary station keeping control of a space elevator during initial cable deployment*. Acta Astronautica, 2012. **70**: p. 85–94.
110 Jorgensen, A., S. Patamia, and B. Gassend, *Passive radiation shielding considerations for the proposed space elevator*. Acta Astronautica, 2007. **60**(3): p. 198–209.
111 Engel, K.A., *Lunar transportation scenarios utilising the Space Elevator*. Acta Astronautica, 2005. **57**(2–8): p. 277–287.
112 Swan, C.W. and P.A. Swan, *Why we need a space elevator*. Space policy, 2006. **22**(2): p. 86–91.
113 Ketsdever, A.D., et al., *Overview of advanced concepts for space access*. Journal of Spacecraft and Rockets, 2010. **47**(2): p. 238–250.
114 Bolonkin, A., *Non-Rocket Space Launch and Flight*, 2010: Elsevier.
115 Quine, B., R. Seth, and Z. Zhu, *A free-standing space elevator structure: a practical alternative to the space tether*. Acta Astronautica, 2009. **65**(3–4): p. 365–375.
116 Liu, Z., et al., *Reflection and absorption contributions to the electromagnetic interference shielding of single-walled carbon nanotube/polyurethane composites*. Carbon, 2007. **45**(4): p. 821–827.
117 Chung, D., *Electromagnetic interference shielding effectiveness of carbon materials*. Carbon, 2001. **39**(2): p. 279–285.
118 Luo, X. and D. Chung, *Electromagnetic interference shielding reaching 130 dB using flexible graphite*. MRS Online Proceedings Library, 1996. **445**(1): p. 235–238.
119 Pande, S., et al., *Improved electromagnetic interference shielding properties of MWCNT–PMMA composites using layered structures*. Nanoscale Research Letters, 2009. **4**(4): p. 327–334.
120 Gupta, T.K., et al., *Improved nanoindentation and microwave shielding properties of modified MWCNT reinforced polyurethane composites*. Journal of Materials Chemistry A, 2013. **1**(32): p. 9138–9149.
121 Singh, B., et al., *Designing of multiwalled carbon nanotubes reinforced low density polyethylene nanocomposites for suppression of electromagnetic radiation*. Journal of Nanoparticle Research, 2011. **13**(12): p. 7065–7074.
122 Zhang, W., et al., *Advances in waterborne polymer/carbon material composites for electromagnetic interference shielding*. Carbon, 2021. **177**: p. 412–426.
123 Wang, Z., et al., *Highly stretchable graphene/polydimethylsiloxane composite lattices with tailored structure for strain-tolerant EMI shielding performance*. Composites Science and Technology, 2021. **206**: p. 108652.
124 Sharma, S., et al., *Improved static and dynamic mechanical properties of multiscale bucky paper interleaved Kevlar fiber composites*. Carbon, 2019.
125 Sharma, S., et al., *Excellent mechanical properties of long length multiwalled carbon nanotube bridged Kevlar fabric*. Carbon, 2018.
126 Shi, J., et al., *Graphene reinforced carbon nanotube networks for wearable strain sensors*. Advanced Functional Materials, 2016. **26**(13): p. 2078–2084.
127 Park, S.J., et al., *Highly flexible wrinkled carbon nanotube thin film strain sensor to monitor human movement*. Advanced Materials Technologies, 2016. **1**(5): p. 1600053.
128 Michelis, F., et al., *Highly reproducible, hysteresis-free, flexible strain sensors by inkjet printing of carbon nanotubes*. Carbon, 2015. **95**: p. 1020–1026.
129 Wang, Y., et al., *Wearable and highly sensitive graphene strain sensors for human motion monitoring*. Advanced Functional Materials, 2014. **24**(29): p. 4666–4670.
130 Liu, Q., et al., *High-quality graphene ribbons prepared from graphene oxide hydrogels and their application for strain sensors*. Acs Nano, 2015. **9**(12): p. 12320–12326.
131 Gong, T., et al., *Highly responsive flexible strain sensor using polystyrene nanoparticle doped reduced graphene oxide for human health monitoring*. Carbon, 2018. **140**: p. 286–295.
132 Li, Y., et al., *Hybrid strategy of graphene/carbon nanotube hierarchical networks for highly sensitive, flexible wearable strain sensors*. Scientific Reports, 2021. **11**(1): p. 1–9.

133 Nandihalli, N., C.-J. Liu, and T. Mori, *Polymer based thermoelectric nanocomposite materials and devices: Fabrication and characteristics*. Nano Energy, 2020. **78**: p. 105186.
134 Lu, Y., et al., *Ultrahigh power factor and flexible silver selenide-based composite film for thermoelectric devices*. Energy & Environmental Science, 2020. **13**(4): p. 1240–1249.
135 Schweicher, G., et al., *Molecular Semiconductors for Logic Operations: Dead-End or Bright Future?* Advanced materials, 2020. **32**(10): p. 1905909.
136 Zhang, D., et al., *Multifunctional Superelastic Graphene-Based Thermoelectric Sponges for Wearable and Thermal Management Devices*. Nano letters, 2022. **22**(8): p. 3417–3424.
137 Sun, X., et al., *Anisotropic electrical conductivity and isotropic seebeck coefficient feature induced high thermoelectric power factor> 1800 $\mu W\ m^{-1}\ K^{-2}$ in MWCNT films*. Advanced Functional Materials, 2022: p. 2203080.
138 Feng, S., et al., *Quasi-industrially produced large-area microscale graphene flakes assembled film with extremely high thermoelectric power factor*. Nano Energy, 2019. **58**: p. 63–68.
139 Zeng, W., et al., *Defect-engineered reduced graphene oxide sheets with high electric conductivity and controlled thermal conductivity for soft and flexible wearable thermoelectric generators*. Nano Energy, 2018. **54**: p. 163–174.
140 Wang, L., et al., *Large thermoelectric power factor in polyaniline/graphene nanocomposite films prepared by solution-assistant dispersing method*. Journal of Materials Chemistry A, 2014. **2**(29): p. 11107–11113.
141 Wang, Y., et al., *Polypyrrole/graphene/polyaniline ternary nanocomposite with high thermoelectric power factor*. ACS Applied Materials & Interfaces, 2017. **9**(23): p. 20124–20131.
142 Schrade, M., et al., *Centimeter-sized monolayer CVD graphene with high power factor for scalable thermoelectric applications*. ACS Applied Electronic Materials, 2022. **4**(4): p. 1506–1510.
143 Parida, K., et al., *A kinetic, thermodynamic, and mechanistic approach toward adsorption of methylene blue over water-washed manganese nodule leached residues*. Industrial & Engineering Chemistry Research, 2011. **50**(2): p. 843–848.
144 Han, Z., et al., *Ultrafast and selective nanofiltration enabled by graphene oxide membranes with unzipped carbon nanotube networks*. ACS Applied Materials & Interfaces, 2021. **14**(1): p. 1850–1860.
145 Nasrollahzadeh, M., et al., *Carbon-based sustainable nanomaterials for water treatment: state-of-art and future perspectives*. Chemosphere, 2021. **263**: p. 128005.
146 Gopinath, K.P., et al., *Environmental applications of carbon-based materials: a review*. Environmental Chemistry Letters, 2021. **19**(1): p. 557–582.

2 Carbon Nanotubes and Graphene for Ballistic Protection

Sushant Sharma, Reena Goyal, Mamta Rani, Sanjay R. Dhakate, and Bhanu Pratap Singh

CONTENTS

2.1 INTRODUCTION

With the rapid development of ammunition and warfare technologies, the dynamics of the arena have changed from world wars to the current terrorist activities, which is the major threat on mankind. Hence, there is a stringent requirement of effective defensive mechanisms which can be used for prolonged duration without any physical fatigue. Therefore, it is observed that modern man-made synthetic fibers and their reinforced composites are frequently replacing metallic armor for high velocity projectile resistance applications. High-performance fibers such as glass fiber [1], highly oriented ultrahigh molecular weight polyethylene (UHMWPE) (e.g., Dyneema, Spectra) [2, 3], aramids (Kevlar 29, Kevlar 49, Kevlar 129, Kevlar KM2) [4], polybenzoxazole (PBO) (e.g., Zylon), polypyridobisimidazole (PIPD) (e.g., M5) [5] are the new class of polymeric fibers which are frequently practiced for protective armor application along with ceramic-based amors. Some advanced designs incorporate both ceramic armors as a front protective plate and fiber-reinforced polymer (FRP) composite plate as a back plate assembled together, which play different roles in

DOI: 10.1201/9781003231943-2

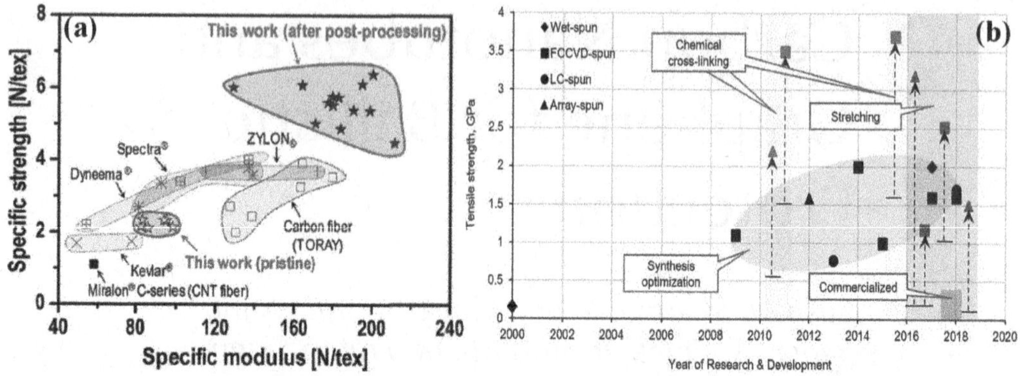

FIGURE 2.1 (a) represents the comparison of strength and modulus of the various synthetic fibers with carbon nanotube (CNT) fibers (Reprinted with permission from Ref. [6]. Copyright (2020) American Chemical Society), and (b) evolution in the strength of CNT fiber prepared by various technique (Reprinted from Ref. [7]. Copyright (2019) with permission from Elsevier).

different positions. Generally, the front ceramic protective plate absorbs a major part of projectile energy and damages it, while the FRP back plate protects the wearer from serious trauma.

Although these materials have high specific strength, there is some scope to cut down the weight and enhance the impact strength of body armor by using high-performance carbon materials including nanotubes (CNTs) and graphene. CNT and graphene are the allotropic forms of carbon, having outstanding mechanical properties. Compared to widely used ballistic fibers, they are rigid, have a high degree of toughness, and can absorb a lot of energy. Figure 2.1(a) represents the mechanical properties of CNT fiber prepared by the chemical vapor deposition (CVD) technique in context with commercially available high-performance fibers. It is a fact that the intrinsic features of CNTs are not completely realized in their microscopic forms, such as sheets, fibers, etc. But still, processed CNT yarns represent a tremendous improvement in the mechanical properties and are capable of replacing synthetic fibers for various engineering applications (Figure 2.1(b)).

Apart from this, various research groups are already working in this direction and have established various facts related to the high velocity impact resistance properties of carbon nanomaterials. They observed that these high aspect ratio carbon nanomaterials, when interacting with high velocity projectiles, dissipate its kinetic energy in compaction, stretching, fibrillation and finally tensile fracture. They immediately transfer the shock wave that spreads uniformly from the first point of contact upon collision and largely absorb the projectile's kinetic energy. These experimental and simulation results motivate the researchers to use these carbon nanomaterials in futuristic personal protection gadgets to improve their specific impact resistance properties. Hence, carbon nanomaterials are used with different combinations of ceramics, polymer matrices, and fibers to enhance the intrinsic impact resistance properties in hybrid composite forms.

According to the ballistic materials, there are three different methods for incorporating the carbon nanomaterials in it. First, reinforcing the ballistic grade fibers like UHMWPE, Kevlar and reinforced composites with carbon nanomaterials to enhance its energy absorption capacity. Secondly, carbon nanomaterials can be transformed into woven and non-woven fabric for achieving, incredible ballistic performance. Third, the carbon nanomaterials are blended into polymer matrix, metals and ceramic to improve their impact resistance properties. Here, selection of an appropriate approach depends on various parameters such as type of materials design, failure mechanism, projectile velocity, and shape. It is very important to develop and design a splendid ballistic material by utilizing the outstanding mechanical properties of carbon nanomaterials to its full extent for futuristic threats.

Hence, this chapter will elaborate the failure mechanism of various ballistic materials such as ceramics, ballistic fibers and reinforced composites, carbon nanomaterials and scaffolds, and finally hybrid composites. Due to the outstanding straining capabilities, carbon nanomaterials reinforcement in hybrid ballistic material improves the specific impact resistance properties and modifies the original failure mechanism. Hence, this chapter will also elaborate the effect of carbon nanomaterial reinforcement on the impact resistance properties of different ballistic grade futuristic materials.

2.2 TYPES OF IMPACT

An impact load is suddenly applied over a short span of time when two or more objects collide. The failure behavior of the colliding objects and therefore their mechanical properties depend on velocities of respective objects. According to the velocity, it is categorized into three types; first low velocity impact (below 10 m/s), second intermediate velocity impact (10–50 m/s), third high velocity impact (50–1000 m/s) and hyper velocity impact (2000–5000 m/s) [8].

2.2.1 Low Velocity Impact

According to the stiffness and material characteristics of the target as well as the mass and stiffness of the projectile material, low velocity impact (LVI) is described as an event that occurs in the 1–10 m/s range [9]. Low velocity impact can occur during in-service or maintenance and it can be treated as the most vulnerable load on any engineering material. Because the impact lasts for a long enough time for the structure to react to it and absorb more energy elastically, dynamic response of the target material is extremely important in low velocity impacts. For low velocity impact, the pendulum type (Charpy and Izod test) and drop weight tests have been used extensively. These tests enable the numerical and theoretical simulation of a broad range of impact conditions and collect detailed information of performance variables [10]. It also enables the wide variety of test geometries, thereby involving more complex components to be tested [11].

Charpy and Izod impact tests are generally used to study the impact response of isotropic material having different compositions and manufacturing conditions. A pendulum of known mass, dropped from a certain height freely to fracture the specimen. After eroding the specimen, height gained by the pendulum gives the energy absorption capacity. The only difference in the Charpy and Izod impact test is that in a Charpy test specimens are clamped as a simple supported beam while in Izod tests specimens are supported as a cantilever beam. Apart from materials' characteristic properties, the damage mechanism, crack propagations and loading capability are analyzed more accurately by using high-speed photography [12].

Delamination of fiber-reinforced laminar composites under low velocity impact results in the interlaminar characteristics at the boundary layer under transverse loading. Low velocity impact does not cause perforation and any surface damage, it only creates delamination between the layers [13]. There are various parameters that govern the failure of composites under low velocity impact i.e., fiber properties, resin properties, lay-up technique, sample thickness, boundary conditions, and type of projectile (shape, size, velocity, and mass) [14]. Unlike ceramics, fibers and their reinforced composites absorb energy in elastic deformation fibrillation and delamination. While in the case of ceramic materials, depending on their volume present, particle size, packing density and distribution, ceramic materials show brittle fracture and absorb the maximum amount of energy during failure.

2.2.2 Intermediate Velocity Impact

Intermediate velocity impact (IVI) is a moderate velocity impact which lies in the intermediate velocity range of 20–100 m/s with some blunt impactor. These impacts are significant in the case of low velocity projectiles, road debris, hail impact or cricket bat striking a cricket ball. In test methodology, it is crucial to find the energy dissipation and pinpoint the failure mechanism over projectile interaction. Here energy absorption is associated with two components E_{damage} and $E_{SL.}$ E_{damage} includes matrix failure, fiber matrix debonding, fiber pull outs, and fibrillation, while E_{SL} is caused due to absorption of energy imparted to the system through elastic response, vibrations, heat, etc. [15]. Split Hopkinson pressure bar (gas gun) is used to determine the ballistic impact and IVI. Up until the gun velocity reaches zero, a specimen's ability to absorb energy is measured.

2.2.3 High Velocity Impact

In a high velocity impact (HVI) test, the contact time of projectile and targeted component systems is a fraction of a second that surface erosion and finally perforation (plastic deformation) take place. The high velocity impact is subjugated by stress wave propagation through ballistic material, and it does not take much time to respond to impact load, leading to localized damage [16]. It is a very complex mechanism in which various parameters are involved. There are basically two types of projectile launching systems which can deliver a velocity near 2000 m/s: first, single-stage gas gun, and second, gunpowder gun. According to their name, a gas gun uses the compressed helium or hydrogen gas while in the other, gunpowder is used as a propellant to transfer the kinetic energy to the projectile [17]. Previous research has indicated the penetration and perforation phenomenon of ceramics and fiber-reinforced laminar composite by means of high velocity flat-faced, round nose, conical tip and truncated cone nose tip projectiles [18, 19]. The energy absorbed by the targeted material is determined by measuring the velocity of the projectile before and after a strike, using optically operated actuators at the front and back of the targeted material, respectively. The loss in the kinetic energy of the projectile is equivalent to the energy absorbed during impacting. This absorbed energy is used in damaging and perforation of the target material and damaging the projectile.

The majority of engineering applications today, including those in the automotive, aircraft, marine, sporting goods, and personal armor industries, include these kinds of impact conditions. High-performance antiballistic materials, including fiber-reinforced composites, are being employed to meet this difficulty. These are used in vehicles to create the body framework, the chassis, the engine, the drive shafts, the leaf springs, the internal and exterior assemblies, etc. These fiber-reinforced composites carry out a variety of tasks, such as reducing weight and enhancing fuel efficiency, while also having high specific stiffness, good damage tolerance, and outstanding surface polish and appearance. In aerospace and aeronautics, these are used in radome, radar antenna, canopy, door skin, intake of the engine nacelle, turbine blade, wings, tail, etc. which reduces the weight by 20–30 percent. In marine hulls, body structures and rotor parts consistently experience the harsh corrosive environment and therefore are designed by using composites. In sporting goods these are used to make racquets, head bicycle frames, golf club shafts, skis, canoe helmets, fishing poles, tent poles, etc. to design weight reduction vibration damping design and have high fatigue resistance properties. In personal armor, they are generally used in protective clothing, combat helmets, spall liner, ballistic shields, etc. due to light weight, high impact resistance, comfortable and easy to transport.

Different ballistic grade materials behave in a different manner upon high velocity impact. Their interaction with projectiles helps in determining the failure mechanism and talks about the scope of improvement in impact resistance properties. When the high velocity projectile interacts with a component a high velocity wave is generated which propagates in materials for dissipation of energy. This interaction governs the impact resistance properties of the material which is discussed in upcoming sections.

2.3 WAVE PROPERTIES OF IMPACT RESISTANT MATERIALS

2.3.1 Wave Propagation in Ceramics

Ceramics are the promising high velocity impact resistance materials due to their high specific hardness and compressive strength. Amid them, alumina, aluminum oxynitride (AlON), magnesium aluminate spinel ($MgAl_2O_3$), silicon carbide, boron carbide, zirconium diborides, hafnium diborides and other ceramics are widely used in high impact resistance materials such as personal armors, space vehicles and launchers, etc. But the impact resistance is the most critical characteristic of ceramic materials for reassuring the suitability to serve various engineering applications.

Whenever the two objects collide, a sudden force applied on the surface generates a strain that can show elastic (temporary or reversible) or plastic (permanent or irreversible) behavior depending on the materials characteristics of the colliding objects. Different materials perform differently upon impact, for example metal represents both types of behavior (elastic and plastic) depending on the intensity impinging force. For small values of forces the proportionality between resultant strain and modulus is maintained, but afterward, on exceeding the certain limit, the strain ceases and loses its proportionality w.r.t. applied force. However, in the case of ceramic materials a rapid transition will take place from elastic to fracture, because of their small mechanical stress relaxation properties and hence lies in the category of brittle materials.

Unlike ductile materials in which fracture is preceded by the plastic strain, in case of ceramic materials impacted by a high velocity projectile, the failure takes place in a different manner. For instance, a steel bullet hitting an aluminum plate at 1000 m/s velocity will produce an impact pressure of magnitude 19.4 GPa, which is larger than the compressive strength of the majority of ceramic materials (σ_c= 2–7 GPa). However, this contact was only sustained for a brief period of time because the relief wave from the ceramic-free surface reduced the pressure to a quasi-steady state with a tenth of the peak value (i.e., ~ 2 GPa). Few ceramics existing at the site of contact will fail in compression during this brief period of intense triaxial compressive stress while abrading the projectile material. No additional compression failure happens because the quasi-steady state value is less than the compressive strength of ceramic. When a fracture conoid is built up at impacting point (striking surface) at the same time a relief wave from the rear end develops a tensile stress having the magnitude of 300–500 MPa, which moves forward in the ceramic. Since the magnitude of tensile stress is greater than the compressive stress, an axial cracking will generate as shown in the Figure 2.2(a). Because of their low tensile strength and high brittleness, ceramic materials fail prematurely at an impacting point, therefore a lot of modifications in micro or nano-scale are employed for improving the impact resistance properties. It can also be bonded over other composite backing made of high-performance fibers [20, 21]. Although, ceramic materials are brittle in nature but elementally their particles are harder than projectile materials, therefore they are capable of deforming, blunting and eroding the projectile material, which enhances the impact resistance efficiency.

2.3.2 Wave Propagation in Fibers

The mechanism of impact resistance is a complicated mechanical process that takes into account a number of different factors, including the density of the material being targeted, its density, thickness, modulus, strength, ductility, and toughness. High-performance fibers are often some of the potential materials for soft or stiff forms of personal protective equipment utilized against several sorts of ballistic and stab impact hazards. Light weight, high impact resistance, good comfort and low cost are the parameters, which are always associated with textile based protective armors. Nowadays, various research groups are progressively working in the direction of improvements of impact resistance properties of these kinds of protective equipment by considering various design parameters such as weaving patterns (i.e., 2D, 3D, interlocked, etc.), laminar stacking with different fashion, reinforcing in polymers, etc. Moreover, development of high tenacity yarn from high modulus, high strength, and superior environmental stability fiber are major priority before developing the impact resistance textiles. Therefore, in order to create any fiber-based protective equipment, it is crucial to take into account the real impact event and comprehend the fiber reaction.

A high velocity impact on a protective material may be thought of as an energy balance between the projectile making the impact and the substance it is intended to hit. The impacting projectile possesses the kinetic energy, which is transmitted to the targeted substance upon impacting. This kinetic energy is transformed into different types of energy that the intended substance will absorb. Particularly, textile-based targeted materials include the kinetic and potential energy of the simplest element of textile i.e., yarns and fibers. The behavior of yarn under impact can give insights into the complex behavior of fabric assemblies. Two different sorts of waves that originate at the site of contact and move away from it are evidence of the distortion caused by the high velocity impact against a yarn. One kind of wave is associated as a longitudinal wave, which causes the yarn to stretch (Figure 2.2(b)). The second wave, known as the transverse wave, moves in the direction of the projectile and causes the yarn to flow outward in that direction (Figure 2.2(b)). Independent of the impact velocity, the longitudinal wave moves outward at the speed of sound. Figure 2.2(b) represents the three different zones in the yarn under the impact of high velocity projectiles. The horizontal deflection of polymer molecules occurs in two locations, one on either side of the impact, when the necking of the yarn components occurs. Third is the middle V-shaped zone, which deflects the bullet in that direction. Once material reaches the V-shaped zone from the nearby necking zone, all horizontal motion stops, and only transverse wave components remain [22]. All of the yarn material in this area of horizontal deflection is under some strain, the amount of which depends on the material's characteristics and the impact velocity (Vp). The vertical component of velocity in the V-shaped zone will stay constant for all yarns and be equal to the velocity of the projectile since all yarns in the horizontal stretching zone move at the same speed and the projectile's velocity is the same for all yarns. The transverse (out of plane) deformation progresses outward, and the longitudinal (in-plane) strain propagates along the yarn in the woven textile with the velocity of sound. Hence, higher the in-plane velocity greater the strain dissipation in the material away from the point of impact and improves the energy absorption capacity. The square root of the elastic modulus and the square root of the yarn density both affect the wave propagation velocity in different ways. But this velocity will never be the same for the constructed fabric, as it depends on the type of construction which introduces crimp at the crossover points.

Another crucial factor is impact velocity; when impact velocity is low and high propagation velocity is present in the fabric, the produced stress is kept below the threshold level of yarn strength needed for rupture. One of the key requirements for energy absorption in this situation is the transfer of energy during deformation in the transverse direction. The projectile penetrates by pulling out the yarn from the fabric in the impact direction and it depends on the inter yarn friction and projectile yarn friction. Therefore, for fabric-based protection equipment tenacity, elongation at rupture, density, wave propagation, inter yarn friction, weaving pattern are the most important properties.

2.3.3 Wave Propagation in Carbon Nanomaterial

It is well known that carbon nanostructures are both lightweight and have excellent mechanical characteristics [23, 24]. Body armor also works by virtue of being both powerful and

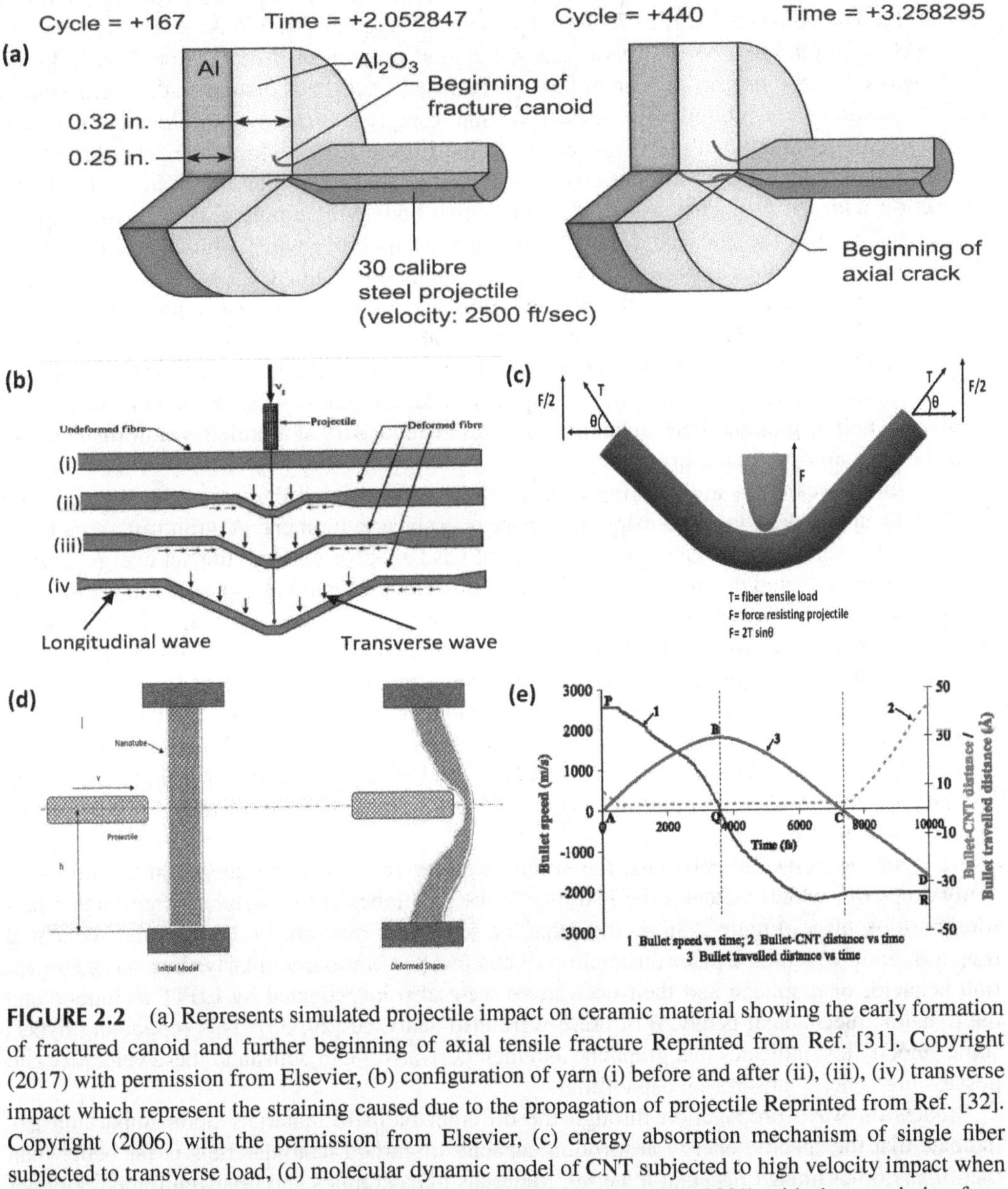

FIGURE 2.2 (a) Represents simulated projectile impact on ceramic material showing the early formation of fractured canoid and further beginning of axial tensile fracture Reprinted from Ref. [31]. Copyright (2017) with permission from Elsevier, (b) configuration of yarn (i) before and after (ii), (iii), (iv) transverse impact which represent the straining caused due to the propagation of projectile Reprinted from Ref. [32]. Copyright (2006) with the permission from Elsevier, (c) energy absorption mechanism of single fiber subjected to transverse load, (d) molecular dynamic model of CNT subjected to high velocity impact when a CNT was fixed at both the ends Reprinted from Ref. [33]. Copyright (2021) with the permission from Elsevier, and (e) variation of bullet speed, CNT-bullet distance and bullet traveled distance w.r.t. time to represent the straining and multiple hit capability of individual CNT (Reprinted with the permission from Ref. [28]. Copyright (2007) with permission from IOP science).

light. One may make a composite material that is even stronger and lighter by combining two strong and lightweight materials. It has excellent strength, light weight, incredible energy absorption capacity, high strain to failure ratio, multiple hit capacity, therefore, has great potential for antiballistic material [25]. The carbon nanomaterials can be used as ballistic textile material [26, 27]. With the perspective of ballistic material, ballistic properties of carbon nanomaterials are investigated experimentally by Mylvaganam et al., [28, 29] and numerically by Koziol et al. [30]. The impact resistance dynamics and energy absorption capacity were examined by performing a molecular dynamic simulation on a single wall carbon nanotube (SWCNT). The constant length ~75 Å and varied diameter of (10.576 Å, 8.606 Å, 7.051 Å) SWCNTs were clamped on both side and a diamond projectile of dimension 35.6 x 35.6 x 7.1 $Å^3$ was used as a projectile. The bullet was released from 15 Å distance at its center point axis; the results were plotted with respect to time to represent the dynamic behavior of projectile when it interacts with CNT as shown in the Figure 2.2(d) and 2.2(e). The three body Tersoff-Brenner potential model was used to describe the atomic interaction in the CNT and interaction with the projectile was described by two body Morse potential. According to the study, the CNT has the maximum ballistic resistance capability when a bullet enters it in the middle, and bigger tubes can tolerate faster bullets. This shows that CNT body armor can have a consistent ballistic resistance and multiple hitting capabilities. On a subsequent impact after a brief pause, CNT could withstand nearly the same speed as in the first impact. However, there is a huge difference in the mechanical properties of individual carbon nanomaterials and their microscopic assemblies in which a number of heterogeneous carbon nanomaterials are involved. Their alignment, heterogeneity, structural irregularity at atomic level limits to attain their theoretically predicted properties.

Therefore, researchers are working progressively to understand the impacting behavior and prove their application in nano-inspired impact resistance equipment. A group of researchers from Rice University, have shown the potential of CNT mat for extreme impact energy absorption experimentally and theoretically [34]. They used a unique technique, laser induced projectile impact test (LIPIT) for determining the impact strength of CNT assembly, In which a micron sized silica sphere was impacted with the hypersonic velocities ranging from 305 m/s to 916 m/s and the interaction between CNTs and projectile is analyzed. It is observed that a major part of the projectile kinetic energy was lost in friction interactions between CNT-CNT and CNT-projectile, adiabatic heating, tube stretching and collapsing and ultimately fracture of taut tube and fibrillation of tubes as depicted in the Figure 2.3. The remarkable energy absorbed per unit mass during this interaction with the 205 nm thin CNT mat ranges from 7–12 Mj/kg which was 5–14 times higher than that of Kevlar.

At higher velocity i.e., 916 m/s, the strain rate was very high and energy absorption suddenly drops off, which indicates the inability of the nanotubes at the peripheral region and perforation took place. Figure 2.3(i–l) indicates the schematic diagram to describe the sequential transformation which took place during impact and mat perforation. Similarly, the energy absorption behavior of graphene and their derivatives were also investigated by LIPIT technique and the dynamic mechanical behavior of these were also analyzed [29, 35]. This replicating hypersonic velocity test indicates that graphene and their derivatives can withstand these velocities and possess high energy absorption capabilities.

Studies on wave propagation through the different ballistic materials upon impact clearly indicate that the specific energy absorption capacity of carbon nanomaterials is far better than that of any other impact resistant material. Materials like ceramics and fiber-reinforced polymer composites in the laminar form are enormously vulnerable to brittle failure, crack initiation and propagation through laminar interface in various impacting conditions. Although there are various

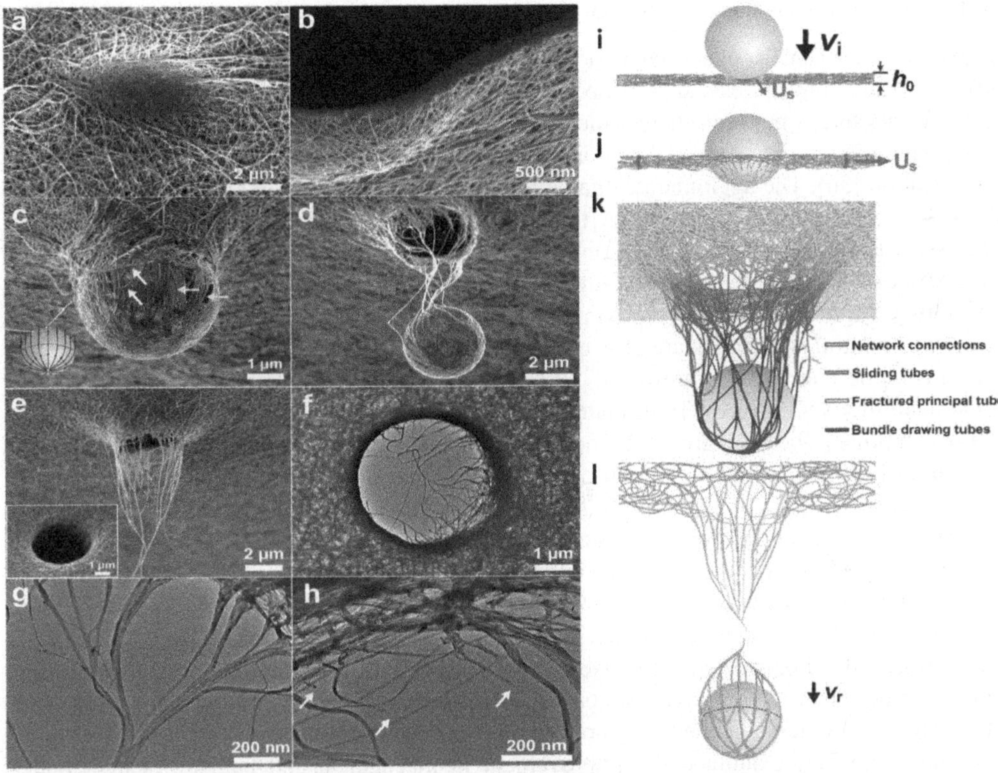

FIGURE 2.3 (a–e) SEM micrographs of back side of CNT assembly subjected LIPIT test for determining the impact resistance properties, it gives morphological changes caused during projectile shock compression, penetration and perforation, (f–h) represents TEM micrographs of perforated surface of CNT assembly which represents the drawn tube bundles in the projectile direction which also get fibrillated and finally fractured. White arrows represent the fractured tip of CNT in (h), and (i–l) represents the schematic diagram to show morphological and material deformation during perforation (Reproduced with permission from Ref. [34]. Copyright 2021, John Wiley and Sons).

techniques available for improving the impact resistance properties, which will be covered in more detail in the sections that follow.

2.4 CARBON NANOMATERIAL REINFORCED CERAMIC (CNRC) COMPOSITES

As was indicated in the previous section, ceramic materials have a high degree of brittleness and need novel design approaches and manufacturing methods to cope with these inherent shortcomings in order to perform effectively in high velocity impact events. Altering the ceramic structures at nanoscale is a leading-edge research; whilst modification via carbon nanofiller reinforcement has become a renowned exercise for advanced engineering application. Strategies like tuning and optimization of doping, new dispersion, and sintering techniques are the most recently adopted ways to improve the properties of carbon nanofiller reinforced ceramic (CNRC) composites. This section will highlight some recent strategies and key difficulties in manufacturing the CNRC composites for the improvement of their related impact resistance properties.

2.4.1 Dispersion of Carbon Nanofillers

Dispersion of carbon nanomaterial is a highly imperative parameter for achieving the better mechanical and impact resistance properties. Carbon nanofillers are associated with strong van der Waals forces between them which restricts in attaining the good dispersion in ceramic matrix, which is possible only at low concentration (<2 wt%), higher than this leads to severe agglomeration [36]. The agglomerated nanofiller gives poor interface and microstructure, which ultimately hamper the mechanical properties. To deal with this problem, a majorly colloidal technique is used, which involves dispersion of nanofiller with the help of ultrasonication energy. This provides better results with less uncertainty compared to ball milling, magnetic stirring, tape casting, high rpm centrifuge mixing, sol-gel, hydrothermal, etc. Not only this, cationic, anionic and neutral surfactants are also used to detangling to nanofiller and further reinforcing them in ceramic matrices uniformly [37, 38]. Several studies also involve the direct growth of carbon nanofiller specially CNTs in matrices by using conventional chemical vapor deposition (CVD) technique [39]. CNT growth in the ceramic matrix is a fascinating technique and various efforts have been made for the growth of ordered CNTs within the pores of ceramic material [40, 41]. Chen et al., impregnate the Si_3N_4 with 10% cobalt acetate tetrahydrate and feed it in the CVD system under the flow of Ar gas and $(CH_3)_2CO$ vapor at 550°C. The growth of CNT interconnects the Si_3N_4 grains and helped in improving load distribution capability [42]. Table 2.1 represents some useful mechanical properties of CNRC composites, prepared by various different techniques having different loading concentration of nanofiller. The reinforcement of nanofiller in ceramic is useful in improving the relative density, hardness, flexural strength, fracture toughness and most importantly impact resistance properties. Recently, Yin et al. [43] have prepared boron carbide (B_4C) ceramic toughened by graphene platelets (GPLs) using the spark plasma sintering technology, and they examined the improvement in mechanical and ballistic characteristics. Additionally, 5 wt% of Al_2O_3/Y_2O_3 were also used to improve the sintering process. Only 1.5 wt% of GPL improves the fracture toughness by ~23% compared to without GPL, which clearly indicates the improvement in pinning effect and packing density caused due to the reinforcement of GPLs. Furthermore, ballistic performance was analyzed by performing conventional depth of penetration (DOP) test, and observed that, GPLs/B4C ceramic armor has the least penetration depth when compared to SiC and B_4C, and its impact resistance has been enhanced by 50% and 12.5%, respectively. Earlier, the effect reinforcement of CNTs on structural and ballistic properties were analyzed by Nepochatov et. al [44]. The nanocomposite of CNT and B_4C were prepared by vacuum sintering at 2100°C and 30 MPa. The reinforcement of CNTs improves the ballistic properties by ~26% as compared to without reinforcement. It was also observed that the fracturing pattern is changed to branched pattern after reinforcement, which causes effective kinetic energy absorption of the armor piercing projectile.

2.4.2 Densification Processes

Although there are various challenges involved while developing the CNRC composites, obtaining full density composite without damaging the structure of nanofiller in the confinement of ceramic matrix is a fundamental requirement, as most of the mechanical properties including impact resistance properties are associated with proper densification techniques. Because carbon nanomaterial is present at the grain borders, coalescence of ceramic grains is hampered, resulting in a poorly modeled microstructure [45]. As a result, pressure aided sintering, which includes simultaneously pressing and heating ceramic powder, is a frequently utilized approach to address this issue. It gives high packing density with excellent mechanical properties, not only in the case of ceramics but also in case of carbon nanofiller reinforced ceramic composites. Coble et al.

TABLE 2.1
Enhancement in Mechanical Properties of CNRC Composites with the Reinforcement of Carbon Nanomaterials

S. No.	Type of Matrix	Type of carbon nanofillers	Concentration of carbon nanofillers	Synthesis Technique	Improvement in impact resistance properties				Year	Ref
					Relative density (%)	Hardness (GPa)	Flexural Strength (MPa)	Fracture toughness ($MPa.m^{1/2}$)		
1	Al_2O_3	MWCNTs	0	Wet mixing	99.5	17.5	222	3.92	2012	[58]
			0.15 vol.%		98.4	21.4	242	5.27		
2	Si_3N_4	Graphene platelet	0		-	15.4	-	6.9	2012	[59]
			1 wt%		-	16.4	-	9.9		
3	Al_2O_3	MWCNTs	0	Slip casting	-	16.9	-	5.5	2013	[60]
			1 vol.%		-	13.5	-	6.0		
4	Al_2O_3	MWCNT	0		97.54	17.05	-	3.34	2014	[61]
		SWCNT	0.1 wt%		96.41	16.66	-	3.00		
		MWCNT+ SWCNT	0.1 wt.%		95.28	15.55	-	2.85		
			0.05 + 0.05 wt%		92.87	12.45	-	1.96		
5	Al_2O_3	MWCNT	0	Freeze drying and hot pressing	99.8	17	-	5.5	2014	[62]
	Al_2O_3 +		1 vol. %		99.5	15.5	-	5.9		
	5 vol. % ZrO_2		1 vol. %		99.4	14.5	-	7.5		
6	Al_2O_3	MWCNTs	0	Spark plasma sintering	-	14.73	-	3.95	2015	[63]
			5.7 vol. %		-	19.18	-	6.03		
7	ZrO_2	MWCNTs	0	Sol gel	87	8.04	-	1.95	2015	[64]
			1 wt%		94	8.88	-	2.48		
			3 wt%		98	9.18	-	4.08		
			5 wt%		89	9.08	-	3.81		
8	Al_2O_3	SiC+ CNT	0		99.3	18.56		3.61	2016	[65]
			10 vol. % SiC +1 vol. % CNT		98.63	20.81		4.58		
			10 vol. % SiC +2 vol. % CNT		98.02	17.50		6.98		
9	ZrB_2	SiC	0	Hot pressed	90.1	11.9	355	2.1	2017	[66]
		SiC+ carbon black	20 vol. %		93.2	13.1	662	3.7		
			20 vol. % SiC +10 vol. % CB		99.8	16.7	687	5.8		

(continued)

TABLE 2.1 (Continued)
Enhancement in Mechanical Properties of CNRC Composites with the Reinforcement of Carbon Nanomaterials

S. No.	Type of Matrix	Type of carbon nanofillers	Concentration of carbon nanofillers	Synthesis Technique	Improvement in impact resistance properties: Relative density (%)	Hardness (GPa)	Flexural Strength (MPa)	Fracture toughness ($MPa.m^{1/2}$)	Year	Ref
10	WC-TiC-Al_2O_3	Multilayer graphene	0		95.3	24.2	962.6	9.2	2017	[67]
			0.1 wt%		99.3	22.9	1021.9	14.1		
11	Al_2O_3	SWCNT	0	Spark plasma sintering	97.9	20.2	700	3.7	2018	[68]
		RGO	0.5 wt%		97.4	18.8	950	4		
			1 wt%		96.7	18.5	725	4.2		
			0.5 wt%		99	19.6	870	4.2		
			1 wt%		98.2	19.6	980	4.7		
12	Al_2O_3	CNT	0	Spark plasma sintering	91	18.21	-	2.4	2018	[69]
		Graphene	1 wt.% CNT		98.9	17.41		6.4		
			1 wt.% CNT+ 0.5 wt.% Graphene		99.9	16.66		8.33		
13	ZrB_2	g-C_3N_4	0	Spark plasma sintering	76.5	10.1	187.6	1.9	2019	[70]
			5 wt%		99.8	16.2	516.4	5.4		
14	TiC	SiC whiskers	0	Spark plasma sintering	99.23	25	540	-	2019	[71]
			10 vol. %		98.73	24.8	511			
			20 vol. %		101.68	26	644			
			30 vol. %		102.95	29.04	610			
15	ZrB_2_25 vol. % SiC	Nano-diamond	0	Spark plasma sintering	99.9	19.5		4.3	2020	[72]
			1 wt%		100	23.4		4.8		
			2 wt%		99.9	24.6		5.4		
			3 wt%		98.8	22.2		5.8		
16	TiC	Nano-graphite	0	Spark plasma sintering	95.5	30.67	504	-	2020	[73]
			5 wt%		97.1	20.31	633			
17	TiC	SiC	0	Spark plasma sintering	98.44		410		2021	[74]
	TiC +		40 vol. %		100		543			
	5 vol% WC		0		99.92		491			
			40 vol. %		100		590			
18	Al_2O_3	graphene graphene	0	Spark plasma sintering	99.7	16.44	415.78	4.9	2021	[75]
		and CNT	0.4 wt%		99.4	17.47	671.8	6.785		
			0.4 wt% & 1 wt%		99.2	15.8	494	7.325		

noticed that, prolonged sintering at elevated temperature and pressure, grain growth takes place which damages the CNT structure, which is a major drawback of high pressure sintering [46, 47]. To avoid the structural damage in carbon nanofiller during sintering, it is important to reduce the temperature and shorten the halt period. Therefore, spark plasma sintering techniques are used, which provides a fully dense composite structure without damaging the nanofiller. Pressureless sintering is also a cheaper densification technique, but reproducibility is a big concern associated with this technique. For example, Zhang et al. [48] and Ahmed et al. [49] reported large variation in the densities of the Al_2O_3/CNTs nanocomposite, in which 1 wt% CNT reinforced composite showed highest density of 99% and lowest density of < 90%. Microwave sintering is also an interesting technique which involves lower densification temperature and reduced processing time. Various industrial ceramics and their composites such as Al_2O_3, ZrO_2, Si_3N_4, etc. have been already prepared by this technique without any agitation [50, 51]. Although, the mixed large and fine grained microstructure is achieved by this technique, but it is very useful for CNRC composites, as it involves localized heating which is advantageous for carbon nanomaterials to get reinforced without any structural deformation [52, 53].

2.4.3 Microstructural Properties

Variation in microstructural properties is caused due to the variation in reinforcement structural properties. 1D and 2D nanofiller act differently upon reinforcement in ceramic matrices and they have different interactions with ceramic grains. For example, 1D CNTs cause pinning effect and restrict the grain growth, therefore in case of CNT reinforced ceramic finer grains are achieved compared to monolithic ceramics with coarser grain. Due to these microstructural properties, fracture mode is also changed from inter-granular in case of monolithic ceramic to trans-granular in CNT reinforced ceramics [54]. The morphological analysis of fractured surface is depicted in the Figure 2.4, which helped in understanding the fracture mechanism in the case of monolithic Al_2O_3 ceramic and CNT reinforced composite. In the case of monolithic Al_2O_3 it shows edged and corned fractured surface caused due to intra-granular fracture mode, while Figure 2.4 represents glazed surface caused by the CNT reinforcement, indicating trans-granular fractured mode. Naturally, in the case of trans-granular fractured mode, load is transferred by grains through interfacial CNTs positioned at grain borders, across grain boundaries, and inside single grains, improving the toughness of the composite materials and strengthening them for high velocity projectile impacts (Figure 2.4) [54, 55]. In case of 2D graphene nanofiller and their derivative fillers, the accessible surface area is larger than 1D CNTs, therefore 2D graphene nanoparticles (GNPs) anchored with large surface area of ceramic grains, improves the interfacial friction, and required large energy during fracture as compared CNTs. Walker et al. [56] reported the improvement in toughness of 235% in Si_3N_4 ceramics in which GNPs wrapped around it to block the crack propagation. This toughening mechanism was discovered for the first time, and ever since then, various carbon nanofillers have been used as reinforcement to boost mechanical qualities. The role and the reinforcement effect of different carbon nanofillers are well documented, however their behavior in same ceramic matrix is more complex. Recently, Yazdani et al. [57] reinforced the GNTs (i.e., MWCNTs and GNPs) in Al_2O_3 matrix and improvement in flexural strength and fracture toughness. The study suggests that GNPs wrapped around Al_2O_3 matrix enhance pullout energy during fracture and strengthen the grain boundaries. While MWCNTs contribute more in bridging effect for load sharing between ceramic grains. It was also concluded that CNTs stretch more compared to GNPs before collapsing during crack propagation.

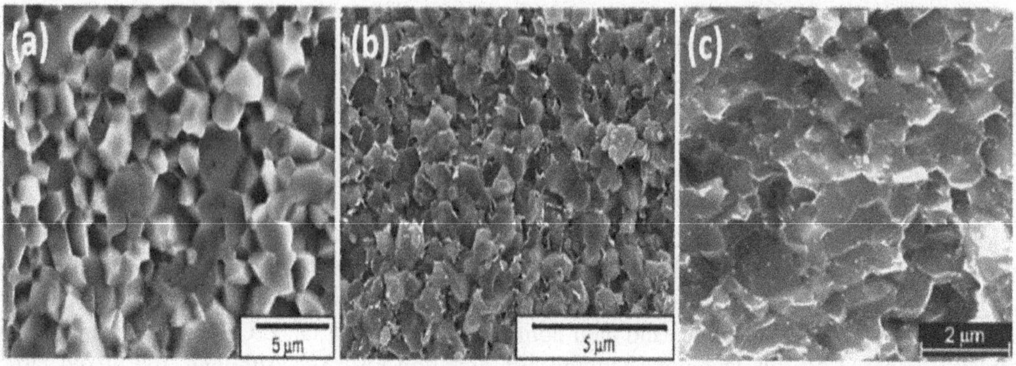

FIGURE 2.4 Morphological features of fractured (a) monolithic Al_2O_3 representing large grain with intergranular fracture Reprinted from Ref. [54]. Copyright (2010) with the permission from Elsevier, (b) CNT/Al_2O_3 ceramic nanocomposite with fine grain Reprinted from Ref. [36]. Copyright (2014) with the permission from Elsevier, and (c) trans-granular fracture mode appears in Al_2O_3 (Reprinted from Ref. [54]. Copyright (2010) with the permission from Elsevier).

2.5 CARBON NANOMATERIAL/FIBER REINFORCED MULTISCALE COMPOSITES

In the current scenario, high-performance fiber like aramid, carbon fiber, glass fiber, ultrahigh molecular weight polyethylene fiber (UHMWPE), nylon fiber, Twaron fiber are the main choice of reinforcement in high velocity impact resistant engineering structures. However, their through-thickness properties under actual impact events lack some most important physical and mechanical property requirements, which cause unpredictable failure. These structural failure challenges raised the demand of improved laminar composites with structural integrity and impact resistance properties subjected to any kind of velocity events.

The fiber design and resin modification are the two most extensively studied aspects that aid in enhancing the impact resistance and damage tolerance qualities of fiber reinforced polymer (FRP) composites [76]. 2D FRP composites (i.e., unidirectional and bidirectional weaved fibers) have been used for the past four decades, they are associated with poor through-thickness (transverse) properties. With the development in weaving technology, 3D weaved FRP composites are now frequently used in high impact resistance composite structures [77]. Delamination in laminar composites is the major problem as it is difficult to detect in multiple layer composite systems and it effectively degrades the residual strength [78]. In addition, traditional 2D (bidirectional) weaves have a high stress concentration that results from the crimping of the strands [79]. Various methods are employed to address this issue, such as stitching, Z-pinning, 3D interlocking, etc., which finally enhances the transverse qualities such as interlaminar fracture toughness and delamination resistance [80, 81]. The impact that fabric architecture has on the qualities of impact resistance is shown in table 2.2. Although these techniques improve the transverse properties, they also limit the in-plane properties of the composites. For example, Tan et al. [82] reported an increase in interlaminar fracture toughness, however the creation of a resin-rich zone as a result of fisheye production had an impact on all in-plane parameters. Similar to that, Z-pinning is a helpful approach for enhancing Mode I and Mode II interlaminar fracture toughness [83, 84], but it significantly affects the in-plane properties (compressive or tensile) [85].

Therefore, in another approach i.e., resin toughening, mechanical properties in both the transverse and longitudinal directions are persevered and interlaminar properties are improved. Various types of nanomaterials are used; among them carbon nanomaterials are ideal candidates for resin

TABLE 2.2
Influence of Fibre Design on Impact Resistance and Damage Tolerance Characteristics

S. No.	Fiber architecture	Effect on Impact resistance properties	Ref.
1	UD Composites	Poor transverse properties under low velocity impact	[9]
2	2D woven composites	Lower in-plane properties	[86]
3	Non crimp fabrics (NCF) composites	Generates fisheye defects and limits in-plane properties	[87]
4	3D composites	2.5 times higher impact energy absorption	[88]
5	Stitched Composites	Generates resin-rich regions and fisheye defects	[89]
6	Z-pinned composites	Causes 25% reduction in the in-plane properties	[85]
7	2D/Epoxy and 3D/Epoxy	3D composites show better impact resistance	[85, 90]
8	2D/Epoxy and 3D/Epoxy	3D composites show impact resistance	[91]
9	Plain weave with seaming	Represents 200% improvements in specific energy absorption capacity	[92]
10	3D angle-interlock woven fabric (3DAWF)	Improves the ballistic impact resistance properties	[93]

toughening [94, 95]. Carbon nanomaterial possess superior specific mechanical properties which are already established theoretically and experimentally. According to the previously mentioned sections, if these mechanically robust fillers can be effectively reinforced in polymer matrix with or without reinforcing ballistic fibers like aramid, UHMWPE, Twaron etc., composites with excellent ballistic properties can be achieved. For example, Hanif et al. [96] analyzed the effect of CNT reinforcement on ballistic performance of Twaron/epoxy based multiscale composite. Using varied weight percentages of MWCNTs to toughen the epoxy resin, he constructed the nine-layer laminar composite by hand lay-up and vacuum bagging. Only 1 wt% MWCNTs in epoxy matrix helped in improving the fracture toughness values and ultimately improves the ballistic resistance properties. The maximum energy absorption obtained was 264.8 J which was superior than without MWCNTs. It was also discovered that the qualities of fracture toughness and energy absorption increase in direct proportion. Due to the inertness of CNTs their reinforcement efficiency is dropped drastically, and it causes agglomeration problem. To overcome this problem, surface deformation and functionalization are majorly used. But this CNT dispersion technique only facilitates overall 10–20% improvement energy absorption capability [97]. Rather than dispersing CNT or f-CNT in matrix, the most sophisticated methods to increase impact resistance capabilities include stacking pre-grown CNT forest spun fiber mat with fabric sheets, transferring pre-grown CNT forest to prepreg, and strengthening freestanding CNT sheets [79, 98–100]. Recently, our team has also prepared the freestanding CNT sheets and reinforced it between the Kevlar layers to improve the static and dynamic impact resistance properties and further ballistic properties were also determined. It was observed that with reinforcement of MWCNT sheets between the Kevlar stacking back face signature drops ~30%, which confirms the improved interlaminar properties [98]. But, nowadays, CNTs are directly grown over the fiber by various techniques and further reinforced into matrix to form multiscale composites. In this approach CNTs behave like that they are anchored on the fiber surface and provides good interfacial properties with matrix which ultimately improves composite interlaminar shear strength, toughness and impact resistance properties. For example, Boddu et al. [101] grow the MWCNTs over the E-glass fiber by using floating catalyst CVD technique and interleaved these layers between other E-glass fabric layers. Both the ballistic V50 test and the split Hopkinson pressure bar test were conducted in the lab. The dynamic impact test revealed that interleaving the CNT produced

E-glass fibers enhanced the specific energy absorption by 106% at high strain rates and the energy density dissipation by 64.3%. The ballistic test reveals that the V50 value increased by ~11% after interleaving CNT grown E-glass fabric. Similarly, Sodano and group use the laser induced graphene (LIG) technique to introduce the graphene over Kevlar fabric for in-situ impact damage monitoring. In that work, they correlated electrical impedance during ballistic test and analyzed the severity of impact with respect to time. Additionally, it will also improve the toughness of the laminar composite. The specific V50 velocity was higher than that of pristine Kevlar reinforced composites having same number of layers and thickness, which was caused due to interlaminar friction caused due to the anchored graphene platelets [102]. Apart from impact resistance properties the resin toughing technique improves the stab resistance properties also, which depends upon the matrix system. Wang et al. [103] developed the novel wearable electronic textile (WET) made up of shear stiffening polymer/MWCNTs/Kevlar composite system. Stab resistance test under dynamic condition was conducted, a knife impactor was impacted on WET and neat Kevlar targets. For WET, 11.76 J energy was required for complete penetration, while in case of neat Kevlar target having same number of layers were penetrated with 7.84 J energy, which clearly showed the improvement of ~50% in energy absorption capacity. In addition to the research listed above, there are several more studies that show potential for enhancing the impact resistance qualities of fiber reinforced composites by using a variety of ways to strengthen the carbon nanostructures (Table 2.3).

From the wave propagation mechanism of different carbon nanomaterials, it is obvious the longitudinal wave propagates in carbon nanomaterial with higher velocity than any other ballistic material and therefore they have excellent energy dissipation properties. Secondly, this material possesses superior mechanical properties and high strain to failure ratio, which are not yet to be fully realized in practical ballistic materials. If compare the specific mechanical properties of these nanomaterial, we found that it is 100 times superior to metallic shields of steel. Therefore, researchers are trying to develop fully carbon nanomaterials-based body armors which are not only light in weight but also sustain many high velocity impact shots without deformation and most importantly reduced back face trauma. In this direction, researchers are preparing the aligned carbon nanomaterial fibers and their fabrics so that they used to make multilayer armor system for high velocity projectiles.

2.6 CONCLUSION

In the current world scenario, where every nation wants to protect their soldiers and law enforcement from lethal weapon attacks, there is a stringent requirement for lightweight, movable, easy to wear and robust personal protective equipment. Carbon nanomaterials such as CNT, graphene and their hybrids and derivatives are the most potential candidates for robust, enormous energy absorber, multiple hit capacity, lightweight, flexible body armor systems. They can be used as; 1. Reinforcement in polymer matrices, ceramics matrices to improve their hardness, toughness and impact resistance properties, 2. Incorporating these nanomaterials along with armor grade synthetic fibers such as Kevlar, Twaron, UHMWPE, Nylon, etc. to improve the elastic modulus and energy dissipation capacity, and 3. Prepare the neat or composite fibers of these nanomaterials for further making the stacked system of woven or non-woven fabrics. These techniques significantly modify the failure behavior of well adopted ballistic grade materials like ceramics and high-performance fibers and improve the ballistic resistance properties. It also inspires the researchers to develop the fully carbon nanomaterial-based body armor system, which not only resist the lethal impact but are also very light and comfortable to wear and move in without any exhaustion.

TABLE 2.3
Impact Resistance Properties of Carbon Nanomaterial Reinforced Multiscale Composites

S. No.	Type of fiber	Type of polymer	Type of filler	Filler conc.	Improvement in impact resistance properties: Impact resistance properties	Improvement in impact resistance properties: Mechanical Property	Year	Ref
1	E-glass	polyester	MWCNT	-	Specific energy absorption increased by 106% & V50 value increased by 11.1%	FS decreased by 47.3% & Interlaminar shear strength decreased by 26%	2016	[101]
2	Kevlar Carbon	Epoxy	MWCNT	1 wt%	Withstand 400 m/s metal projectile impact without penetration	FS = 500 MPa	2016	[104]
3	UHMWPE	Epoxy	SWCNT	0.5 wt%	Withstand hypervelocity of range 6.5–7 km/s		2013	[105]
4	Aramid	Epoxy	Graphene oxide Graphene nanoplatelets	0.3 wt%	The ballistic limit and energy absorption capacity of composite (HVI) are increased by 23 and 52%, respectively		2020	[106]
5	Carbon	Epoxy	MWCNT	2 wt%	At 100 m/s impact shock energy absorption improved by 21%	TS = 780 MPa Strain to failure =1.5 (improved by 12%)	2013	[97]
6	Kevlar	Polystyrene ethyl acrylate	CNT	1 wt%	Ballistic limit velocity of Kevlar was improved from 84.6 m/s to 96.5 m/s after impregnation		2020	[107]
7	Aramid	High-density polyethylene	Graphene nanoplatelets	0.5 wt%	Energy absorption efficiency improved by 32% & Ballistic protection level Type III-A can with stand	Storage modulus increased to 114%	2022	[108]
8	Carbon	Epon 862 Epoxy	CNT	1 wt%	Absorb highest amount of energy of hardened steel projectile at low velocity (7 m/s) & Lower deformation area		2017	[109]
9	Aramid	-	Graphene oxide	2 mg/ml aqueous suspension filtration	Coating GO on two side improves the energy absorption capacity to 50%		2020	[110]
10	Kevlar	Epoxy	MWCNT	0.3 wt%		Tensile strength increased by 93% & Storage modulus increased by~193%	2018	[79]
11	Kevlar	Epoxy	MWCNT, Buckypaper	0.3 wt%	Static energy absorption increased by 49% & Average back face scattering improved by 30%	Storage modulus increased by ~233%	2019	[98]

ACKNOWLEDGMENT

One of the authors, Sushant Sharma (SS), is grateful to the Council of Scientific Industrial Research (CSIR), India for providing the financial award (No. 13(9199-A)/2021) for conducting this study.

REFERENCES

1 Sevkat, E., et al., *A combined experimental and numerical approach to study ballistic impact response of S2-glass fiber/toughened epoxy composite beams*. Composites Science and Technology, 2009. **69**(7): p. 965–982.
2 Grujicic, M., et al., *A ballistic material model for cross-plied unidirectional ultra-high molecular-weight polyethylene fiber-reinforced armor-grade composites*. Materials Science and Engineering: A, 2008. **498**(1): p. 231–241.
3 Langston, T., *An analytical model for the ballistic performance of ultra-high molecular weight polyethylene composites*. Composite Structures, 2017. **179**: p. 245–257.
4 Bandaru, A.K., S. Ahmad, and N. Bhatnagar, *Ballistic performance of hybrid thermoplastic composite armors reinforced with Kevlar and basalt fabrics*. Composites Part A: Applied Science and Manufacturing, 2017. **97**: p. 151–165.
5 Cunniff, P.M., et al. *High performance "M5" fiber for ballistics/structural composites*, in *23rd. Army Science Conference*, 2002.
6 Oh, E., et al., *Super-strong carbon nanotube fibers achieved by engineering gas flow and postsynthesis treatment*. ACS Applied Materials & Interfaces, 2020. **12**(11): p. 13107–13115.
7 Mikhalchan, A. and J.J. Vilatela, *A perspective on high-performance CNT fibres for structural composites*. Carbon, 2019. **150**: p. 191–215.
8 Vaidya, U.K., *Impact response of laminated and sandwich composites*, in *Impact Engineering of Composite* Structures, 2011. Springer. p. 97–191.
9 Richardson, M. and M. Wisheart, *Review of low-velocity impact properties of composite materials*. Composites part A: Applied Science and Manufacturing, 1996. **27**(12): p. 1123–1131.
10 Mathivanan, N.R. and J. Jerald, *Experimental investigation of low-velocity impact characteristics of woven glass fiber epoxy matrix composite laminates of EP3 grade*. Materials & Design, 2010. **31**(9): p. 4553–4560.
11 Cantwell, W.J. and J. Morton, *The impact resistance of composite materials—a review*. Composites, 1991. **22**(5): p. 347–362.
12 Ujihashi, S., *An intelligent method to determine the mechanical properties of composites under impact loading*. Composite Structures, 1993. **23**(2): p. 149–163.
13 Meola, C. and G.M. Carlomagno, *Impact damage in GFRP: new insights with infrared thermography*. Composites part A: Applied Science and Manufacturing, 2010. **41**(12): p. 1839–1847.
14 Ghasemnejad, H., A. Furquan, and P. Mason, *Charpy impact damage behaviour of single and multi-delaminated hybrid composite beam structures*. Materials & Design, 2010. **31**(8): p. 3653–3660.
15 Bartus, S. and U. Vaidya, *Performance of long fiber reinforced thermoplastics subjected to transverse intermediate velocity blunt object impact*. Composite structures, 2005. **67**(3): p. 263–277.
16 Kinslow, R., *High-Velocity Impact* Phenomena, 2012: Elsevier.
17 Chhabildas, L.C., L. Davison, and Y. Horie, High-pressure shock compression of solids VIII: The science and technology of high-velocity impact. 2004: Springer Science & Business Media.
18 Wen, H., *Predicting the penetration and perforation of FRP laminates struck normally by projectiles with different nose shapes*. Composite Structures, 2000. **49**(3): p. 321–329.
19 Ahmad, I., B. Yazdani, and Y. Zhu, *Recent advances on carbon nanotubes and graphene reinforced ceramics nanocomposites*. Nanomaterials, 2015. **5**(1): p. 90–114.
20 Akella, K. and N.K. Naik, *Composite armor—A review*. Journal of the Indian Institute of Science, 2015. **95**(3): p. 297–312.
21 Crouch, I., *The Science of Armour Materials*. 2016: Woodhead Publishing.

22 Abtew, M.A., et al., *Ballistic impact mechanisms–a review on textiles and fibre-reinforced composites impact responses*. Composite Structures, 2019. **223**: p. 110966.

23 Takakura, A., et al., *Strength of carbon nanotubes depends on their chemical structures*. Nature Communications, 2019. **10**(1): p. 1–7.

24 Papageorgiou, D.G., I.A. Kinloch, and R.J. Young, *Mechanical properties of graphene and graphene-based nanocomposites*. Progress in Materials Science, 2017. **90**: p. 75–127.

25 Baughman, R.H., A.A. Zakhidov, and W.A. De Heer, *Carbon nanotubes—the route toward applications*. Science, 2002. **297**(5582): p. 787–792.

26 Qian, L. and J.P. Hinestroza, *Application of nanotechnology for high performance textiles*. Journal of textile and apparel, technology and management, 2004. **4**(1): p. 1–7.

27 Patra, J.K. and S. Gouda, *Application of nanotechnology in textile engineering: An overview*. Journal of Engineering and Technology Research, 2013. **5**(5): p. 104–111.

28 Mylvaganam, K. and L. Zhang, *Ballistic resistance capacity of carbon nanotubes*. Nanotechnology, 2007. **18**(47): p. 475701.

29 Lee, J.-H., et al., *Dynamic mechanical behavior of multilayer graphene via supersonic projectile penetration*. Science, 2014. **346**(6213): p. 1092–1096.

30 Koziol, K., et al., *High-performance carbon nanotube fiber*. Science, 2007. **318**(5858): p. 1892–1895.

31 Crouch, I., et al., *Glasses and ceramics*, in *The Science of* Armour Materials, 2017: Elsevier. p. 331–393.

32 Naik, N., P. Shrirao, and B. Reddy, *Ballistic impact behaviour of woven fabric composites: Formulation*. International Journal of Impact Engineering, 2006. **32**(9): p. 1521–1552.

33 Prabha, P.S., et al., *FEA analysis of ballistic impact on carbon nanotube bulletproof vest*. Materials Today: Proceedings, 2021. **46**: p. 3937–3940.

34 Hyon, J., et al., *Extreme energy dissipation via material evolution in carbon nanotube mats*. Advanced Science, 2021. **8**(6): p. 2003142.

35 Xie, W., et al., *Extreme mechanical behavior of nacre-mimetic graphene-oxide and silk nanocomposites*. Nano Letters, 2018. **18**(2): p. 987–993.

36 Ahmad, I., et al., *Investigation of yttria-doped alumina nanocomposites reinforced by multi-walled carbon nanotubes*. Ceramics International, 2014. **40**(7): p. 9327–9335.

37 Fan, J., et al., *Preparation and microstructure of multi-wall carbon nanotubes-toughened Al2O3 composite*. Journal of the American Ceramic Society, 2006. **89**(2): p. 750–753.

38 Sharma, S., et al., *Structural and mechanical properties of free-standing multiwalled carbon nanotube paper prepared by an aqueous mediated process*. Journal of Materials Science, 2017. **52**(12): p. 7503–7515.

39 Xia, Z., J. Lou, and W. Curtin, *A multiscale experiment on the tribological behavior of aligned carbon nanotube/ceramic composites*. Scripta Materialia, 2008. **58**(3): p. 223–226.

40 Fan, Y., et al., *Preparation and electrical properties of graphene nanosheet/Al2O3 composites*. Carbon, 2010. **48**(6): p. 1743–1749.

41 Kyotani, T., L.-f. Tsai, and A. Tomita, *Preparation of ultrafine carbon tubes in nanochannels of an anodic aluminum oxide film*. Chemistry of Materials, 1996. **8**(8): p. 2109–2113.

42 Chen, M., et al., *Electromagnetic interference shielding properties of silicon nitride ceramics reinforced by in situ grown carbon nanotubes*. Ceramics International, 2015. **41**(2): p. 2467–2475.

43 Yin, Z., et al., *Mechanical property and ballistic resistance of graphene platelets/B4C ceramic armor prepared by spark plasma sintering*. Ceramics International, 2019. **45**(17): p. 23781–23787.

44 Nepochatov, Y., et al. *Effect of carbon nanotubes on ballistic armour performance of ceramics from boron carbide*. in *IOP Conference Series: Materials Science and Engineering*. 2020. IOP Publishing.

45 Gao, L., L. Jiang, and J. Sun, *Carbon nanotube-ceramic composites*. Journal of Electroceramics, 2006. **17**(1): p. 51–55.

46 Coble, R.L., *Diffusion models for hot pressing with surface energy and pressure effects as driving forces*. Journal of Applied Physics, 1970. **41**(12): p. 4798–4807.

47 Ahmad, I., et al., *Multi-walled carbon nanotubes reinforced Al2O3 nanocomposites: mechanical properties and interfacial investigations*. Composites Science and Technology, 2010. **70**(8): p. 1199–1206.
48 Zhan, G.-D. and A. Mukherjee, *Processing and characterization of nanoceramic composites with interesting structural and functional properties*. Reviews on Advanced Materials Science, 2005. **10**(3): p. 185–196.
49 Ahmad, I. and M.A. Dar, *Structure and properties of Y2O3-doped Al2O3-MWCNT nanocomposites prepared by pressureless sintering and hot-pressing*. Journal of Materials Engineering and Performance, 2014. **23**(6): p. 2110–2119.
50 Katz, J.D., *Microwave sintering of ceramics*. Annual Review of Materials Science, 1992. **22**(1): p. 153–170.
51 Xie, Z., J. Yang, and Y. Huang, *Densification and grain growth of alumina by microwave processing*. Materials Letters, 1998. **37**(4–5): p. 215–220.
52 Adam, S.F. Microwave Theory and Applications. 1969: Prentice-Hall.
53 Fujitsu, S., M. Ikegami, and T. Hayashi, *Sintering of partially stabilized zirconia by microwave heating using ZnO–MnO2–Al2O3 plates in a domestic microwave oven*. Journal of the American Ceramic Society, 2000. **83**(8): p. 2085–2087.
54 Ahmad, I., et al., *Carbon nanotube toughened aluminium oxide nanocomposite*. Journal of the European Ceramic Society, 2010. **30**(4): p. 865–873.
55 Brosnan, K.H., G.L. Messing, and D.K. Agrawal, *Microwave sintering of alumina at 2.45 GHz*. Journal of the American Ceramic Society, 2003. **86**(8): p. 1307–1312.
56 Walker, L.S., et al., *Toughening in graphene ceramic composites*. ACS nano, 2011. **5**(4): p. 3182–3190.
57 Yazdani, B., et al., *Graphene and carbon nanotube (GNT)-reinforced alumina nanocomposites*. Journal of the European Ceramic Society, 2015. **35**(1): p. 179–186.
58 Sarkar, S. and P.K. Das, *Microstructure and physicomechanical properties of pressureless sintered multiwalled carbon nanotube/alumina nanocomposites*. Ceramics International, 2012. **38**(1): p. 423–432.
59 Dusza, J., et al., *Microstructure and fracture toughness of Si3N4+ graphene platelet composites*. Journal of the European Ceramic Society, 2012. **32**(12): p. 3389–3397.
60 Michálek, M., et al., *Alumina/MWCNTs composites by aqueous slip casting and pressureless sintering*. Ceramics International, 2013. **39**(6): p. 6543–6550.
61 Aguilar-Elguézabal, A. and M. Bocanegra-Bernal, *Fracture behaviour of α-Al2O3 ceramics reinforced with a mixture of single-wall and multi-wall carbon nanotubes*. Composites Part B: Engineering, 2014. **60**: p. 463–470.
62 Michálek, M., et al., *Mechanical properties and electrical conductivity of alumina/MWCNT and alumina/zirconia/MWCNT composites*. Ceramics International, 2014. **40**(1): p. 1289–1295.
63 Yi, J., et al., *A novel processing route to develop alumina matrix nanocompositesreinforced with multi-walled carbon nanotubes*. Materials Research Bulletin, 2015. **64**: p. 323–326.
64 Almeida, V.O., et al., *Enhanced mechanical properties in ZrO2 multi-walled carbon nanotube nanocomposites produced by sol–gel and high-pressure*. Nano-Structures & Nano-Objects, 2015. **4**: p. 1–8.
65 Saheb, N. and K. Mohammad, *Microstructure and mechanical properties of spark plasma sintered Al2O3-SiC-CNTs hybrid nanocomposites*. Ceramics International, 2016. **42**(10): p. 12330–12340.
66 Farahbakhsh, I., Z. Ahmadi, and M.S. Asl, *Densification, microstructure and mechanical properties of hot pressed ZrB2–SiC ceramic doped with nano-sized carbon black*. Ceramics International, 2017. **43**(11): p. 8411–8417.
67 Sun, J., et al., *Multilayer graphene reinforced functionally graded tungsten carbide nanocomposites*. Materials & Design, 2017. **134**: p. 171–180.
68 Shin, J.-H., et al., *Comparative study on carbon nanotube-and reduced graphene oxide-reinforced alumina ceramic composites*. Ceramics International, 2018. **44**(7): p. 8350–8357.

69 Rahman, O.A., et al., *Synergistic effect of hybrid carbon nanotube and graphene nanoplatelets reinforcement on processing, microstructure, interfacial stress and mechanical properties of Al2O3 nanocomposites*. Ceramics International, 2018. **44**(2): p. 2109–2122.
70 Ahmadi, Z., et al., *A novel ZrB2–C3N4 composite with improved mechanical properties*. Ceramics International, 2019. **45**(17): p. 21512–21519.
71 Asl, M.S., et al., *Spark plasma sintering of TiC–SiCw ceramics*. Ceramics International, 2019. **45** (16): p. 19808–19821.
72 Fattahi, M., et al., *Nano-diamond reinforced ZrB2–SiC composites*. Ceramics International, 2020. **46**(8): p. 10172–10179.
73 Fattahi, M., et al., *Strengthening of TiC ceramics sintered by spark plasma via nano-graphite addition*. Ceramics International, 2020. **46**(8): p. 12400–12408.
74 Foong, L.K. and Z. Lyu, *Sintering and mechanical behavior of SiC and WC co-added TiC-based composites densified by hot-pressing*. Ceramics International, 2021. **47**(5): p. 6479–6486.
75 Shah, W., X. Luo, and Y. Yang, *Microstructure, mechanical, and thermal properties of graphene and carbon nanotube-reinforced Al2O3 nanocomposites*. Journal of Materials Science: Materials in Electronics, 2021. **32**(10): p. 13656–13672.
76 Dubary, N., et al., *Influence of temperature on the impact behavior and damage tolerance of hybrid woven-ply thermoplastic laminates for aeronautical applications*. Composite Structures, 2017. **168**: p. 663–674.
77 Tong, L., A.P. Mouritz, and M. Bannister, *3D Fibre Reinforced Polymer* Composites, 2002: Elsevier.
78 Wisnom, M., *The role of delamination in failure of fibre-reinforced composites*. Philosophical Transactions of the Royal Society A: Mathematical, Physical and Engineering Sciences, 2012. **370**(1965): p. 1850–1870.
79 Sharma, S., et al., *Excellent mechanical properties of long multiwalled carbon nanotube bridged Kevlar fabric*. Carbon, 2018. **137**: p. 104–117.
80 Mouritz, A., *Ballistic impact and explosive blast resistance of stitched composites*. Composites Part B: Engineering, 2001. **32**(5): p. 431–439.
81 Shah, S., et al., *Impact resistance and damage tolerance of fiber reinforced composites: A review*. Composite Structures, 2019. **217** : p. 100–121.
82 Tan, K.T., N. Watanabe, and Y. Iwahori, *Impact damage resistance, response, and mechanisms of laminated composites reinforced by through-thickness stitching*. International Journal of Damage Mechanics, 2012. **21**(1): p. 51–80.
83 Tan, K.T., N. Watanabe, and Y. Iwahori, *Experimental investigation of bridging law for single stitch fibre using Interlaminar tension test*. Composite Structures, 2010. **92**(6): p. 1399–1409.
84 Wood, M.D., et al., *A new ENF test specimen for the mode II delamination toughness testing of stitched woven CFRP laminates*. Journal of Composite Materials, 2007. **41**(14): p. 1743–772.
85 Yasaee, M., et al., *Influence of Z-pin embedded length on the interlaminar traction response of multi-directional composite laminates*. Materials & Design, 2017. **115**: p. 26–36.
86 Baker, A.A., *Composite Materials for Aircraft* Structures, 2004: AIAA.
87 Greve, L. and A. Pickett, *Delamination testing and modelling for composite crash simulation*. Composites Science and Technology, 2006. **66**(6): p. 816–826.
88 Seltzer, R., et al., *X-ray microtomography analysis of the damage micromechanisms in 3D woven composites under low-velocity impact*. Composites part A: Applied Science and manufacturing, 2013. **45**: p. 49–60.
89 Francesconi, L. and F. Aymerich, *Numerical simulation of the effect of stitching on the delamination resistance of laminated composites subjected to low-velocity impact*. Composite Structures, 2017. **159**: p. 110–120.
90 Hart, K.R., et al., *Comparison of compression-after-impact and flexure-after-impact protocols for 2D and 3D woven fiber-reinforced composites*. Composites part A: Applied Science and manufacturing, 2017. **101**: p. 471–479.
91 Umer, R., et al., *The mechanical properties of 3D woven composites*. Journal of Composite Materials, 2017. **51**(12): p. 1703–1716.

92 Li, H. and Y. Zhou, Parametric study on the ballistic performance of seamed woven fabrics. Defence Technology, 2022.

93 Wei, Q., et al., *Numerical analysis of strain rate effect on ballistic impact response of multilayer three dimensional angle-interlock woven fabric*. International Journal of Damage Mechanics, 2021. **30**(6): p. 923–944.

94 Sharma, S. and B.P. Singh, *Mechanical Properties of Graphene–Carbon Nanotube Reinforced Hybrid Polymer Nanocomposites*, in *All-carbon Composites and Hybrids*, 2021: Royal Society of Chemistry. p. 278–316.

95 Sharma, S., et al., *Mechanical Properties of CNT Network-Reinforced Polymer Composites*, in *Organized Networks of Carbon Nanotubes,* 2020, CRC Press. p. 75–105.

96 Hanif, W.W., M. Risby, and M.M. Noor, *Influence of carbon nanotube inclusion on the fracture toughness and ballistic resistance of twaron/epoxy composite panels*. Procedia Engineering, 2015. **114**: p. 118–123.

97 Tehrani, M., et al., *Mechanical characterization and impact damage assessment of a woven carbon fiber reinforced carbon nanotube–epoxy composite*. Composites Science and Technology, 2013. **75**: p. 42–48.

98 Sharma, S., et al., *Improved static and dynamic mechanical properties of multiscale bucky paper interleaved Kevlar fiber composites*. Carbon, 2019. **152**: p. 631–642.

99 Garcia, E.J., B.L. Wardle, and A.J. Hart, *Joining prepreg composite interfaces with aligned carbon nanotubes*. Composites part A: Applied Science and Manufacturing, 2008. **39**(6): p. 1065–1070.

100 Zeng, Y., et al., *Design and reinforcement: vertically aligned carbon nanotube-based sandwich composites*. ACS Nano, 2010. **4**(11): p. 6798–6804.

101 Boddu, V.M., et al., *Energy dissipation and high-strain rate dynamic response of E-glass fiber composites with anchored carbon nanotubes*. Composites Part B: Engineering, 2016. **88**: p. 44–54.

102 Steinke, K., L. Groo, and H.A. Sodano, *Laser induced graphene for in-situ ballistic impact damage and delamination detection in aramid fiber reinforced composites*. Composites Science and Technology, 2021. **202**: p. 108551.

103 Wang, S., et al., *Smart wearable Kevlar-based safeguarding electronic textile with excellent sensing performance*. Soft Matter, 2017. **13**(13): p. 2483–2491.

104 Micheli, D., et al., *Ballistic and electromagnetic shielding behaviour of multifunctional Kevlar fiber reinforced epoxy composites modified by carbon nanotubes*. Carbon, 2016. **104**: p. 141–156.

105 Khatiwada, S., C.A. Armada, and E.V. Barrera, *Hypervelocity impact experiments on epoxy/ultra-high molecular weight polyethylene fiber composites reinforced with single-walled carbon nanotubes*. Procedia Engineering, 2013. **58**: p. 4–10.

106 Safamanesh, A., et al., *On the low-velocity and high-velocity impact behaviors of aramid fiber/epoxy composites containing modified-graphene oxide*. Polymer Composites, 2021. **42**(2): p. 608–617.

107 Cao, S., et al., *The CNT/PSt-EA/Kevlar composite with excellent ballistic performance*. Composites Part B: Engineering, 2020. **185**: p. 107793.

108 de Tomasi Tessari, B., et al., *Influence of the addition of graphene nanoplatelets on the ballistic properties of HDPE/aramid multi-laminar composites*. Polymer-Plastics Technology and Materials, 2022. **61**(4): p. 363–373.

109 El Moumen, A., et al., *Dynamic properties of carbon nanotubes reinforced carbon fibers/epoxy textile composites under low velocity impact*. Composites Part B: Engineering, 2017. **125**: p. 1–8.

110 da Silva, A.O., et al., *Effect of graphene oxide coating on the ballistic performance of aramid fabric*. Journal of Materials Research and Technology, 2020. **9**(2): p. 2267–2278.

3 Carbon Nanotubes and Graphene-Based Sensors for Human and Structural Health Monitoring

Tejendra K Gupta, Deepshikha Gupta, Rohit Verma, and Gaurav Kumar

CONTENTS

3.1 INTRODUCTION

Nowadays, sensing devices and systems have become a crucial part of human needs for their involvement in all types of daily life activities and in all job sectors. Therefore, the development of these sensing devices is highly required. The essential feature of a sensor is to transform the input energy or signals into measurable electrical energy or signals. Because of the rapid

DOI: 10.1201/9781003231943-3

growth of electronic device miniaturization, convenient and detailed health monitoring is now possible. Wearable sensors are already on the market and can be incorporated into smartwatches and bracelets. These devices can collect and interpret data any time of day or night in terms of improving the individual's health in terms of various parameters such as: heartbeat [1], blood oxygen level [2], and relaxing patterns [3].

The next generation of these devices is expected to be created as smart attire because of its low power and micro size. As of now, some apparel-based products have been scientifically validated and tested for feasibility, giving a reason to believe that this milestone may be attained soon. Nanocarbon or other nanomaterials have been used in the creation of a significant portion of these e-textiles [4, 5].

Carbon nanotubes (CNTs) and graphene nanostructures are appropriate to use for biosensing applications due to its flexible structure due to light in-weight, electrically conducting, and highly reacting on stimulation. Although there are other exclusive materials which are used for wearable sensors, however, these devices are prone to failure when pulled or twisted due to their discomfort and poor mechanical properties. Considering the foregoing, it is not unexpected that in recent years there has been a growing interest in using CNTs and graphene in smart clothing (Figure 3.1 (a, b)).

Recently, several research studies have demonstrated the use of nanomaterials, such as two-dimensional materials [6, 7], for sensing applications. But the comprehensive summary of the use of CNT and graphene is still required. This chapter describes and covers up-to-date research on the most illustrative advancement in the field of sensors for structural and human health monitoring systems using carbon nanotubes (CNTs) and graphene-based stretchable polymer nanocomposites. We start this chapter with the primary and most feasible methods of synthesis of CNTs and graphene and polymer composite systems, for sensing applications. Detailed discussions and work carried out by other researchers have also been explored in this chapter to discuss the mechanism of strain sensing, and sensitivity. The several applications of these innovative polymer nanocomposites in structural and human health monitoring are also highlighted. Finally, the conclusions and future perspectives of this work with future research directions are also given in order to get the immediate attention for the development in this field.

3.1.1 Structural and Human Health Monitoring

The functionality of a part during operation is monitored in real time using structural and human health monitoring. The structural health monitoring (SHM) system is a tool for assessing and tracking the functioning conditions which are employed in a huge variety of engineering products. These sensors have detected promptly adverse structural changes, that increase the reliability, and govern the stability of structures. In composite materials, the structures are heterogeneous in nature, and damages such as micro-level damages, macro-level damages, and coupled micro- and macro-level damages are generally occurring in composite materials, but these such damages are below the surface of the part or structure. These damages cannot be monitored by visual inspection.

Ultrasonic lamb waves, X-rays, and acoustic emission are some of the technologies used to monitor structural health. Lamb waves are commonly employed in structural health monitoring systems to reflect specific properties in solid materials such as fatigue cracks, voids, and de-bonding. These characteristics necessitate a significant amount of effort in order to reliably and efficiently reconstruct the detected signals. Acoustic emission is more for post-failure diagnosis and determining failure modes, and it entails translating acoustic emissions into specific failure modes. However, separating noise from the actual signal and identifying the number of sensors required to detect sound waves have made acoustic emission a study subject that is still being

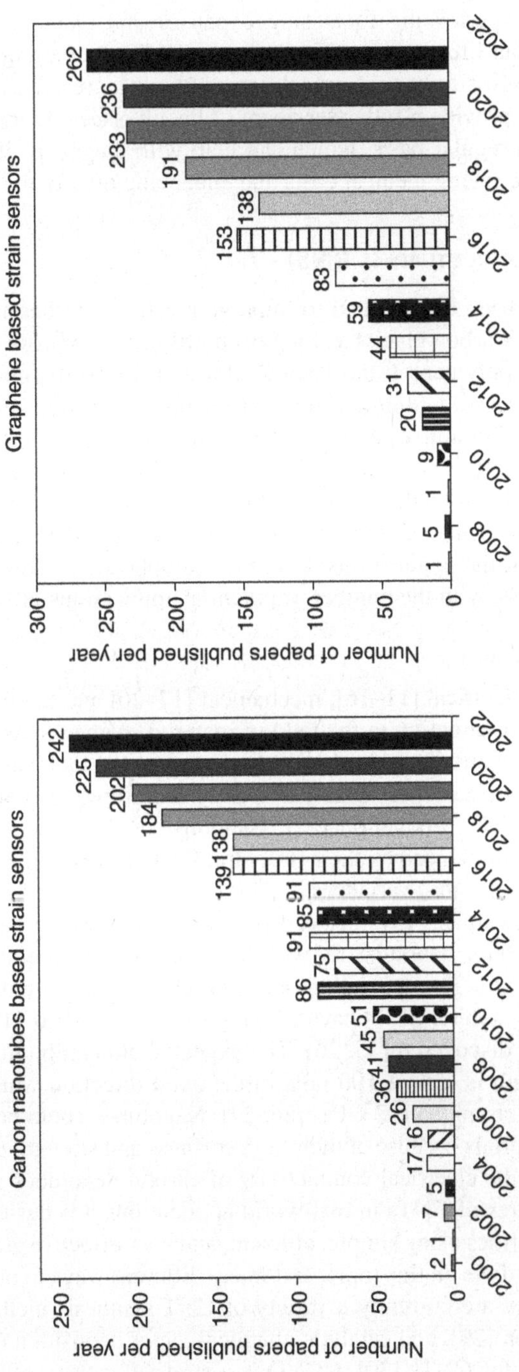

FIGURE 3.1 The number of articles published per year indexed by Scopus database under terms (a) carbon nanotubes and strain sensor, and (b) graphene and strain sensor.

explored. These techniques are both expensive and time-consuming, therefore there is a need for materials and techniques which can detect early-stage damages in structures without losing the inherent required properties.

The healthcare system is experiencing increasing expenses and challenges as the world population grows fast and human life expectancy rises dramatically, prompting governments to discover realistic solutions to provide basic medical care without increasing healthcare costs. Health monitoring systems can provide a full assessment of health issues by tracking important indicators and biomarkers on a regular basis, which can help with diagnosis, illness treatment, and postoperative rehabilitation, saving medical costs and enhancing quality of life.

3.1.2 Carbon-Based Nanostructures (CBNS)

Carbon-based fillers have long been investigated to improve mechanical, thermal and electrical properties in polymer matrices. Carbon black has long been utilized to improve these properties of thermoplastic and thermoset polymers. It has been used to reduce the thermal degradation in racing tires [8]. Carbon fiber is also a common material for producing high-strength and conductive PNCs for aerospace and automotive applications [9, 10]. Carbon nanofibers (CNF), carbon nanotubes (CNTs), graphene, and their derivatives have a special place in nanoscience because of their extraordinary electrical, thermal, chemical, and mechanical capabilities. Composite materials, sensors, energy storage, medication delivery, field emission devices, and nanoscale electronic parts are just a few of the applications. CNTs and graphene are growing rapidly and influencing a wide range of fields, with the number of potential applications growing by the day.

3.1.2.1 Carbon Nanotube (CNT)

Because of their extraordinary electrical [11–16], mechanical [17–20] and thermal [21] capabilities, CNTs have received a lot of attention in the field of material sciences. CNTs have already been marketed in a variety of athletic items, including high-end tennis racquets, golf clubs and baseball bats. CNTs are becoming easier to make these days, therefore they could become one of the main reinforcements in PNC production [22]. Nanotube features such as conductivity, density, and lattice structure, are influenced by their structure. CNTs have distinct band structures and thus variable band gaps, which cause variances in their conducting capabilities [23]. Rolling a single sheet graphene into a seamless cylinder can produce single-walled carbon nanotubes (SWCNTs). A multi-walled carbon nanotube (MWCNT) is a coaxial assembly of SWCNT cylinders, one inside the other [24]. The separation between tubes is almost equal to the separation between layers in natural graphite. CNTs have a Young's modulus of 1.0 TPa, making them the stiffest known material ever discovered [25, 26]. The expected elongation to failure of CNT is 20%–30%, and tensile strength is almost 100 GPa in the axial direction which is far higher than high-strength steel (tensile strength: 1–2 GPa) [26, 27]. Nanotubes could be an appropriate reinforcement in composite materials because of their high stiffness and strength paired with their low density [26]. In this study, the electrical conductivity of carbon nanotubes is the most crucial aspect. To employ the features of CNTs in real-world applications it is desirable to produce high-quality CNTs in large quantities using simple, efficient, and cost-effective growing methods. There has been a lot of research done on this topic, and many different ways to make CNTs have been examined; as a result, there are currently a variety of CNT synthesis methods accessible. Arc discharge [28], laser ablation [29], and catalytic chemical vapor deposition (CCVD) are the three main production methods for CNTs [30]. CCVD is a straightforward, efficient, and low-cost process for producing high-quality carbon nanotubes. A metal catalyst is used to achieve thermal decomposition of a hydrocarbon vapor in this procedure. As a result, it is sometimes called thermal CVD or catalytic CVD. This is a simple process for making CNTs that involves pyrolysis of organometallic compounds like metallocenes in a reactor furnace. Many researchers have already employed various hydrocarbons, catalysts, and inert gases to create CNTs using the CVD process [26, 31, 32].

3.1.2.2 Graphene

Graphene can be considered the generation of all the CBNS due to the sheet like structure, variety of size, morphology and extraordinary properties that are similar to CNTs [33, 34]. Recently, graphene and its derivatives are widely used as 2D CBNS in several applications such as strain sensing, EMI shielding, solar cells, energy efficient devices, lithium-ion batteries, biomedical applications and super capacitors [35–40]. Graphene is a 2D CBNS which is composed of sp^2 bonded carbon atoms with a hexagonal packed lattice structure having many unique properties such as high carrier mobility of ≈10,000 $cm^2 V^{-1}S^{-1}$ at room temperature [41], large specific surface area of 2630 m^2g^{-1} [42], good optical transparency of ≈97.7%, high Young's modulus of ≈1 TPa [43] and high thermal conductivity of 3000–5000 $Wm^{-1}K^{-1}$ [34]. There are various methods for the synthesis of graphene such as mechanical exfoliation, chemical vapor deposition, microwave exfoliation, thermal exfoliation and other chemical routes. However, another chemical route, Hummer's method, is one of the best nominated routes for large-scale synthesis of graphene in the form of graphene oxide (GO), which can be reduced into reduced graphene (RG) or reduced graphene oxide (RGO) via thermal as well as microwave exfoliation [35, 38–40]. The improved Hummer's method is the best suited method for the synthesis of GO [44], in which oxidation reaction of natural graphite flakes is done in the presence of a mixture of conc. H_2SO_4/H_3PO_4. This approach is quite easy, and no hazardous gases are released throughout the preparation process. To avoid an explosion, the $KMnO_4$ was added slowly into sol with constant magnetic stirring [38, 44–46]. The material was ultrasonicated for 8 hours in ethanol to exfoliate graphite oxide into graphene oxide (GO) sheets before being washed and filtered with HCl and water, and then dried in a vacuum oven at 120^{0}C. The exfoliation of the stacked graphene sheets occurs via the emission of carbon dioxide created by heating in a thermal-mediated approach to reduce GO (RGO) at 1000^{0}C. Some vacancies and topological flaws are left in the plane of the RGO platelets during reduction, which are beneficial for good interaction with the polymer matrix [47]. PNC manufacture could benefit from graphene's unique features, such as its flexibility and adjustable structural qualities.

3.2 FABRICATION AND CHARACTERIZATION OF CNT AND GRAPHENE-BASED COMPOSITES SENSORS

Structural health monitoring systems are designed to give an automated and real-time assessment of a structure's ability to perform its intended function. As a result, the advantages of structure health monitoring systems for civil, military, and aerospace applications are required. There are two types of structural health monitoring (SHM): passive and active responses. One of the passive reaction approaches in SHM is condition monitoring. The obtained periodical response will be compared to the regular (default) data to check if any structural integrity-related changes in the signal exist [48–50]. Meanwhile, the active response will monitor the signal and notify the system of these signals changes from the normal signal. Acoustic emission, ultrasonic, fiber optic, laser Doppler vibrometer, and thermal imaging are some of the methods available for detecting damage or failure in a structure using SHM [51]. There are a few common methods which are generally used to synthesize carbon nanostructures reinforced polymer nanocomposites. These methods are discussed below.

3.2.1 Hand Layup Method

Hand layup is one of the simple, open molding, and cost-effective technique for the synthesis of CNT and graphene-reinforced polymer nanocomposite laminates. In this technique, fibers in the form of stitched/woven/bonded fabric are placed in the mold and the CNT/graphene reinforced polymer resin is impregnated in the fiber fabric using a roller. The laminates of polymer nanocomposites fabricated using the hand layup method are cured under the standard given conditions. The polymer resins such as polyester, epoxy, vinyl ester, and phenolic resin are used to prepare the laminates. Large- and small-scale items, such as storage tanks, boats, bathing tubs, and car bumpers can be prepared via this method [52].

3.2.2 Compression Molding Method

In the compression molding method, the CNT/graphene-modified polymer resin was impregnated in fiber fabric to get the prepregs. A vacuum drying oven was used to remove the entrapped air and solvent (if any) from the prepregs. These prepregs were stacked in a desired angle in the metallic mold, and then hot pressed with a hot-pressing machine [53]. This method is used to prepare hybrid polymer nanocomposite laminates, which is shown in Figure 3.2.

Figure 3.2 is showing the dispersion of CNT or graphene in the appropriate solvent and then transferring the dispersion into the polymer solution. A roller is used to coat the polymer resin on the surface of the fiber fabric and the fabric is then cut into small equal size pieces and then stacked and compressed to form a final composite with the desired thickness.

3.2.3 Solution Mixing and Evaporation Method

Solution mixing method is an effective and laboratory-scale method to synthesize CNT/graphene-reinforced polymer nanocomposites. In this method, initially, we dissolve the polymer granules in a suitable solvent, and CNT/graphene are also dispersed in the compatible solvent

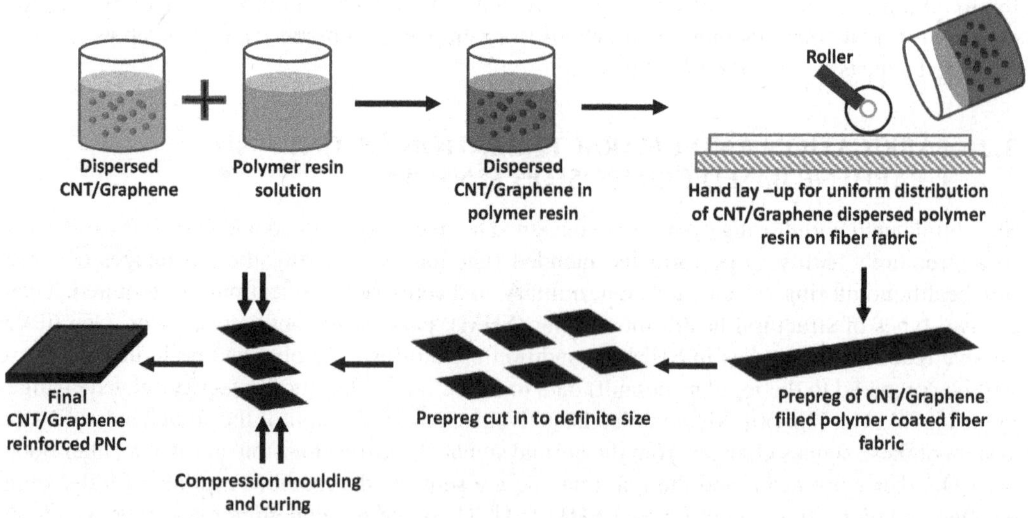

FIGURE 3.2 Schematic of compression molding method of CNT/Graphene reinforced polymer nanocomposite laminates.

separately. Dispersed CNTs/graphene solution is then mixed into the polymer solution. The mixed solution of CNT/graphene-polymer is transferred into the petri dish and evaporates the solvent under vacuum to form a CNT/graphene-reinforced polymer nanocomposite film. Vigorous high-energy ultrasonication, magnetic stirring and high-speed homogenization techniques were tried to disperse the CNT/graphene uniformly into the polymer solution. This method is advantageous to achieve good dispersion and de-bundling of nanosized fillers in the polymer matrix [54]. However, few disadvantages of this method are that the long-time use of ultrasonication and shear mixing can degrade the fillers morphology which may reduce the properties of polymer nanocomposites, and that the large number of solvents can contaminate the environment. Therefore, this technique is not well-suited to a large-scale production, and not good for insoluble polymers.

3.2.4 Extrusion Method

The melt mixing and processing method is the most favorable methodology to synthesize polymer nanocomposites on an industrial level. This method is also suitable for insoluble thermoplastic polymers that cannot be processed via solution mixing methods because they are insoluble in common solvents. In this method, the polymers are melt blended with CNT/graphene by applying the intense shear forces using a twin screw extruder at elevated temperatures, and composites are molded using injection molding into desired shapes [55–58]. High-intensity shear force de-agglomerates the CNT bundles and stacked graphene layers and increases the level of dispersion in polymer matrices. The screws and channels slits of the extrusion system were cleaned properly via passing some amount of polymer formulation, and then, the micro-compounder was set to "cycle" mode, in such a way that all material can be injected into the slit channel. The melt blending of polymers with CNT/graphene can be done in a continuous operation or in batches using a high-speed shear mixer via a twin screw extruder. As compared to a single screw extruder, a twin screw extruder is more effective to blend the fillers into the polymer melts uniformly. The polymer granules are put into the extruder during the operation, and the rotating screw catches and pushes them forward. The filler materials are shear mixed with the help of these two screws, and the temperature inside the heating zone melts the polymer granules. After mixing of the reinforcement such as CNTs and graphene in polymer melts, it reached to the homogenization zone [26]. Finally, the CNT/graphene-reinforced polymer semisolid composite materials are transferred into the desired size and shaped die mold, and cooled by an air-drying process, before being chopped into the granules for further use. The shear force and temperature must be properly optimized to protect the damages in polymer chains and fillers morphology. This is simple, fast, environmentally friendly and an industrial viable technique for the large production of composite materials without the use of solvent. The preparation of nanocomposites by this method is not as effective as the solution-mixing method in the breaking of filler bundles. Also, this technique is not suitable for those materials which degrade with temperature [54].

3.2.5 Vacuum-Assisted Resin Transfer Molding Method

Vacuum-assisted resin transfer molding (VARTM), also called vacuum assisted resin infusion molding (VARIM) [59], is a closed mold method to synthesize the polymer nanocomposites. This is mainly an extension version of the hand layup method, where the pressure is applied to the polymer laminates to improve its consolidation. The wet laid-up laminates are sealed using a plastic film or by taking a sealed polybag. The air is removed from the plastic bag by using a vacuum pump, and the pressure is applied on the wet composite laminates to consolidate it as

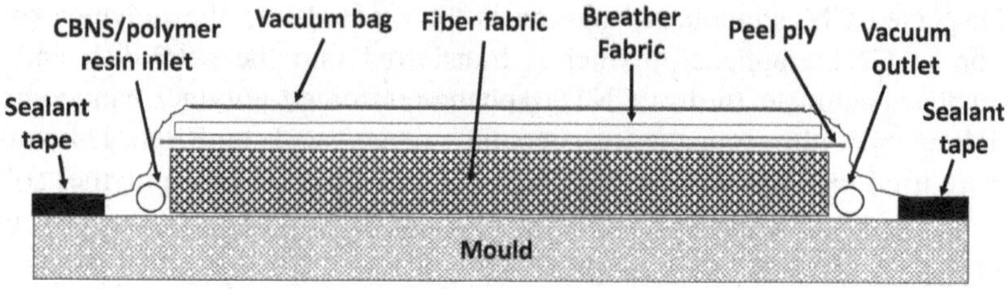

FIGURE 3.3 Schematic of vacuum assisted resin transfer molding method for the synthesis of CNT/graphene reinforced polymer nanocomposites.

shown in Figure 3.3. VARTM is different from the conventional RTM technique, where a vacuum is used to remove the air and other impurities from the mold. A vacuum is also used to pull the reinforced polymer resin through the fabric. The VARTM process includes the use of a vacuum to ease the CNTs/graphene-reinforced polymer resin flow into a stacked fabric placed on a mold and covered by a vacuum bag. The CNTs/graphene reinforced polymer resin is passed into the mold under the pressure difference created by the vacuum. The CNT/graphene-reinforced polymer resin is infused between the fabric fibers until the mold is entirely filled, at which point the infusion process is stopped and the part is cured at a high temperature. Epoxy and phenolic type resins can be used in this process. Polyesters and vinylesters cannot be used due to extraction of styrene from the resin.

3.3 BIOCOMPATIBILITY OF CNT AND GRAPHENE

Graphene, CNTs, and their derivatives are some of the most advanced materials for development of medical devices. Carbon-based nanomaterials (CBNS) are becoming attractive materials due to their inorganic semiconducting capabilities and organic-stacking properties. Due to its efficient interaction with biomolecules and reactivity to light, recent studies have shown the production of carbon-based nanomaterials for the applications in targeting medication delivery, cancer therapy, and biosensors. Before graphene is merged with human skin, particularly when implanted into the human body, its impact such as biocompatibility, toxicity, and potential environmental concerns, is investigated. CBNS' inherent physicochemical features, such as dosage, shape, purity, surface chemistry, layers, thickness, and lateral size, are mostly influenced by their toxicity and have an impact on their biodistribution, translocation to secondary organs, accumulation, degradation, and clearance. The toxicity of CBNS on the immune system and cardiovascular system and organs like skin, has been studied thoroughly. Skin proteins may react with CBNS, causing irritation and allergy symptoms. In vitro assays on cytotoxicity in human skin keratinocytes and fibroblasts revealed that aggregated graphene sheets have higher cytotoxicity over individual graphene sheets. Long term and high concentration exposure of graphene may lead to penetration through keratinocytes leading to plasma membrane damage indicating cytotoxicity to some extent. Graphene-based implants are mainly used in neuroscience, so their effect on neural cells and neuronal networks has been studied. To increase biocompatibility of CBNS, graphene-biomacromolecule hybrids like graphene-biopolymer hybrid, graphene-polysaccharide hybrid has been synthesized to meet growing demands of stable, low toxic material. Also GO and rGO were found to have excellent antibacterial activity too [60].

CNTs have been employed in tissue engineering, drug delivery, gene therapy, and as a biosensor in a variety of medicinal applications. It has a wide range of applications because of its unique properties, such as increased surface area, high stability, and the ability to conjugate with a variety of therapeutic antibodies, DNA, enzymes, and release loaded pharmaceuticals near the targeted cells. Particles with larger surface area to mass ratio with have greater contact area with the cellular membrane, greater absorptivity, and transport across membranes. Long-term exposure to pristine CNT, SWCNT or MWCNT may cause lung inflammation, granuloma formation, skin irritation and cytotoxicity in comparison to carbon black [61]. Cytotoxicity of MWCNT can be assessed by measuring lactase dehydrogenase released at 24 h of incubation, while inflammation potential was assessed by measuring mRNA expression of TNF-α at 6 h of incubation [61]. CNTs have been shown to increase the specificity and sensitivity of antibody-based responses by inducing class II major immune-compatibility complex (MHC). CNT is important for antibacterial activity because it activates the antioxidant glutathione, causing greater oxidative stress in bacterial cells, which causes pathogenic germs to die [62].

3.4 STRUCTURAL HEALTH MONITORING

Due to the stringent safety and performance requirements in aerospace industries for public aircrafts, the current components and structures of composite materials are typically overdesigned, and such designs create the chances of sudden internal failures, which works against the weight reduction goal of using them in the first place. The ability to recognize and monitor the inner damage status, and hence set suitable safety thresholds and limit the danger of catastrophic failures, is a promising technique for better using the potential of FRPs. Despite the fact that structural health monitoring (SHM) is not a new concept in the composites industry, it has recently gained traction. Many concepts have been offered for monitoring the health, which includes revolutionary non-destructive testing methodologies, embedded sensors, and self-sensing systems. The goal of a SHM system, as shown in Figure 3.4, is to mimic the function of the human nervous system, resulting in an in-situ monitoring network of health of the future generation of aircrafts, bridges, and railway tracks.

3.4.1 Carbon Nanotubes for Structural Health Monitoring

Kang et al. [63] has synthesized SWCNTs bucky paper and SWCNTs reinforced polymethyl methacrylate (PMMA) by the solution mixing method, and their strain-sensing properties were

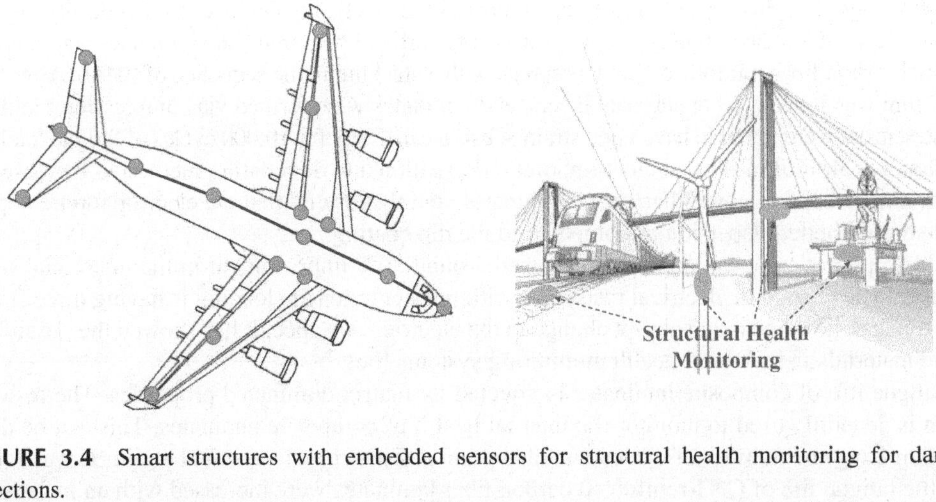

FIGURE 3.4 Smart structures with embedded sensors for structural health monitoring for damage detections.

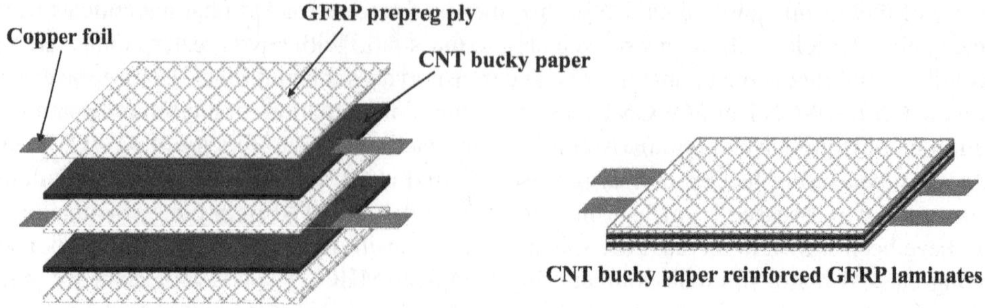

FIGURE 3.5 Schematic of layup sequence of CNT bucky paper in between GFRP sheets, and copper foil is attached on the sides for electrical connections.

measured. In their first study, they dispersed SWCNTs in dimethyl formamide (DMF) solvent at low power bath sonicator, filtered via P8 filter paper, and then dried in a vacuum oven to convert it into the freestanding bucky paper film of 20 μm thickness. In the second study, PMMA polymer was added into the SWCNT dispersion. After complete dissolution, the solvent was evaporated into the vacuum oven, and finally freestanding composite film of SWCNTs/PMMA of ~85 μm was formed. In both samples, piezoresistive strain-sensing properties were measured with respect to the increase in weight % of SWCNTs, and they found that the gauge factor of bucky paper composites was higher as compared to the other sensors in the linear tensile bending and compressive ranges. However, bucky paper may not be suitable to measure the piezoresistive strain response in whole elastic range, while, the composite sensors show the lower value of gauge factor than bucky paper, but in both compressive and bending ranges, they demonstrate a better linear symmetric strain response [63].

Wang et al. [64] has sandwiched CNT bucky paper in between glass fiber-reinforced plastics (GFRP) to make bucky paper reinforced GFRP laminates (see Figure 3.5) and their tensile tests and piezoresistive performance were studies.

According to the findings, bucky paper reinforced GFRP laminates exhibit two separate linear changes in bucky paper resistance as a function of applied strain, and this linear relationship is robust and recoverable in the first sensing stage of 0 to 30,000 με. The bucky paper-reinforced GFRP laminates appear to be suitable for strain-sensing applications in the low strain regime based on these findings [64]. Boztepe et al. [65] has prepared the CNT-film-sandwiched carbon fiber epoxy laminates, in which they initially, prepared cross-ply carbon fiber composite laminates using unidirectional carbon fiber-reinforced epoxy prepregs with a stacking in the sequence of (0/90). After that, CNT film was introduced in alternate layers, and laminates were formed via compression molding. These sensors were found to have good strain-sensing capability for 10,000 cycles of fatigue loading.

Zhang et al. [66] have used electrophoretic deposition and dip coating method to increase the dispersion of CNT in glass fiber/epoxy laminates, and they found that the electrophoretic deposition showed better dispersion as compared to the dip coating.

Electrophoretic deposition method showed around 30% improvement in the interfacial shear strength. The change in electrical resistance with respect to tensile loading is having three stages with a linear, non-linear and abrupt change in the electric resistance. It has showed the possibility of the materials in structural health monitoring systems [66].

Fatigue life of composite laminates is affected by matrix dominated properties. The residual strain is generally used to monitor the internal health of composite laminates. This can be done by comparing the number of cycles and change in resistance with residual strain. It was found that the fatigue life of CNT-reinforced carbon fiber laminates were increased with an increase in the number of cycles as compared to the reference materials [67]. Delamination is also a common failure in composite laminates because laminated behavior of composites has weakness in out-of-plane properties. To improve the resistance during delamination, and the health monitoring of structures, several studies have been done with the addition of CNTs in composite laminates.

Zhang et al. [68] used a direct spray method to deposit CNTs into glass fiber-reinforced plastics through interleaves and monitored the damage sensing. In this method, very low concentration of CNT around ~0.01 wt% was introduced to create a percolation network to monitor the internal health of structures. Fine CNT networks on the surface of glass fibers and in between the glass fibers were observed. Internal propagation of cracks was correlated to measure the changes in electrical signals and reports a good correlation.

Zhang et al. [68] has also used this method with carbon fiber prepregs, and they found good sensing properties. As compared to the carbon fiber laminates, the CNTs reinforced at the interfacial regions of carbon fiber laminates improved the stability of the sensing signals. Such reduction in the scattering of sensing signals is essential for various industrial applications. The sensing mechanism in this scenario shifts from physical contacts between carbon fibers to tunnel current between CNTs in percolated networks in the resin regions, which improves the sensing results' stability and consistency.

3.4.2 Graphene for Structural Health Monitoring

Graphene-based strain sensors are also in demand nowadays. The principle vibrational frequency and electrical conductivity of graphene strongly depend upon its topological structure. This topological structure can be modified by the applied strain, making its usefulness for stretchable sensors with high sensitivity [69]. Nie et al. [69] have prepared a flexible strain sensor, where they initially prepared the graphene film mesh by vacuum filtration of graphene oxide solution, and then, this graphene film mesh is attached on the surface of liquid crystal polymer film and connected via copper wires for electrical contacts. The schematic of the experimental setup is shown in Figure 3.6, where the strain sensor was placed between the gauges. The change in the

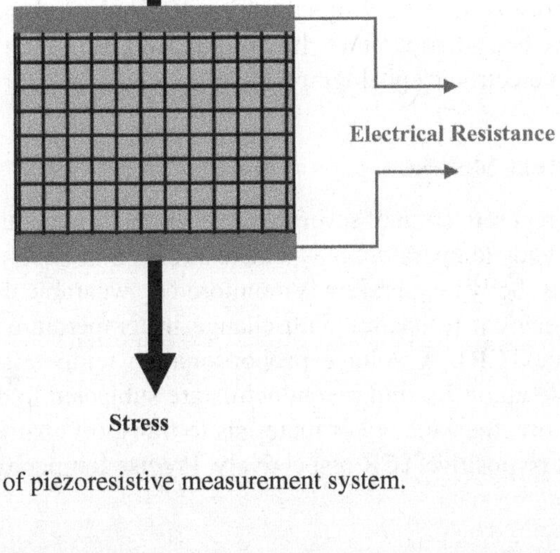

FIGURE 3.6 Schematic of piezoresistive measurement system.

length of sensor with respect to the tensile stress was measured by force gauge, and the relative change in resistance was measured at the same time by the electrical resistance measurement system. The relative change in resistance and strain of graphene film mesh on liquid crystal polymer showed extraordinary performance under tensile loading. To determine the sensitivity, gauge factor (GF) was calculated by taking the slope of the linear curve. The GF was found to be in the range of 375 to 473 in whole strain range, which was much higher compared to previously reported strain sensors [69].

The long cyclic test of graphene film mesh-based strain sensor up to 1000 cycles were performed with 0.16% strain, and it was found that the relative changes in the resistance were always near to 75% with an interval of 100 cycles. This showed the good stability, and reproducibility of the strain sensor. These strain sensors could be useful in monitoring the structural health, such as buildings, bridges, and other infrastructures.

Fiber-reinforced composites have extensively been used to replace the traditional metals in several fields. In the last decade, the use of CNTs and graphene to develop the strain sensors that can detect the early stages of internal damages and the failure of composite structures has been widely studied. The electrons can pass through the structure or materials, due to percolated networks of electrically conductive CNTs and graphene. These movements of electrons provide the possibility to monitor the internal damages and deformation in composite materials using relative change in resistance. The different dispersion methods of CNTs in composite laminates, and synthesis process can provide a significant effect on the feasibility of industrial scale-up. Each synthesis route has its benefits and drawbacks, which might have certain applications in many fields.

3.5 CNTS AND GRAPHENE FOR HUMAN HEALTH MONITORING

CNT is known for its exceptional current carrying capacity, excellent thermal conductivity, tunability and wide band gap, with extraordinary tensile strength and antibacterial properties. These features make it an excellent material to be used in different innovative platforms. Both CNT and graphene are ideal for wearable sensor application due to their flexibility, lightweight, low power input, and wide spectrum to stimuli response. CNT based wearable sensors can be obtained by solubilizing CNT in a medium overcoming Van der Waals interaction between individual CNT. Surface active functional groups on CNT can establish stronger interaction between CNT and solvent. Use of ionic or polymeric surfactants for water-based solutions or solutions in organic solvents may exhibit selective dispersion of CNTs. Due to high current carrying capacity, CNTs can be used to record bioelectrical signals quite efficiently and can be used for monitoring of health parameters like body temperature, blood glucose level, heart rate, electromyography, electrocardiography, and electroencephalography [70].

3.5.1 Body Temperature Sensing

Body temperature is a direct reflection of several health conditions requiring immediate attention. Persistence of very high body temperature may lead to stroke, internal bleeding, and similar emergency conditions. Usually, body temperature is monitored by wearable thermocouple devices that senses the change in electrical resistance with change in temperature (known as Temperature Coefficient of Resistance-TCR). A voltage proportional to temperature difference is formed when two junctions made up of dissimilar conductors are subjected to different temperatures. If during a rise in temperature, the capacity of materials to transport charge increases or decreases, they have negative TCR or positive TCR respectively. Precise temperature measurement is very important in many industrial as well as scientific fields, especially in cryogenics. In the technical field, precise temperature measurement can be done by direct contact of a thermocouple device to the surface or indirectly using IR sensors or fluorescence imaging using temperature sensitive material [70]. CNT-based temperature sensors present good sensitivity at a varied range of temperature. CNT-reinforced PNC structures incorporated with dispersed Fe_3O_4 nanoparticles or gallium, and carbon nanotube wires (CNWs) hybrids can be used for producing CNT-based temperature sensors [70]. Surface type CNT film thermal sensors were developed in which CNT was assumed to be a bulk hetero system, leading to high responsiveness of the CNTs layers due to the effect of temperature, where inter-particles contact areas and intrinsic conductivity of the nanoparticles increase, resulting in increased conductivity and decreased resistance of the samples as temperature rises. [71]. Silica based supports with lithographically printed Au or Pt electrodes may provide platform for CNT based sensors. Textile based temperature sensor using multiwalled CNT and Nylon 6 with polypyrrole coating were synthesized by Blasdel and Monty [72]. Such sensors were water resistant and skin unfriendly. To overcome this problem multiwalled CNT and cotton-based wearable sensors were fabricated to produce e-Textile having good mechanical and electrical conductivity. Another silk-based biodegradable and breathable and durable fabric with CNT and an ionic liquid was reported by Wu et al. having temperature sensitivity of 1.23% $°C^{-1}$ [73].

3.5.2 Non-Invasive Blood Glucose Sensing

Diabetic patients need continuous monitoring of blood glucose levels for which the invasive needle method is commonly practiced. Such a method is not comfortable with a large proportion of the population; therefore a non-invasive detection method is welcomed. In this regard, CNT-based non-invasive glucose sensors are being developed that can be simply implanted or placed on the skin to check the glucose level in body fluids. Non-invasive glucose sensor type devices can check glucose levels in body fluids like tears, sweat, saliva or interstitial fluid. Lower concentrations of glucose can also be detected by using glucose oxidase (Gox) immobilized on CNT enzymatic biosensors [74]. Changes in electrochemical characteristics of the sensor is observed upon oxidation of glucose to gluconic acid [5]. Gox oxidizes glucose to yield gluconic acid and hydrogen peroxide (H_2O_2), as mentioned in the equation below [75].

$$Gluose + H_2O + O_2 \xrightarrow{GOx} Gluconic\ acid + H_2O_2$$

Kang, Park, and Ha [76] have reported a SWCNT-based wearable glucose sensor, which can detect glucose down to 50 μM with a response time of 5 s, sensing mechanism through molecular oxidation-reduction reactions in (Gox) with glucose [75].

3.5.3 ECG/EMG/EEG Interface

The electrocardiograph (ECG) can detect the DC component up to 300 mV (due to electrode-skin contact) and it can also detect very weak signals from 0.5 mV to 5.0 mV and a common-mode component up to 1.5 V (due to the potential between the electrodes and the ground) [77]. CNTs have applications in this industry because of their excellent electrical conductivity high flexibility and low mass density, which reduces the device's weight substantially. To make the device more user-friendly, ensuring proper skin adhesion, the nanocarbon component is mixed with a flexible polymer matrix like polyurethane or polydimethylsiloxane. Chi and co-workers [78] showed that traditional Ag/AgCl electrodes can be outperformed by CNT-based sensors for daily monitoring of ECG signals.

Electroencephalography (EEG) is employed to record real-time brain activity through scalp-based electrodes for measurement of biopotential. It can also track brain disorders like epilepsy or brain tumors. Hairs on the head decrease the signal of the EEG. Multi-walled CNT electrodes or Ag/AgCl blocks receive X-rays, thereby creating artifacts on Computed Tomography (CT) images during surgery, complicating the procedure [79]. CNT-based dry electrodes do not have any side effects such as itching, redness, or irritation six months after their application when directly exposed to skin.

The well-being of muscles and nerve cells controlling them can be monitored by electromyography (EMG), a non-invasive technique. Even in the presence of wrinkles in the skin, a combination of carbon nanotubes (for monitoring) and polydimethylsiloxane (for adhesion) can reliably detect bioelectrical signals. There was no need for gel or glue to electrically link the skin to the electrodes, allowing for comfortable, continuous, and long-term monitoring. CNT sensors on the forearm performed similarly to Ag/AgCl electrodes, which required the use of a conductive gel sheet. To avoid injury, ECG textile sensors allow you to monitor the status of your muscles as you exercise. Yarns composed of CNT-coated polyurethane nanofibers which may be woven effectively to make a piezoresistive sensor, are used to sense wrist bending, cheek bulging, swallowing, and breathing [80].

3.5.4 Graphene for Human Health Monitoring

Every second, the human body creates a slew of physiological signals that might reveal the state of the body. The amount and quality of physiological signals are critical for diagnosing and administering treatments. The signals obtained by rigid silicon-based sensors do not bind effectively with the body due to the poor contact between the sensors and the skin, lowering the signal quality. Wearable and implanted sensors have higher sensitivity and specificity, making them more efficient. Electronic transducers that are ultrathin, flexible, conformal, and skin-like are emerging as attractive alternatives for non-invasive and nonintrusive human health monitoring. However, to obtain an affordable sensor few constraints like biocompatibility, biofouling, stability, comfort and convenience, low cost, small size real-time monitoring must be kept into mind. Monolayer, multilayers, and graphene-like structures such as graphene oxide, and reduced graphene oxide are the most common types of graphene. Due to its good mechanical qualities, wide surface area, and excellent electrochemical activity, graphene is an attractive substrate, which can be used in making high sensitivity sensing and biosensing devices. The analytical performances of graphene-based materials as sensors and biosensors for glucose, cholesterol, dopamine (DA), ascorbic acid (AA), uric acid (UA), bisphenol A, cancer biomarkers, and heavy metal ions were studied. Because of its great sensitivity relative to silicon, broad working range, extraordinary flexibility, and inexpensive cost, graphene offers a substantial advantage over traditional materials in producing flexible electronics and wearable sensors [81]. The quality and quantity of the physiological signals can be improved by using wearable graphene sensors that have great potential for healthcare and telemedicine in the future. The simple assembly of graphene structures such as 1D fiber-like, 2D film, and 3D foam-like structures can ensure that E-skins can function as flexible touch sensors [82]. Also, the ultrathin dimensions of graphene, high optical transparency and good electrical conductivity make graphene an excellent material for wearable sensor applications. The cytotoxicity of graphene-based nanomaterials cannot be ruled out for long-term usage in the form of wearable or implantable sensors due to reaction with skin proteins or fibroblasts. But the use of graphene-biopolymer nanohybrids with biocompatible polysaccharides can minimize toxicity and enhance solubility and stability, making the material biodegradable too.

Graphene-based sensors can be designed in both invasive and non-invasive devices. Researchers developed a graphene-based transparent tattoo that can be applied to the skin directly with water, similar to a temporary tattoo. Graphene-based electronic sensors have a thickness of 450–490 nm with 40% stretchability and excellent skin adhesion. The graphene electronic tattoo is like the commercially available electronic health and fitness trackers in that it can monitor heart rate and detect bioimpedance (the body's response to an electric current). However, because the ultrathin graphene tattoos can fully adhere to the skin, they provide medical-grade data quality, as opposed to the rigid electrode sensors put on bands and fastened to the wrist or chest, which have a lesser performance. The researchers believe that due to the high-quality sensing, graphene tattoos could be a feasible replacement for present pharmaceutical sensors in contact mode with skin which requires a lubricant for smooth functioning. The bilayer of graphene/PMMA as a small portion is transformed into regular tattoo paper during production. This layer is subsequently imprinted into various serpentine ribbon forms to create several bunches of sensors. Then, the final tattoo is placed on any region of the human skin by touching the graphene side of the tattoo paper and releasing the tattoo with water on the reverse of the tattoo paper. The tattoos last for two days or longer and even pull off. The graphene-PMMA tattoo has a thickness of 460 nm with 85% optical transparency and more stretchability than the human body. Thus, the graphene tattoos are just appreciable due to their mechanical and optical strength. These graphene-based electronic tattoos are used to measure the electrical pulse of the heart and activities of muscles and brain, and also bring in various pharmaceutical measurements as an electrocardiogram (ECG), and electroencephalogram (EEG) [60, 83]. The graphene-based composite materials for human health monitoring such as implantable devices, for the spontaneous measurement of temperature of body, pulse rate and monitoring the heartbeat, pressure, glucose level monitoring, ECG, EMG, and EEG signal detection are shown in Figure 3.7.

3.5.5 Wearable Strain Sensors

Wearable strain sensors are critical in the development of artificially intelligent devices in the future. It is challenging to strike a suitable balance between required performance and ease of manufacture for strain sensors. Nowadays, CNTs and graphene are the fast developing materials for biosensing applications in human movement detection, stretchable wearable sensors, health care, robotics, etc. [84–86]. In this part of research, we provide a flavor of how these CNTs and graphene can shape the current research in these areas. These kinds of wearable sensing devices provide a connection between humans and technology. Kedambaimoole et al. [87] prepared a simple and low-cost reduced graphene oxide tattoo for proximity sensors. In this work, they have synthesized graphene oxide by the chemical route and then the freestanding GO film was prepared via the vacuum filtration method, which is coated with PDMS polymer. The detailed synthesis method has been discussed elsewhere [87]. The electrostatic potential is applied on the sensor using a point probe through an electrostatic voltage source in a controlled manner. The contact probes of the sensor were connected to the electrometer to monitor the variation in resistance. The potential was applied in vice-versa direction (positive and negative) to examine the relative change in resistance. The sensor showed the high response in resistance of up to 208% at 6 kV while applying negative potential with the step height of 1 kV. It was found that the sensor presented the higher response for negative potential than the positive potential, which may be due

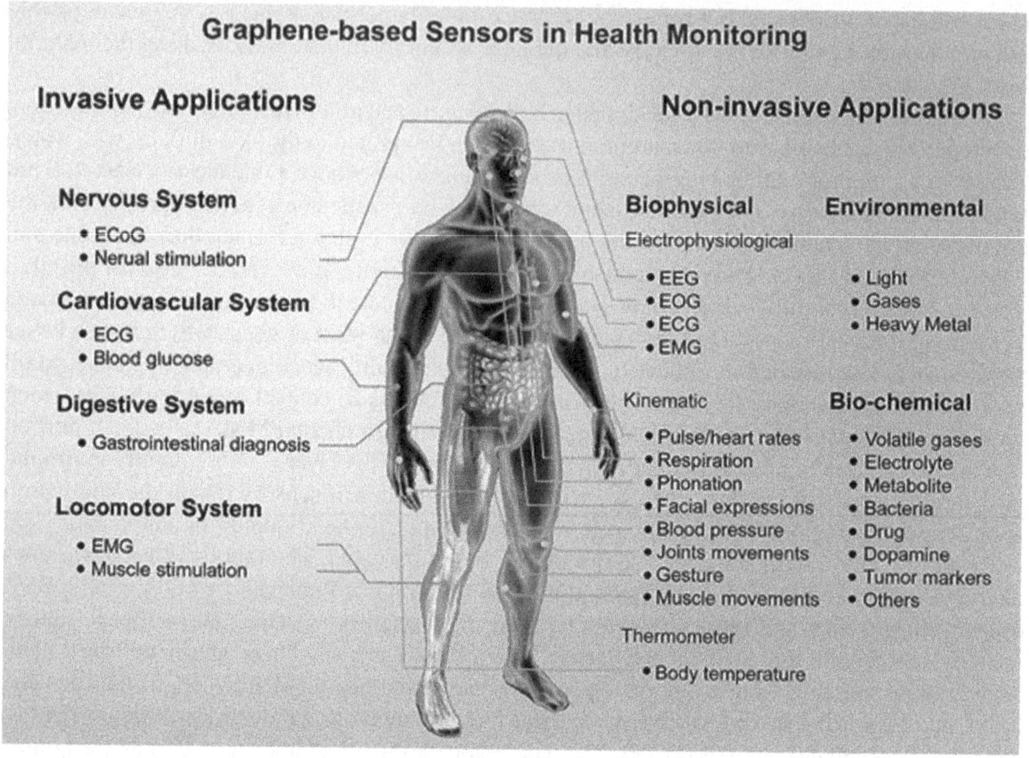

FIGURE 3.7 Graphene-based composite material sensors for human motion detection in invasive, and non-invasive applications. Reproduced with permission from [60] (Copyright under under the terms of the Creative Commons Attribution License (CC BY)).

to the presence of holes as majority carriers. The sensor has also showed excellent reproducibility at −2 kV potential with respect to 100 cycles repetition of static voltage, when the distance is fixed at 1 cm between the sensing layer and probe tip [87].

These flexible pressure sensors could be used in touch panels, human motion detection, artificial e-skins, etc. Due to low sensitivity and rigidness, uses of metal-based sensors are limited. Therefore, graphene-based strain sensors can be used to overcome such challenges. Chen et al. [87] have developed the percolated graphene film on flexible polyethylene terephthalate (PET) substrate via the spray deposition method, and they recorded the change in resistance with respect to applied bending force. The gauge factor of each percolated graphene film was calculated using $\frac{\Delta R}{R.\varepsilon}$, where $\frac{\Delta R}{R}$ was the relative change in electrical resistance, and ε was the mechanical strain [87].

From this study, it was found that the size of graphene sheets affected the gauge factor [87]. The gauge factor was found to be larger with a smaller size of graphene flakes because of presence of dominated out-of-plane conducting paths in the percolated graphene network. Furthermore, the two-fold enhancement in the gauge factor was the result of the addition of surfactant molecules in the graphene solution. The surfactant molecules enhance the out-of-plane resistance and also improve the uniformity of percolated graphene film. The array of percolated graphene film showed excellent flexibility, transparency, and good pressure response, and 0 to

60 kPa pressure with good reproducibility was monitored using the percolated graphene array network. This range shows the pressure from very low (gentle touch) to high (normal person's weight). The details of the fabrication process and the measurement of change in resistance with respect to applied pressure are given in [87].

Arif et al. [86] has reported the strain sensors of carbon nanostructure (CNS)-reinforced PDMS nanocomposites. The composites of PDMS with CNS were prepared via the high-speed homogenization technique, and the electrical percolation of 0.05 wt% was achieved. The tensile specimens were cut into the desired size, and the experimental setup for piezoresistive measurement of the sample specimen was designed as shown in Figure 3.8 (a).

The classical percolation theory was used to determine the electrical percolation, which is given as [86];

$$\sigma = \begin{cases} 0 & if\ \phi < \phi_c \\ \sigma_0 (\phi - \phi_c)^t & if\ \phi \geq \phi_c \end{cases}$$

Where, σ represents electrical conductivity, ϕ is the wt% of CNS, ϕ_c is the electrical percolation threshold, and t is the universal critical exponent, that shows the dimensional state of the conductive path. The values of ϕ_c and t can be determined by the non-linear fitting of electrical conductivity curve using power low. CNS-reinforced PDMS nanocomposites showed strong linear piezoresistive response up to 110% strain with a high gauge factor of 8 to 47, which is shown in Figure 3.8 (b).

The resistance change in nanocomposite samples was increased during stretching and decreased during releasing of the tensile load at 30% strain up to 100 cycles. The stress relaxation occurs as the tensile load increased and decreased, however, the decrease in stress was very small. The hysteresis loop was constant during cyclic loading and unloading, and the composite samples showed the hyper-elastic behavior of CNS/PDMS nanocomposites with good stability

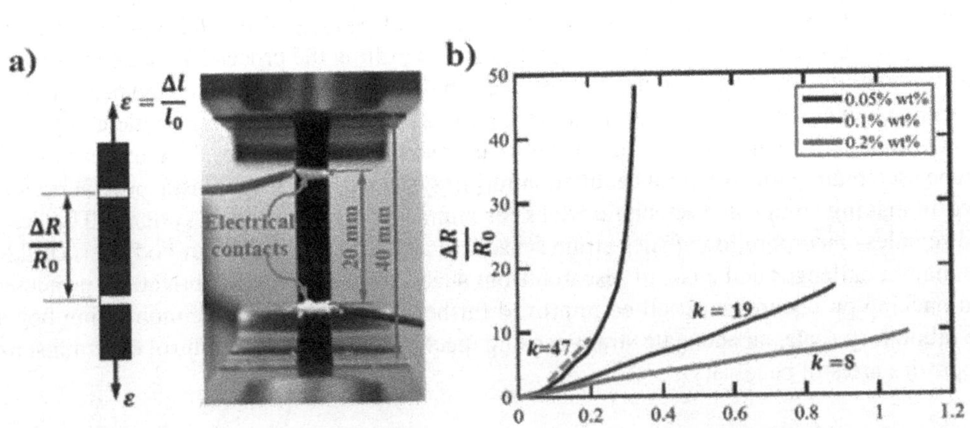

FIGURE 3.8 (a) Experimental setup for tensile testing and piezoresistive measurement, and (b) Curve of normalized change in resistance versus strain with respect to different wt% of CNS loading. Reprinted from [86] Copyright (2021), with permission from Elsevier.

and reproducibility up to 100 cycles for 30% strain. These composites found applications in making wearable and health fitness devices [86].

In a study of Kumar et al. [84], stretchable and ultrasensitive nanocomposites of thermoplastic polyurethane (TPU) with MWCNTs were prepared via the solution mixing method, and their relative change in resistance with respect to tensile loading was evaluated. The ultra-high stretchable TPU nanocomposite films up to 200% strain showed low electrical percolation network at 0.1 wt% MWCNT loading, and excellent piezoresistive response in elastic and inelastic regimes. These TPU nanocomposite sensing materials demonstrated the ultra-high sensitivity with stretchability in both small and large strain regimes up to 200% strain, and such strain-sensing materials could be useful in making sensing devices for human motion detection [84].

Gupta et al. [35] has synthesized biomedical grade ultra-high molecular-weight polyethylene (UHMWPE), reinforced with CNT and graphene nanoplatelets (GNP). Their self-sensing performance was evaluated. The hybrid fillers increased the dispersion in polymer matrix and lowered the probability of CNT entanglement, because GNP acts as a bridge between CNTs to separate the entanglement between CNTs. The synergistic effect of CNT/GNP created the more conductive network at low electrical percolation and improved the mechanical and electrical performance of these nanocomposites. These CNT/GNP-reinforced UHMWPE nanocomposites showed extraordinary piezoresistive response with a high gauge factor in elastic and plastic regimes. Such nanocomposite materials could be useful in orthopedic implant materials.

3.6 CONCLUSIONS AND FUTURE PERSPECTIVES

The present chapter highlights the synthesis methods of CNT and graphene, preparation of strain-sensing materials, measurement of strain sensing performance of these materials, and the usefulness of these materials in structural and human health monitoring. The important features of the perfect strain sensor are the gauge factor, stretchability, linearity, biocompatibility, reproducibility, and hysteresis. The gauge factor can be optimized by the incorporation of an appropriate amount of conductive fillers, controlling the conductive networks, and using the best processing methodologies. Stretchability could be achieved by selecting the stretchable materials such as polymers; conductive networks could be created by addition of conductive fillers; and reproducibility could be controlled by controlling the processing parameters. From the above study, it is found that the carbon nanotubes and graphene could be the perfect choice for making such strain-sensing devices, because CNT and graphene can provide a good electrically conductive path along with extraordinary mechanical properties. Thermoset polymers can be used mainly for structural health monitoring systems, and thermoplastic polymers can be used in making stretchable sensing devices for human health monitoring systems. The design and seamless incorporation of such strain sensors in any structure, fabric, or body part could be the major challenges and areas of research from such sensing devices. Fabrication, protection, and packing procedures must all be improved further. To enable real-time monitoring beyond the laboratory scale, an adequate strain-sensing mechanism and proper control of strain sensor properties are still necessary.

ACKNOWLEDGMENTS

The authors are grateful to Amity Institute of Applied Sciences, Amity University Uttar Pradesh, Noida, India for encouragement. During this work, the support of the family members and kids is highly appreciable.

REFERENCES

1 Marín-Morales, J., et al., *Affective computing in virtual reality: emotion recognition from brain and heartbeat dynamics using wearable sensors*. Scientific Reports, 2018. **8**(1): p. 13657.

2 Davies, H.J., et al., *In-ear SpO2: A tool for wearable, unobtrusive monitoring of core blood oxygen saturation*. Sensors, 2020. **20**(17): p. 4879.

3 Saleem, K., et al., *IoT healthcare: Design of smart and cost-effective sleep quality monitoring system*. Journal of Sensors, 2020. **2020**: p. 8882378.

4 Shim, B.S., et al., *Smart electronic yarns and wearable fabrics for human biomonitoring made by carbon nanotube coating with polyelectrolytes*. Nano Letters, 2008. **8**(12): p. 4151–4157.

5 Rdest, M. and D. Janas, *Carbon nanotube wearable sensors for health diagnostics*. Sensors, 2021. **21**(17): p. 5847.

6 Ma, C., et al., *Flexible MXene-based composites for wearable devices*. Advanced Functional Materials, 2021. **31**(22): p. 2009524.

7 Pang, Y., et al., *Wearable electronics based on 2D materials for human physiological information detection*. Small, 2020. **16**(15): p. 1901124.

8 Hochi, K. and K. Uesaka, *Rubber composition and tire having tread comprising thereof*, 2006, Google Patents.

9 Das, T.K., P. Ghosh, and N.C. Das, *Preparation, development, outcomes, and application versatility of carbon fiber-based polymer composites: a review*. Advanced Composites and Hybrid Materials, 2019: p. 1–20.

10 Rajak, D.K., et al., *Fiber-reinforced polymer composites: Manufacturing, properties, and applications*. Polymers, 2019. **11**(10): p. 1667.

11 Peng, L.-M., Z. Zhang, and S. Wang, *Carbon nanotube electronics: recent advances*. Materials Today, 2014. **17**(9): p. 433–442.

12 Nilsson, J., et al., *Electronic properties of graphene multilayers*. Physical Review Letters, 2006. **97**(26): p. 266801.

13 Li, X., et al., *Structural and electrical properties tailoring of carbon nanotubes via a reversible defect handling technique*. Carbon, 2018. **133**: p. 186–192.

14 Pokharel, P., et al., *A hierarchical approach for creating electrically conductive network structure in polyurethane nanocomposites using a hybrid of graphene nanoplatelets, carbon black and multi-walled carbon nanotubes*. Composites Part B: Engineering, 2019. **161**: p. 169–182.

15 Han, S., et al., *Mechanical and electrical properties of graphene and carbon nanotube reinforced epoxy adhesives: experimental and numerical analysis*. Composites Part A: Applied Science and Manufacturing, 2019. **120**: p. 116–126.

16 Jyoti, J., et al., *Dielectric and impedance properties of three dimension graphene oxide-carbon nanotube acrylonitrile butadiene styrene hybrid composites*. Polymer Testing, 2018. **68**: p. 456–466.

17 Ruoff, R.S., and D.C. Lorents, *Mechanical and thermal properties of carbon nanotubes*. Carbon, 1995. **33**(7): p. 925–930.

18 Han, Z., and A. Fina, *Thermal conductivity of carbon nanotubes and their polymer nanocomposites: A review*. Progress in Polymer Science, 2011. **36**(7): p. 914–944.

19 Cha, J., et al., *Comparison to mechanical properties of epoxy nanocomposites reinforced by functionalized carbon nanotubes and graphene nanoplatelets*. Composites Part B: Engineering, 2019. **162**: p. 283–288.

20 Milowska, K.Z., et al., *Carbon nanotube functionalization as a route to enhancing the electrical and mechanical properties of Cu–CNT composites*. Nanoscale, 2019. **11**(1): p. 145–157.

21 Zare, Y., and K.Y. Rhee, *Following the morphological and thermal properties of PLA/PEO blends containing carbon nanotubes (CNTs) during hydrolytic degradation*. Composites Part B: Engineering, 2019. **175**: p. 107132.

22 Dai, H., *Carbon nanotubes: synthesis, integration, and properties*. Accounts of Chemical Research, 2002. **35**(12): p. 1035–1044.

23 Mittal, V., *Polymer Nanotube Nanocomposites Synthesis, Properties, and Applications*. First ed2010: Scrivener Publishing LLC.

24 Rao, C.N.R., R. Voggu, and A. Govindaraj, *Selective generation of single-walled carbon nanotubes with metallic, semiconducting and other unique electronic properties*. Nanoscale, 2009. **1**(1): p. 96–105.

25 Yu, M.-F., et al., *Tensile Loading of Ropes of Single Wall Carbon Nanotubes and their Mechanical Properties*. Physical Review Letters, 2000. **84**(24): p. 5552–5555.

26 Gupta, T.K., and S. Kumar, *Fabrication of carbon nanotube/polymer nanocomposites*, in *Carbon Nanotube-Reinforced* Polymers, 2018: Elsevier. p. 61–81.

27 Yakobson, B.I., C.J. Brabec, and J. Bernholc, *Nanomechanics of carbon tubes: Instabilities beyond linear response*. Physical Review Letters, 1996. **76**(14): P. 2511–2514.

28 Journet, C., et al., *Large-scale production of single-walled carbon nanotubes by the electric-arc technique*. Nature, 1997. **388**(6644): p. 756–758.

29 Rinzler, A., et al., *Large-scale purification of single-wall carbon nanotubes: process, product, and characterization*. Applied Physics A: Materials Science & Processing, 1998. **67**(1).

30 Ren, Z., et al., *Synthesis of large arrays of well-aligned carbon nanotubes on glass*. Science, 1998. **282**(5391): p. 1105–1107.

31 Garg, P., et al., *Effect of dispersion conditions on the mechanical properties of multi-walled carbon nanotubes based epoxy resin composites*. Journal of Polymer Research, 2011. **18**(6): p. 1397–1407.

32 Gupta, T.K., et al., *Improved nanoindentation and microwave shielding properties of modified MWCNT reinforced polyurethane composites*. Journal of Materials Chemistry A, 2013. **1**(32): p. 9138–9149.

33 Muschi, M. and C. Serre, *Progress and challenges of graphene oxide/metal-organic composites*. Coordination Chemistry Reviews, 2019. **387**: p. 262–272.

34 Balandin, A.A., et al., *Superior thermal conductivity of single-layer graphene*. Nano Letters, 2008. **8**(3): p. 902–907.

35 Gupta, T.K., et al., *Self-sensing and mechanical performance of CNT/GNP/UHMWPE biocompatible nanocomposites*. Journal of Materials Science, 2018. **53**(11): p. 7939–7952.

36 Gupta, T.K., et al., *Multi-walled carbon nanotube-graphene-polyaniline multiphase nanocomposite with superior electromagnetic shielding effectiveness*. Nanoscale, 2014. **6**(2): p. 842–851.

37 Gupta, T.K., et al., *MnO2 decorated graphene nanoribbons with superior permittivity and excellent microwave shielding properties*. Journal of Materials Chemistry A, 2014. **2**(12): p. 4256–4263.

38 Gupta, T.K., et al., *Superior nano-mechanical properties of reduced graphene oxide reinforced polyurethane composites*. RSC Advances, 2015. **75**(22): p. 16921–16930.

39 Jyoti, J., et al., *Synergetic effect of graphene oxide-carbon nanotube on nanomechanical properties of acrylonitrile butadiene styrene nanocomposites*. Materials Research Express, 2018. **5**(4): p. 045608.

40 Umrao, S., et al., *Microwave-Assisted Synthesis of Boron and Nitrogen co-doped Reduced Graphene Oxide for the Protection of Electromagnetic Radiation in Ku-Band*. ACS applied materials & interfaces, 2015. **7**(35): p. 19831–19842.

41 Novoselov, K.S., et al., *Electric field effect in atomically thin carbon films*. Science, 2004. **306**(5696): p. 666–669.

42 Stoller, M.D., et al., *Graphene-based ultracapacitors*. Nano Letters, 2008. **8**(10): p. 3498–3502.

43 Lee, C., et al., *Measurement of the elastic properties and intrinsic strength of monolayer graphene*. Science, 2008. **321**(5887): p. 385–388.

44 Marcano, D.C., et al., *Improved synthesis of graphene oxide*. ACS Nano, 2010. **4**(8): p. 4806–4814.

45 Dreyer, D.R., et al., *The chemistry of graphene oxide*. Chemical Society Reviews, 2010. **39**(1): p. 228–240.

46 Garg, B., T. Bisht, and Y.-C. Ling, *Graphene-based nanomaterials as heterogeneous acid catalysts: A comprehensive perspective*. Molecules, 2014. **19**(9): p. 14582–14614.

47 Nakajima, T., A. Mabuchi, and R. Hagiwara, *A new structure model of graphite oxide*. Carbon, 1988. **26**(3): p. 357–361.

48 Tandon, N., and A. Choudhury, *A review of vibration and acoustic measurement methods for the detection of defects in rolling element bearings*. Tribology International, 1999. **32**(8): p. 469–480.

49 Betz, D.C., et al., *Structural damage location with fiber bragg grating rosettes and lamb waves*. Structural Health Monitoring, 2007. **6**(4): p. 299–308.
50 Li, D., et al., *A review of damage detection methods for wind turbine blades*. Smart Materials and Structures, 2015. **24**(3): p. 033001.
51 Ciang, C.C., J.-R. Lee, and H.-J. Bang, *Structural health monitoring for a wind turbine system: A review of damage detection methods*. Measurement Science and Technology, 2008. **19**(12): p. 122001.
52 Elkington, M., et al., *Hand layup: understanding the manual process*. Advanced Manufacturing: Polymer & Composites Science, 2015. **1**(3): p. 138–151.
53 Han, X., et al., *Effect of graphene oxide addition on the interlaminar shear property of carbon fiber-reinforced epoxy composites*. New Carbon Materials, 2017. **32**(1): p. 48–55.
54 Kim, H., Y. Miura, and C.W. Macosko, *Graphene/polyurethane nanocomposites for improved gas barrier and electrical conductivity*. Chemistry of Materials, 2010. **22**(11): p. 3441–3450.
55 Lee, J.-H., S.K. Kim, and N.H. Kim, *Effects of the addition of multi-walled carbon nanotubes on the positive temperature coefficient characteristics of carbon-black-filled high-density polyethylene nanocomposites*. Scripta Materialia, 2006. **55**(12): p. 1119–1122.
56 Kalaitzidou, K., H. Fukushima, and L.T. Drzal, *A new compounding method for exfoliated graphite–polypropylene nanocomposites with enhanced flexural properties and lower percolation threshold*. Composites Science and Technology, 2007. **67**(10): p. 2045–2051.
57 Wang, W.-P., and C.-Y. Pan, *Preparation and characterization of polystyrene/graphite composite prepared by cationic grafting polymerization*. Polymer, 2004. **45**(12): p. 3987–3995.
58 Cooper, C.A., et al., *Distribution and alignment of carbon nanotubes and nanofibrils in a polymer matrix*. Composites Science and Technology, 2002. **62**(7): p. 1105–1112.
59 Sánchez, M., et al., *Effect of the carbon nanotube functionalization on flexural properties of multiscale carbon fiber/epoxy composites manufactured by VARIM*. Composites Part B: Engineering, 2013. **45**(1): p. 1613–1619.
60 Huang, H., et al., *Graphene-based sensors for human health monitoring*. Frontiers in Chemistry, 2019. **7**(399).
61 Smart, S.K., et al., *The biocompatibility of carbon nanotubes*. Carbon, 2006. **44**(6): p. 1034–1047.
62 Rajakumar, G., et al., *Current use of carbon-based materials for biomedical applications—A prospective and review*. Processes, 2020. **8**(3): p. 355.
63 Kang, I., et al., *A carbon nanotube strain sensor for structural health monitoring*. Smart Materials and Structures, 2006. **15**(3): p. 737–748.
64 Wang, X., et al., *Tensile strain sensing of buckypaper and buckypaper composites*. Materials & Deslgn, 2015. **88**: p. 414–419.
65 Boztepe, S., et al., *Novel carbon nanotube interlaminar film sensors for carbon fiber composites under uniaxial fatigue loading*. Composite Structures, 2018. **189**: p. 340–348.
66 Zhang, J., et al., *Functional interphases with multi-walled carbon nanotubes in glass fibre/epoxy composites*. Carbon, 2010. **48**(8): p. 2273–2281.
67 Böger, L., et al., *Improvement of fatigue life by incorporation of nanoparticles in glass fibre reinforced epoxy*. Composites Part A: Applied Science and Manufacturing, 2010. **41**(10): p. 1419–1424.
68 Zhang, H., et al., *Improved fracture toughness and integrated damage sensing capability by spray coated CNTs on carbon fibre prepreg*. Composites Part A: Applied Science and Manufacturing, 2015. **70**: p. 102–110.
69 Nie, M., Y.-h. Xia, and H.-s. Yang, *A flexible and highly sensitive graphene-based strain sensor for structural health monitoring*. Cluster Computing, 2019. **22**(4): p. 8217–8224.
70 Monea, B.F., et al., *Carbon nanotubes and carbon nanotube structures used for temperature measurement*. Sensors, 2019. **19**(11): p. 2464.
71 Karimov, K.S., M.T.S. Chani, and F.A. Khalid, *Carbon nanotubes film based temperature sensors*. Physica E: Low-dimensional Systems and Nanostructures, 2011. **43**(9): p. 1701–1703.
72 Blasdel, N.J., et al., *Fabric nanocomposite resistance temperature detector*. IEEE Sensors Journal, 2015. **15**(1): p. 300–306.

73 Wu, R., et al., *Silk composite electronic textile sensor for high space precision 2D combo temperature–pressure sensing*. Small, 2019. **15**(31): p. 1901558.
74 Kim, J., et al., *Wearable biosensors for healthcare monitoring*. Nature Biotechnology, 2019. **37**(4): p. 389–406.
75 Lee, H., et al., *Enzyme-based glucose sensor: from invasive to wearable device*. Advanced Healthcare Materials, 2018. **7**(8): p. 1701150.
76 Kang, B.-C., B.-S. Park, and T.-J. Ha, *Highly sensitive wearable glucose sensor systems based on functionalized single-wall carbon nanotubes with glucose oxidase-nafion composites*. Applied Surface Science, 2019. **470**: p. 13–18.
77 Shahbeigi-Roodposhti, P. and S. Shahbazmohamadi, *Tensile Strength of Resorbable Biomaterials*. JoVE Science Education Database. Biomedical Engineering. Acquisition and Analysis of an ECG (electrocardiography) Signal. 2022.
78 Chi, M., et al., *Flexible carbon nanotube-based polymer electrode for long-term electrocardiographic recording*. Materials, 2019. **12**(6): p. 971.
79 Awara, K., et al., *Thin-film electroencephalographic electrodes using multi-walled carbon nanotubes are effective for neurosurgery*. Biomedical Engineering Online, 2014. **13**: p. 166–166.
80 Qi, K., et al., *Weavable and stretchable piezoresistive carbon nanotubes-embedded nanofiber sensing yarns for highly sensitive and multimodal wearable textile sensor*. Carbon, 2020. **170**: p. 464–476.
81 Qiao, Y., et al., *Graphene-based wearable sensors*. Nanoscale, 2019. **11**(41): p. 18923–18945.
82 Chen, S., et al., *Recent developments in graphene-based tactile sensors and E-skins*. Advanced Materials Technologies, 2018. **3**(2): p. 1700248.
83 Nurazzi, N.M., et al., *Mechanical performance and applications of CNTs reinforced polymer composites-A review*. Nanomaterials (Basel, Switzerland), 2021. **11**(9): p. 2186.
84 Kumar, S., T.K. Gupta, and K. Varadarajan, *Strong, stretchable and ultrasensitive MWCNT/TPU nanocomposites for piezoresistive strain sensing*. Composites Part B: Engineering, 2019. **177**: p. 107285.
85 Reddy, S.K., et al., *Strain and damage-sensing performance of biocompatible smart CNT/UHMWPE nanocomposites*. Materials Science and Engineering: C, 2018. **92**: p. 957–968.
86 Arif, M.F., et al., *Strong linear-piezoresistive-response of carbon nanostructures reinforced hyperelastic polymer nanocomposites*. Composites Part A: Applied Science and Manufacturing, 2018. **113**: p. 141–149.
87 Kedambaimoole, V., et al., *Reduced Graphene Oxide Tattoo as Wearable Proximity Sensor*. Advanced Electronic Materials, 2021. **7**(4): p. 2001214.

4 Development of Carbon Nanotube and Graphene-Based Gas Sensors

Manish Pal Chowdhury

CONTENTS

4.1 INTRODUCTION

4.1.1 Usefulness of Gas Sensors

Earth's atmosphere is composed of approximately 78% nitrogen, 21% oxygen and 1% other gases. Argon is the main component amongst other gases, along with carbon dioxide, inert gases, and traces of other gases. Oxygen is the life sustaining essential gas for humans and animals, and carbon dioxide for plants. Deviation from the natural composition of the atmosphere poses danger to humans, animals, and plants. Natural phenomena like volcanic eruptions release toxic gases into the atmosphere. The main sources of gaseous forms of air pollution are manufactured, e.g., emission from factories, vehicle exhaust, using fossil fuels, etc. Gas sensors are critical for the identification and monitoring of hazardous gases. For the safety of individuals working in the hazardous environment of factories, industrial applications, agriculture and food industry, packaging and transportation industry, research applications, indoor air quality, aerospace, etc.

DOI: 10.1201/9781003231943-4

4.1.2 Why CNT and Graphene-Based Gas Sensors?

A good sensor has the characteristics of high sensitivity, analyte selective, fast response and recovery time, stability and accurate measurement, low cost, low power consumption, small and portability. Active sensing materials with high surface to volume ratio are ideal for resistive gas sensing application. In this respect, carbon nanotube, graphene, and their composites are promising candidates for such applications. Carbon nanotube and graphene alone cannot be used for a desired wide range of gas detection processes, as they lack selectivity and sensitivity. This led to the concept of CNT and graphene-based composite sensors. Upon adsorption of analyte molecules by carbon nanotube/graphene electron exchange occurs resulting in a change in resistance. Single-wall, as well as multi-wall, CNTs were used as gas sensors [1, 2].

4.2 SENSORS

4.2.1 Sensor Architecture

Gas sensor devices may consist of single CNT [3], aligned multiple CNTs [4], CNT network [5] and vertically aligned CNTs [6]. Single CNT has the advantage of high sensitivity over devices with multiple CNT. The widely adopted device fabrication process involves photolithography or e-beam lithography steps. The pristine nature of the CNTs and its composites is hindered by the resist reminiscent of the lithography processes [7]. Alternatively, CNTs are deposited on prefabricated electrodes [8] to reduce contamination but with poor contact quality. Many researchers also adopted the shadow mask electrode deposition technique [9] to have good electrical contacts and to maintain pristine nature of the sensing material. Several researchers have fabricated CNT devices by growing CNTs on platinum electrodes to maintain pristine nature of the CNTs [10–12]. The disadvantage is complex and lengthy device fabrication processes. Various research groups have integrated CNTs into different device architecture for gas sensing applications. In all the architectures, CNTs are exposed to atmosphere for obvious reason. Schematics of sensor architecture is shown in Figure 4.1.

A two electrode CNT-based sensor is the most widely used gas sensor architecture. In this configuration, CNTs are electrically connected by source-drain electrodes of gold, platinum, palladium, etc. [1, 4–6, 8, 9]. In this configuration a bias voltage (V_{sd}) is applied and drain current (I_{sd}) is measured. Single-walled and multi-walled carbon nanotubes both are reported for gas detection [10–14] applications. Figure 1.1(b) shows the schematics of field effect transistor (FET) configuration sensor. In this configuration, the active sensing layer is separated from the conducting back gate electrode by a dielectric layer [4, 9, 13–16]. However, due to the requirement of an additional electrode not many reports can be found. Necessity of semiconducting sensing material is also a limiting factor in this architecture. In the third configuration, two different electrode materials are used to form Schottky contact. The sensing mechanism is based on the modulation of reverse saturation current [17–21]. In the fourth configuration, sensing the chemiresistor is operated in the presence of UV or visible light. Such photo-assisted sensors show enhanced response properties [22–25]. Pristine SWCNTs lack potential to detect nitrogen dioxide (NO_2), sulfur dioxide (SO_2), hydrogen(H_2), hydrogen sulfide (H_2S), and ammonia (NH_3) with high sensitivity, but may sense the analyte gases – with high selectivity with suitable functionalization. Polyethyleneimine-coated multiple SWNT devices grown by the CVD technique on molybdenum electrode showed excellent sensitivity toward NO_2 sensing [26]. Xiao et al. demonstrated that palladium (Pd)-decorated SWNT show good sensitivity and selectivity for H_2 sensing. SWCNT noncovalent composite with poly (4-vinylpyridine) and platinum-polyoxometalate showed ppm level sensitivity toward CH_4 [27] sensing. The sensing property of SWCNT-Fe_2O_3 composite thin film was tuned to detect toxic gases like NH_3, NO, NO_2 and H_2S [28]. Functionalized MWCNT is also a

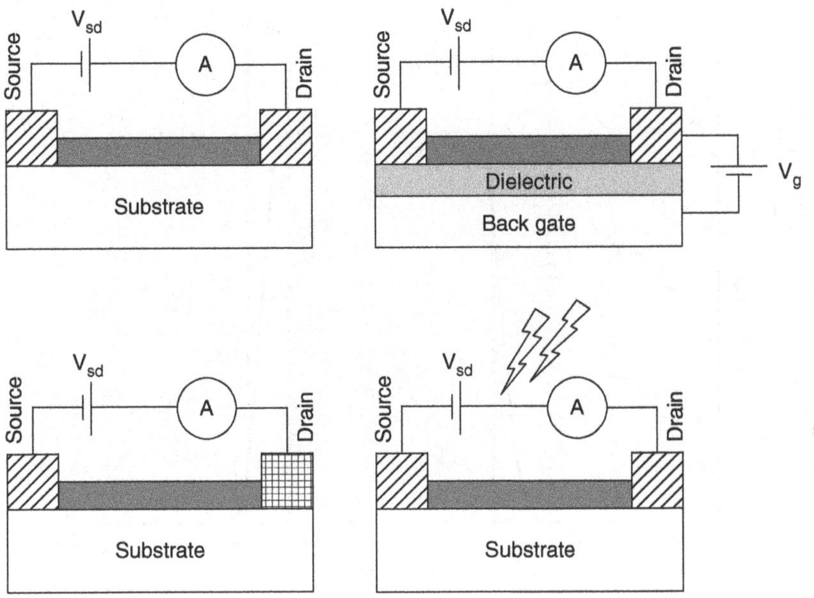

FIGURE 4.1 Schematics of different sensor architecture, (a) two terminal chemiresistor, (b) Field Effect Transistor (FET) architecture, (c) Schottky and (d) Photo assisted chemiresistor.

very potent toxic gas sensor. SnO_2-Pt-decorated MWCNT hybrid system was designed to detect CH_4 [29] at room temperature. Le et al. showed that tungsten oxide (WO_3)-decorated MWCNTs are efficient NH_3 sensors at elevated temperature [30].

Properties of gas sensors can be improved by exploiting FET architecture. FET sensor is operated in the subthreshold regime of the transfer characteristics to have maximum drain current change by adjusting gate voltage. FET parameters such as gate voltage, subthreshold voltage, subthreshold swing, etc. can be tuned to enhance sensing property. Floating gate CNTFET showed 90 ppb H_2 sensing [16] LOD at room temperature. Pd-decorated carbon nanotube FET reported to sense H_2 [31] as low as sub ppm concentration. Sacco et al. reported fabrication of ultrasensitive CNT field effect transistor as NO_2 detector.

The oxidation or reduction process taking place at the surface of graphene by the analyte gas molecules is the key point of the sensing mechanism and is reflected in the source-drain current of the graphene-FET device. Most of the reported GFET gas sensors are homo-junction in the bottom gate configuration. Schedin et al. [32] first reported the gas-sensing property of GFET devices. The authors demonstrated distinct change in drain current in NH_3, CO, H_2O and NO_2 gases. As reported by Kim et al. [33], GFET is capable of sensing NO_2 and NH_3 gases as low as 10 ppm and 50 ppm respectively, however devices showed strong humidity dependent sensing property. A slow response behavior of rGO-based FET in NO_2 gas was observed by Li and his co-workers [34]. The detection limit of the rGO-FET sensors was found to be 0.4 ppm. Wu et al. [35] had reported that micropatterned rGO could be used to sense NO_2 gas at very low concentration, as low as ~6 ppb. Similar sensing property of GFET sensors was reported by Ren et al. [36] in SO_2 environment. Selectivity of the sensor was controlled by the device operation temperature (100°C). Improved NO_2 sensing property [37] of GFET sensors was observed by replacing conventional Si/SiO_2 substrate with 6H-SiC substrates. Compared with the CNTFET sensing limit of 100 ppb, the authors reported a lower detection limit of 10 ppb. An ultrasensitive and fast Schottky GFET oxygen sensor was demonstrated by Cheng et al. [38]. An ultrafast response time of 60 ms and ppm level sensitivity was reported by the authors.

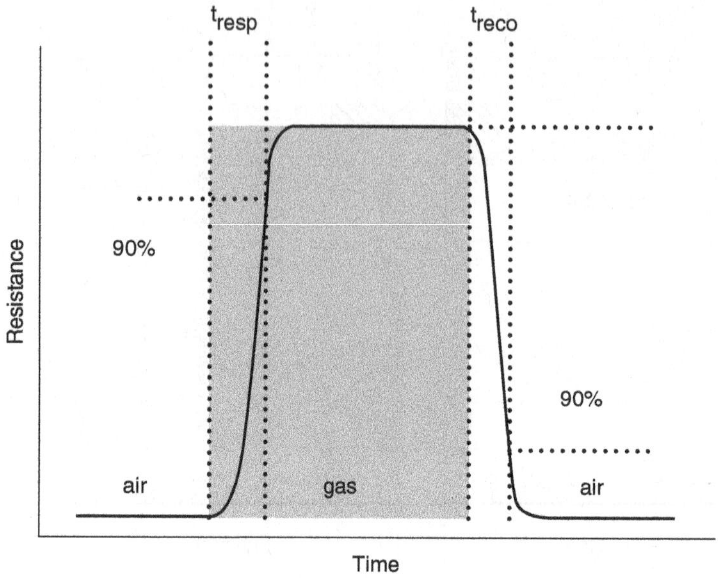

FIGURE 4.2 Schematic diagram of dynamic resistance curve.

4.2.2 Sensing Parameters Calculation

Three parameters are calculated from the $I_{sd} - time$ data to characterize resistive gas sensors. By knowing I_{sd}, resistance (R) of the sensor may be obtained for a given bias voltage. The critical parameters are i) sensitivity, ii) response and recovery time and iii) limit of detection (LOD). Sensitivity of a gas sensor is calculated using the following formula.

$$S = \frac{R_g - R_{air}}{R_{air}}$$

where, R_g and R_{air} are the saturation resistances of sensors in gas and in air.

Response time is estimated from the dynamic resistance plot ($R - Time$) (Figure 4.2), and is the time from the instant of gas discharge in the sensing chamber to 90% of the resistance at saturation. Recovery time is estimated from the instant of air injection into the sensing chamber filled analyte gas to 90% of the resistance value in air. S is calculated for a particular concentration (ppm) of an analyte gas. From the $S - Concentration$ data, LOD is obtained using the formula $LOD = 3\sigma / sl$, where σ and sl are the standard deviation and slope of the plot [39, 40].

4.3 SENSING PROPERTY OF CARBON NANOTUBE AND GRAPHENE-BASED GAS SENSORS

4.3.1 Nitrogen Dioxide (NO_2)

The primary source of nitrogen dioxide (NO_2) in air is combustion of petroleum fuel and industrial emission. Exposure of NO_2 in the long term may increase mortality. NO_2 gas is toxic for our cardiovascular and respiratory system [41, 42]. According to the World Health Organization (WHO), annual mean safety value of NO_2 concentration may be limited to 40 ug/m^3 [43]. Therefore, constant monitoring is essential.

Nitrogen dioxide gas is a potentially toxic gas – it reacts with water molecules to produce nitric acid and forms acid rain. Breathing NO_2-contaminated air for a prolonged time causes irritation in the airways and may result in human respiratory diseases. Carbon nanotube, graphene-based chemiresistor and FET sensors [4, 44–54] could sense NO_2 from ppb to ppm concentration. MWCNT is not responsive to NO_2 gas, but composite with polypyrrole (PPy) [44] showed ~25% response in 5 ppm NO_2 concentration at room temperature. The devices showed 65 s response time with <0.25 ppm LOD. Upon adsorption of NO_2 on PPy surface, PPy oxidizes by releasing electrons, resulting in an increase in resistance of the devices.

Liu et al. [45] demonstrated rGO-CNT-SnO_2 hybrid NO_2 sensing at room temperature. Sensors showed response (R_a / R_g) of 2.5 with 5 ppm NO_2 with 8 s and 77 s response and recovery time. A selectivity test of the sensor with NO_2, Cl_2, CO and NH_3 indicated superior response for NO_2 sensing. NO_2 sensor based on poly (9,9-dioctylfluorene) sorted SWCNT TFT devices [46] display 96% sensitivity at 60 ppm. The sensor showed a response time of 30 s and low detection limit at room temperature. The sensor detects several other gases like H_2, H_2S, and formaldehyde, but with inferior sensitivity. Adsorption of gas molecules at the metal-semiconductor junction modifies the Schottky barrier height [4]. In a SWCNT FET architecture, CNTs are directly connected between the source and drain electrodes. CNTFET devices under exposure of NO_2 molecules display ultrasensitive response ($\frac{I_{NO_2} - I_{N_2}}{I_{N_2}}$). NO_2 concentration in the range from 100 ppb to 10 ppm could be detected in this configuration using nitrogen (N_2) as reference. CNTFET devices exhibit shift in the gate threshold voltage upon exposure to analyte gas concentration [47]. Pulsed voltage gated CNTFET devices showed shift in threshold voltage in the presence of NO_2 and showed substantial shift upon analyte gas concentration in the range from 200 to 1000 ppb. Helbling et al. [48] investigated the influence of back gate dielectric of CNTFET sensor on the response of NO_2 analyte. Two types of devices were fabricated: on the surface, and suspended CNTFET. It was observed that both type of devices produced qualitatively similar results, therefore surface assisted indirect sensing mechanism was ruled out.

In graphene with defect states, doped [55, 56] graphene showed superior sensitivity toward gas sensing. Hydrothermally grown SnO_2-CuO/rGO ternary composite [51] was used to detect NO_2. The ternary composite showed 8–15 times more response than a rGO-based single metal oxide semiconductor sensor. At 50 ppm of NO_2 concentration, the sensor showed 250% response with 150 ppb limit of detection at room temperature. γ-Bi_2MoO_6/graphene composite [57] exhibit NO_2 sensing at room temperature. The synergetic effect of graphene and γ-Bi_2MoO_6 nanoparticles led to superior NO_2 gas sensing property. The devices showed ~95% response at room temperature. Low detection limit of 10 ppb was estimated. SnO_2 nanoparticles loaded on CVD-grown graphene [54] showed remarkable NO_2 sensing property at room temperature. Recovery time was dramatically reduced from ~70 s to <10 s by exposing UV during recovery period. Selectivity test with 10 ppm of SO_2, NH_3, CO, acetone and NO_2 proved to be dominant with NO_2 gas. Hong et al. [53] demonstrated that MoS_2 nanoflakes embedded on graphene is a fast NO_2 gas sensor with 74% response at 100 ppm concentration. The response and recovery time of the sensor was less than 1 s at 200°C sensing temperature. Ethylenediamine modified rGO [58] exhibited 4 to 16 times higher response than rGO alone toward NO_2 sensing. Adsorbed by the functional group

NO_2 gets oxidized by capturing the lone pair electrons. The sensing mechanism led to high selectivity to NO_2 analyte.

4.3.2 Sulfur Dioxide (SO_2) Detection

Sulfur dioxide causes irritation to respiratory tract, nose, eyes, and skin. In combination with water, SO_2 produces sulfuric acid. It is crucial to monitor SO_2, as it is an industrial biproduct and very toxic to humans, animals, and plants [59]. ZnO hollow sphere and MWCNT composite [60] show selective SO_2 sensing property at 300°C in the range of 2–100 ppm concentration. The composite system showed good selectivity compared to CO, CO_2, methanol, toluene, hexane, and xylene. The composite showed a response of 85 at 70 ppm concentration. Near 300°C ZnO surface is dominated by O^{2-} species. SO_2 acts as a reducing gas, upon exposure to the gas, electrons are released, and resistance of the composite system is reduced. SnO_2 nanoparticles alone show very low response (~1.2) as SO_2 sensor. MWCNT composite [61] with SnO_2 nanoparticles showed enhanced response of ~5% at 220°C operating temperature. Adsorbed oxygen on the surface of SnO_2 creates a space charge region. The MWCNT network provides a conducting path in between the grain boundaries of SnO_2 nanoparticles, modifying the response characteristics of the composite sensor.

rGO/SnO_2 composite SO_2 sensors [61] display superior sensing property (response ~22) as compared to bare SnO_2 and MWCNT/SnO_2 composite sensors. The superior sensing property of rGO/SnO_2 composite was attributed to larger surface area of the 2D material for uniform dispersion of the SnO_2 nanoparticles. Metal oxide semiconductors are widely used for gas and volatile organic compounds. The addition of rGO into the system promotes superior response and fast response and decay time. ppb level SO_2 detection performance of WO_3/rGO nanocomposites was reported by Su et al. [62]. The composite sensor showed superior response (~30%) compared to pristine WO_3 sensor (response < 10%) at 300 ppb SO_2 concentration.

4.3.3 Hydrogen Sulfide (H_2S)

H_2S is a toxic gas and can be identified with its pungent smell of rotten egg [63]. H_2S is naturally found in volcanic gases, fossil fuel, hot spring, etc. [64]. It is also produced in the bacterial degradation process of human and animal waste in anaerobic conditions, swamps, sewers [63], and landfill [65] areas. In the industrial processes, such as food processing, paper mills, coke ovens, petroleum refineries, H_2S is produced. When released into air, H_2S oxidized to SO_2 and sulfuric acid, and it is a flammable gas. H_2S concentration in the range 500–1000 ppm is considered as fatal [66], therefore detection of H_2S is an absolute necessity.

Copper nanoparticle/SWCNT composite [67] displays H_2S gas sensing property at room temperature with the exposure of H_2S gas concentration from 5–150 ppm sensor response varied between ~10–70. Sensor stability was tested over the period of 30 days at 5 ppm H_2S exposure. The devices showed fast response and recovery time of ~10 s and ~15 s respectively. The composite showed good selectivity with H_2S over H_2, ethanol, acetone, and methane. Carbocxylated SWCNTs showed poor response to H_2S sensing, while electrochemically deposited gold nanoparticle SWCNT composite [68] showed dramatic improvement in response. When exposed to 20–1000 ppb H_2S concentration the Au-SWCNT network ~ *delta R/R* (%) value changed in the range ~ 5–25. Response time was in the range 6–8 min for all concentrations. A highly effective H_2S detection by redox reaction by functionalized SWCNTs was reported by Jung [69] et al. 2,2,6,6-tetramethylpiperidine-1-oxyl (TEMPO) acts as a catalyst in functionalized SWCNT for detection of H_2S gas. Humidity plays an important role in the sensing of H_2S by TEMPO functionalized SWCNT. At 60% relative humidity, the sensor showed 420% sensitivity which is 17 times higher than the sensitivity measured in dry condition. Cu_2O/CuO nanoparticles showed promising response for H_2S gas sensing. MWCNTs sensors decorated with Cu_2O/CuO nanoparticles showed 1244% response to 1 ppm H_2S gas. Response and recovery time of the sensors varied in the range ~200–600 s and ~80–300 s respectively. Co_3O_4-SWCNT composite [70] synthesized by the arc discharge method showed potential application for H_2S gas sensing property. At 250°C working temperature, the composite sensor showed its highest response of ~500% at 100 ppm H_2S concentration. The devices showed good selectivity toward H_2S as compared with NH_3, CH_4 and H_2 gases. The composite sensors also showed good selectivity for H_2S sensing. The sensors were insensitive to environment humidity at the optimal temperature.

NiO/nitrogen-doped reduced graphene oxide (N-rGO) [71] synthesized by the solvothermal method, showed H_2S gas response (R_g/R_a) >30 at an optimal working temperature of 92°C. The gas response was linear up to 100 ppm concentration with a lower detection limit of 100 ppb. Selective sensing property of H_2S gas by SnO_2 quantum wire/reduced graphene oxide composites was demonstrated by Song et al. [72] at room temperature. Devices showed fast response time (2 s) as compared to recovery time (292 s) at 50 ppm concentration. H_2S gets oxidized on adsorption on SnO_2 nanowire surface, and the electron collection is enhanced through rGO nanosheets. Graphene quantum dot (GQD) functionalized porous and hierarchical SnO_2/ZnO nanostructures [73] showed high response (R_a/R_g) ~16 at 0.1 ppm H_2S concentration. The composites showed remarkably rapid response and recovery time of 14 s and 13 s respectively. The combined effect of GQD/SnO_2/ZnO nanocomposite heterostructure leads to the superior sensitivity of the sensor. Selectivity property of the sensor was studied using H_2S, methanol, iso-propanol, ether, acetone, formaldehyde and toluene and its response in H_2S was found to be 6 times larger than the rest of the analyte gases under similar condition. The response and recovery times were 14 s and 13 s respectively.

4.3.4 Hydrogen (H_2) Detection

Hydrogen has three isotopes, namely hydrogen, deuterium, and tritium. Natural abundance of stable hydrogen and deuterium are 99.9844% and 0.0156% [74]. Hydrogen is used in petroleum, metal, fertilizer, and the food industry. It is also used in rockets as cryogenic fuel, hydrogen fuel cell for clean energy harvest and car fuel [75, 76]. H_2 has an ignition energy of 0.017 mJ and 142 kJ/g heat of combustion. It acts as a strong reducing agent in chemical reactions. It is flammable in a wide concentration range (4–75% v/v) [76]. Hydrogen is a colorless and odorless gaseous substance and cannot be detected by human sensory organs. To detect such a combustible gas, an accurate concentration measurement is necessary.

Palladium nanoparticle (PdNP)-decorated SWCNT-based devices are promising candidates for H_2 detection applications. Li et al. [77] reported H_2 sensing property of the hybrid in wide concentration range (10 ppm-4%). The devices showed relatively low response and recovery time of 62 s and 72 s respectively. Several other researchers [78–81] reported H_2 sensing property of Pd/SWCNT hybrid devices, but with inferior sensor characteristics in all respect. Xiao et al. [30] published on H_2 sensing property of Pd nanoparticle-decorated SWCNT network and rope-CNT network. SWCNT network showed superior response (~1000) at 311 ppm compared to the response (~1) of rope-CNT network devices. The response time of ~7 s is much faster than the earlier reports [77]. A new process of UVC-activated Pt-decorated SWCNT/graphene composite H_2 sensor was reported by Alamri et al. [82]. Up to 10 mins of UVC exposure time, an increase in response (~4.3-fold higher), and fast response time (~3.6 times faster) was observed. The authors claimed that the enhanced sensor property is caused by the desorption of residual molecules from the device surface upon UVC exposure.

Palladium (Pd) in conjunction with other H_2 sensing materials like SnO_2 was investigated by Dhall et al. [83] on the CVD-grown graphene layer. Pd/graphene showed better response (8.2%)

at room temperature. Individual contribution of SnO_2 in sensing H_2 at room temperature and at 200°C operating temperature remained small. At elevated working temperature response of Pd/SnO_2/graphene sensor showed ~15% response in contrary to drastic reduction of response in the case of Pd/graphene devices. NiO nanomaterials show to sense H_2 gas and its morphological effect on the sensing property [84, 85]. Sensing characteristics of NiO/graphene composite [86] measured at 200°C operating temperature in the H_2 concentration range 400–2000 ppm showed ~1.4 times higher response than NiO alone. The higher and relatively fast response of the NiO/graphene composite sensors may have been related to the larger surface area of the composite. Pristine ZnO [87] showed a response in the range 40–70 with H_2 gas flow from 100–1000 ppm at 150°C. The response and recovery time of the composite was 21 s/47 s compared to 46 s/75 s for pristine ZnO sensors indicating improved response property.

4.3.5 Ammonia (NH_3) Detection

Common use of NH_3 is in the agriculture as fertilizer, a refrigerant gas, food industry, manufacturing industry, etc. Despite its many uses, NH_3 is harmful to humans and animals. If inhaled in large quantities, it can cause irritation and death [88–90].

MoS_2/MWCNT composite showed good ammonia-sensing property [39] at room temperature down to 12 ppm concentration. The composite also demonstrated faster response and recovery time (65 s and 70 s respectively) compared to MoS_2 ammonia sensor (400 s and 280 s). Selectivity of MoS_2/MWCNT composite sensor was investigated with 200 ppm of formaldehyde, benzene, methanol, chloroform, and acetone and proved to be superior in sensing NH_3. Polyaniline (PANI) reported to detect flexible polyaniline/carbon nanotube nanocomposite film-based electronic gas sensors [91, 92] NH_3 by several researchers. Synergistic effect [91] in PANI/CNT composite was reflected in the NH_3 sensing property. Composites showed higher response compared to the response of its constituents. The major advantage is the selectivity of the composite for NH_3 compared to other volatile compounds. Other conduction polymer such as polypyrrole (PPy) conjugated with phenylalanine (PA) and SWCNT [92] showed improved sensitivity of NH_3 gas sensor. The composite demonstrated PPy/PA/SWCNT composite is selective to NH_3 gas.

Tin–titanium dioxide/reduced graphene/carbon nanotube nanocomposites [93] synthesized by solvothermal method tested for toluene, DMF, acetone, ammonia, formaldehyde, ethanol, methanol, isopropanol, thinner, carbon dioxide, hydrogen, and acetylene. The sensor was able to identify NH_3 as its relative response is higher compared to other volatile organic compounds (VOCs). The response at room temperature was ~85% at 250 ppm NH_3 concentration.

NH_3 sensing property of mesoporous and nanoporous polypyrrole composites (mPPy/rGO and nPPY/rGO) [94] are shown to have dependency on its microstructure. Both the composites showed non-linear responsivity with NH_3 and displayed maximum responsivity ~40 at 40 ppm concentration. However, wormlike m-PPy/rGO composites are shown to have improved responsivity of ~70% at 40 ppm NH_3 concentration. Selectivity of the composites were tested with NH_3, water, ethanol, methanol, acetone, acetonitrile, ethyl acetate, toluene, CCL_4, H_2S, and CO. Responsivity of the rest of the analyte gases were less than 10%.

NH_3 gas sensing property of WO_3/rGO [95] synthesized by solvothermal method was tested at 20–500 ppm concentration at operating temperature of 300–450°C. The composite showed optimum sensing property at 300°C operating temperature with responsivity 35% with 138 ppb LOD. Ternary nanocomposits such as Pd/SnO_2/rGO are shown to display NH_3 sensing property at room temperature. Presence of reducing gas, resistance of the oxide semiconductor gets reduced and fast response is observed due to the rGO component. Maximum response of 35% was reported for 500 ppm NH_3 concentration. Palladium (Pd) is a versatile and promising material in sensor applications. Sensor based on Pd/SnO_2/RGO ternary composite [96] proved to be an effective NH_3 gas sensor operated at room temperature. The devices showed response of 7.6–35 in the NH_3 concentration 5–300 ppm respectively at room temperature. However, the devices showed slow response and recovery time of 7 min and 50 min respectively. Self-sustaining PPy-graphene-based NH_3 sensor was developed by Gao et al. [94]. The devices showed 45% response at 10 ppm with LOD of 41 ppb at room temperature. Responsivity of the devices was an order higher than other common VOCs. The sensors showed good stability up to 65% relative humidity, beyond which response was reduced. As PANI proved to be a good candidate for NH_3 sensor [97–101], enhanced NH_3 sensing property of PANI-rGO composite was demonstrated by Hadano et al. [102]. The devices showed an improved sensitivity of 250% at 100 ppm NH_3 concentration with response and recovery time of 40 s and 191 s respectively.

4.3.6 Methane (CH_4)

Methane is a highly flammable greenhouse gas. It is an odorless colorless gas, and in high concentration replaces breathable oxygen in air, resulting in breathing trouble. In 2000, the global concentration of CH_4 stabilized at ~1800 ppb [103], and since then has been rising continuously. The primary source of methane emission in the atmosphere is agriculture, followed by fossil fuel, natural gas, and biofuels. Continuous monitoring of CH_4 levels in the mines and industry is of utmost importance for the safety of the workers.

PANI/polymer/CNT composites [104] displayed very fast response and recovery time (<1 s) at 5 ppm CH_4 concentration at room temperature. Functionalized CNT, CNT-COOH composites showed poor responsivity and slower response and recovery time. However, at 60°C the sensitivity of PANI/polymer/CNT-COOH composites is 10 times higher compared to the room temperature value. An UV assisted recovery technique of ZnO-MWCNT chemoresistor CH_4 sensor was described by Humayun et al. [105]. UV (390 nm) exposure reduces the CH_4 desorption barrier. Effect of reduced barrier is reflected in the faster recovery time of the sensor. A recovery time of 3 min was reported which is 10 times faster compared to the normal recovery time [106]. Oxygen plasma and UV-Ozone (UVO) treatment shown to have effect on the responsivity of metal oxide-MWCNT composite based CH_4 sensor [107]. Untreated ZnO-decorated MWCNT samples showed negligible response toward CH_4 sensing. O_2 plasma treated ZnO-MWCNT sensors showed responsivity of 10 in 10 ppm CH_4 flow, while responsivity < 2 for UVO-treated sensors. It indicates that the O_2 plasma-treated activation process is more efficient than the UVO-mediated activation process.

Resistance of graphene/PANI composites [108, 109] increases in the presence of CH_4 gas due to its reducing nature. Wu et al. [109] demonstrated CH_4 sensing property of graphene/PANI composite in the range 1–1600 ppm CH_4 flow with 20 ppm LOD. Response and recovery time estimated from the response curve to be 85 s and 45 s respectively. It was proposed that the $\pi - \pi^*$ conjugated system of graphene/PANI composites have more active sites than PANI alone for the π electron gas molecule interaction [108]. Due to high catalytic activity of palladium (Pd), it has proven to be a good sensor component [2, 110–112]. A comparative study on CH_4 gas sensor by Pd-doped SnO_2/partially reduced graphene oxide (PRGO) and Pd-doped SnO_2/reduced graphene oxide(rGO) was reported by Nasresfahani et al. [113]. Pd-doped SnO_2/rGO demonstrated faster response compared to Pd-doped SnO_2/PRGO sensors with 9.5% sensitivity at room temperature in 12000 ppm CH_4 concentration. However, the materials showed poor detection limit with 250 ppm and 75 ppm for Pd-doped SnO_2/rGO and Pd-doped SnO_2/PRGO sensors respectively. Led sulfide (PbS)/rGO composites also showed response to CH_4 gas at room temperature [109]. Response of the composite increased with the increase of rGO content and attained a maximum ~45% with 6% CH_4 concentration. Response time decreased with increasing rGO content, however, recovery time had little effect on rGO content. Comparison of CNT and graphene-based sensors under different analyte gas ambience are summarized in Table 4.1.

TABLE 4.1
Comparison of sensing performance of carbon nanotube and graphene based devices

Sensing material	Analyte gas	Response calc.	Detection range(p pm)	Concentration (ppm)	Response	Working temperature e	Time (s)		References
							Response	Recovery	
Polypyrrole/MWCNT	NO_2	$\frac{R_g - R_a}{R_a} \times 100$	0.25-9	5	40	RT	65	668	[44]
Polypyrroles rGO-CNT-SnO_2		$\frac{R_a}{R_g}$	5-100	5	2.53	RT	8	77	[45]
isoindigo-based copolymer/SWCNT		$\frac{R_g - R_a}{R_a} \times 100$	10-40	-		RT	-		[46]
MWCNT Schottky barrier		$\frac{I_{NO_2} - I_{N_2}}{I_{N_2}} \times 100$	0.1-10	0.1	100	RT	-		[4]
Chemically modified Graphene		$\frac{G_g}{G_a}$	1-50	50	~25	RT	-		[58]
Graphene nanosheet		$\frac{I_{NO_2} - I_{N_2}}{I_{N_2}} \times 100$	2.5-100	5	30	RT	-		[55]
Doped graphene nanosheet		$\frac{R_g - R_a}{R_a} \times 100$	1-21	21	25	RT	68	635	[114]
SnO_2-CuO/graphene		$\frac{R_a - R_g}{R_g} \times 100$	5-50	50	250	RT	-		[51]
ZnO/MWCNT	SO_2	$\frac{R_a}{R_g}$	2-100	100	270	300°C	84	78	[60]
SnO_2-MWCNT		$\frac{R_a - R_g}{R_g}$	10-500	500	5	60°C	2.4	3.5	[61]

rGO-WO_3 MWCNT-WO_3		$\frac{R_a - R_g}{R_g} \times 100$	0.05-0.30	0.30	~30 ~25	RT	66 61	298 slow	[62]
CuNP-SWCNT	H_2S	$\frac{R_g - R_a}{R_a} \times 100$	5-150	20	25	RT	10	20	[67]
AuNP-SWCNT		$\frac{R_g - R_a}{R_a} \times 100$	0.02-1	1	23	RT	480	1200	[68]
Functionalized SWCNT		$\frac{I_a - I_g}{I_g} \times 100$	10-100	100	~420	RT	-		[69]
Co-SWCNT		$\frac{R_g - R_a}{R_a} \times 100$	10-100	100	500	250°C	-		[70]
NiO-rGO		$\frac{R_g}{R_a}$	0.1-100	50	~32	92°C	-		[71]
SnO_2-rGO		$\frac{R_g}{R_a}$	10-100	50	33	RT	2	292	[72]
GQD-SnO_2/ZnO		$\frac{R_g}{R_a}$	0.025-5	0.1	~16	RT	14	13	[73]
Pd-SWCNT	H_2	$\frac{R_g - R_a}{R_a} \times 100$	10ppm-4%	1000	78	RT	62	67	[77]
Pd-SWCNT		$\frac{R_g - R_a}{R_a} \times 100$	0.89-311	311	1000	RT	7	89	[31]
Pt-/SWCNT /graphene		$\frac{R_g - R_a}{R_a} \times 100$	1-10%	10%	20	RT	~4min	-	[82]
Pd- graphene		$\frac{I_{H_2} - I_{N_2}}{I_{N_2}} \times 100$	2%	2%	~8	RT	27	10	[83]

(continued)

TABLE 4.1 (Continued)
Comparison of sensing performance of carbon nanotube and graphene based devices

Sensing material	Analyte gas	Response calc.	Detection range(p pm)	Concentration (ppm)	Response	Working temperature e	Time (s)		References
							Response	Recovery	
NiO-graphene		$\frac{R_{g-}R_a}{R_a}\times 100$	400-2000	2000	~70	200°C	-		[86]
ZnO-rGO		$\frac{R_g-R_a}{R_a}\times 100$	100-10000	100	484	RT	21	47	[87]
MoS_2-SWCNT	NH_3	$\frac{R_{NH_3}-R_a}{R_a}\times 100$	25-600	475	-80	RT	65	70	[39]
PANI-CNT		$\frac{R_{NH_3}-R_a}{R_a}\times 100$	0.2-50	10	6	RT	85	20	[91]
PPy/phenylalanine-SWCNT		$\frac{R_g-R_{N_2}}{R_{N_2}}\times 100$	0.1-4	1	2	RT	600	-	[92]
rGO/WO_3		$\frac{R_a}{R_g}$	20-500	500	35	300°C	60	711	[95]
Pd/SnO_2/rGO		$\frac{R_a-R_{NH_3}}{R_a}\times 100$	5-300	5	7.6	RT	7-50 min		[96]
PPy-graphene		$\frac{R_g-R_a}{R_a}\times 100$	0.2-40	5	31	RT	85	600	[94]
PANI-rGO		$\frac{R_a-R_g}{R_a}\times 100$	5-600	100	250	RT	97	680	[102]

PANI-MWCNT	CH_4	$\frac{R_a - R_g}{R_a + R_g} \times 100$	5-15	5	3.1	RT	<1	-	[104]
PANI-graphene		$\frac{R_g - R_a}{R_a} \times 100$	1-1600	100	~3.2	RT	85	45	[109]
Pd-doped SnO_2/rGO		$\frac{R_a - R_g}{R_a} \times 100$	800-12000	12000	~9.5	RT	5min	7min	[113]
PbS-rGO		$\frac{R_g - R_a}{R_a} \times 100$	0.5-6%	6%	45	RT	92	70	[115]

4.4 SENSING MECHANISM

The metal oxide(MO) gas sensors such as ZnO [60, 73, 87], SnO_2 [45, 51, 72, 73, 96, 113], CuO [51, 116, 117], TiO_2 [118, 119], MnO_2 [120–122], WO_3 [62, 95, 123], In_2O_3 [124–126], NiO [86, 127, 128], MoS_2 [17, 62, 129, 130], etc. show response toward analyte gases. Pure MO gas sensors and CNT-MO/rGO-MO composite gas sensors have slightly different sensing mechanism.

However, exact gas sensing mechanism is still debatable. The accepted concept in the community is the oxidation reduction by the adsorbed analyte molecules. In some cases, the sensor surface gets chemically modified by the analyte gases. The possible electron conduction path and band model for MO sensor and graphene and CNT based devices is presented in Figure 4.3(a–c) [131–133]. Ambient oxygen molecules get adsorbed at the trap states on the surface of metal oxides. The oxygen molecules accept electrons from the metal oxide thereby creating a depletion layer, which in turn reduces the surface conductivity of the materials. Upon exposure to oxidizing analyte gases, the conduction band electrons will be trapped near the surface region and would lead to increase in depletion width. Whereas, when exposed to reducing gases, adsorbed oxygen at the surface will be released resulting in decrease in depletion width. In MO-decorated/coated CNT and graphene, the n-type metal oxide and p-type CNT/graphene creates p-n junction. In air, the depletion width is larger than in the presence of reducing gases. In turn, the depletion width of the MO-CNT is modified, and the overall resistance of the composite sensor is altered. The CNT and graphene acts as a conduction channel and may influence the electronic property of

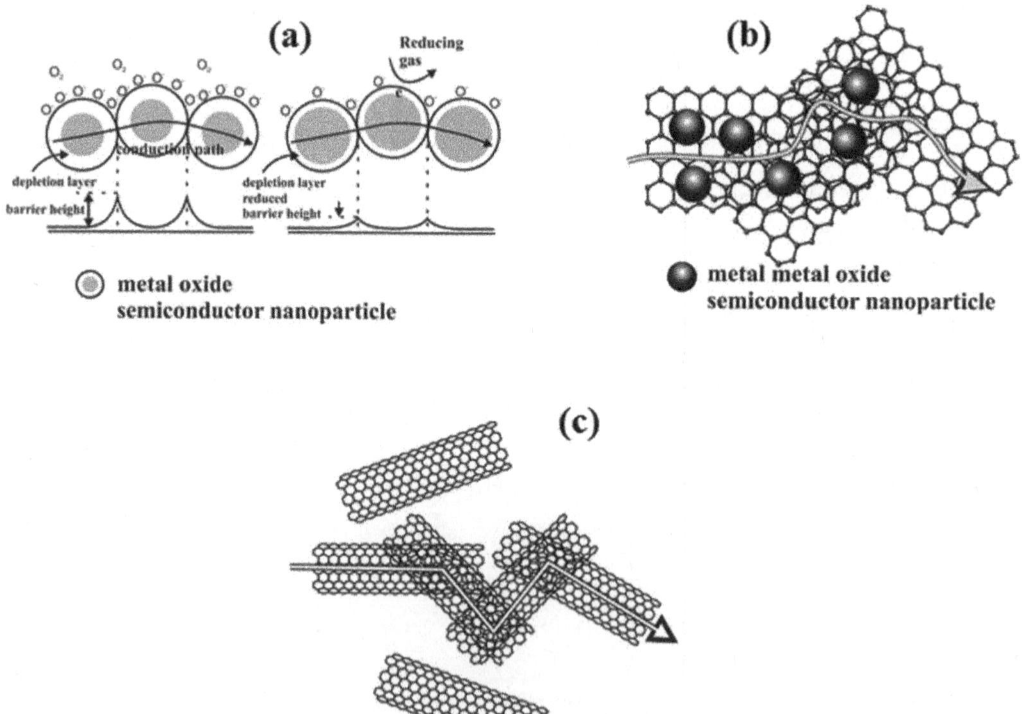

FIGURE 4.3 Schematics of conduction path (a) in a polycrystalline metal oxide gas sensor, depletion region is presented as white and gray region as conduction channel, (b)graphene composites and (c) carbon nanotube device.

the MO [134, 135]. Schematics of depletion region in MO/CNT and MO/graphene is presented in Figure 4.3 [136, 137]. As proposed by Van Duy et al. [136], adsorbed ethanol molecules on SnO_2-TiO_2 grain boundary reduces negatively charged oxygen, as a result grain boundary barrier height is reduced, and the sensor resistance is decreased. SnO_2-TiO_2/CNT composite forms a MO-CNT-MO p-n heterostructure junctions. Generalized schematic diagram of grain boundary barrier height is indicated in figure 4.4(a). On ethanol exposure (reducing gas), it reacts with the initially adsorbed negatively charged oxygen molecules and electron is released to the conduction band. Wu et al. [137] suggested the effective conduction path in graphene-ZnO composites is ZnO nano particle conduction channel and graphene-ZnO heterojunction. In an inert atmosphere, ZnO-ZnO junction barrier height is small and graphene-ZnO heterojunction is Ohmic in nature. In ammonia ambient, the potential barrier height at the graphene-ZnO and ZnO interfaces increases. As a result, increase in responsivity was observed. Change in band gap at the graphene-ZnO interface due to ammonia adsorption may cause enhanced responsivity. Figure 4.4(b) shows the band diagram of MO-MO junction and MO-graphene junction in MO/graphene composite material in ambient air. The depletion regions of the MO particles are indicated by the gray

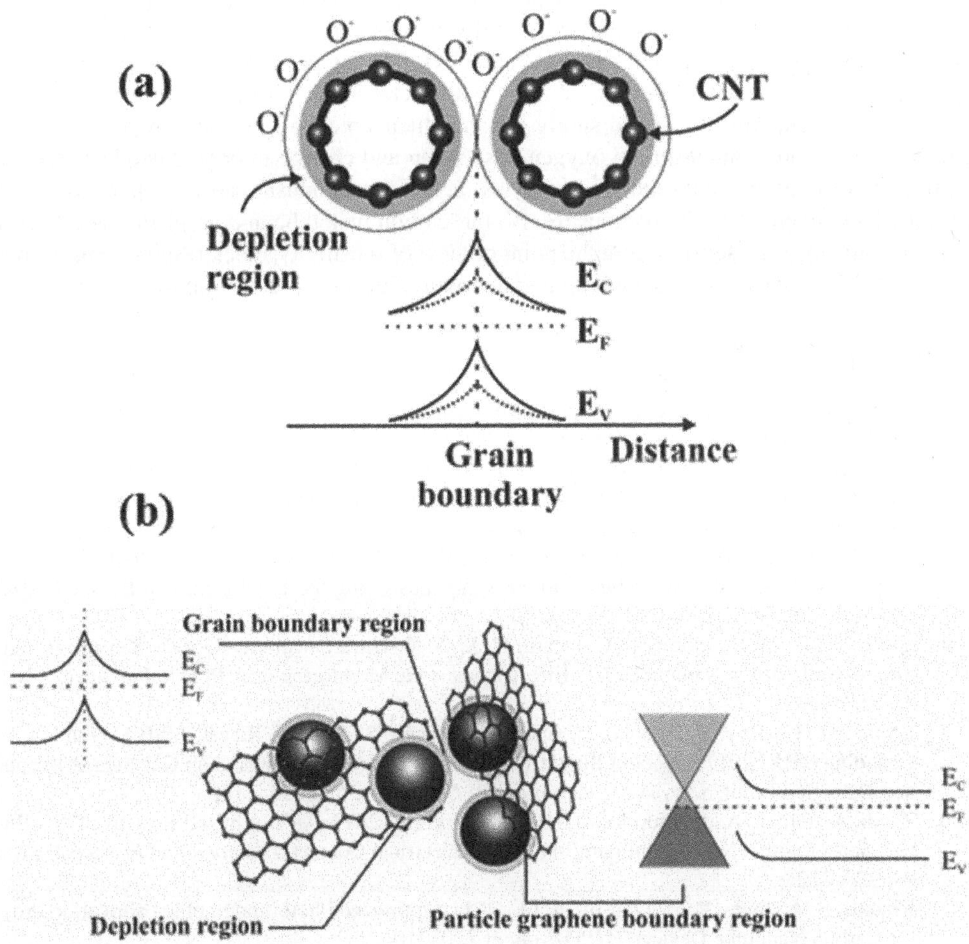

FIGURE 4.4 Schematic diagram of potential barrier modification in air and in reducing gas of (a) MO-CNT composites and (b) MO-graphene composites.

area around it. With the exposure of reducing and oxidizing gas, the barrier height modification occurs, resulting in change in device resistance. n-type and p-type MO materials in the presence of analyte gases behaves in opposite sense. In the presence of reducing gases, the depletion layer of n-type MO is reduced, and resistance is decreased whereas for p-type MO, hole accumulating depletion layer is reduced resulting increase in resistance [138]. A similar trend was observed by other researchers. Reducing analyte gases releases electrons to the p-type conducting polymers like PANI and PPy thereby increasing the electrical resistance. Resistance of PANI and PPy is reduced in the presence of oxidizing analyte gases, as electrons are captured from the polymer. Resistance of PPy hybrid material in the presence of oxidizing NO_2 gas is reduced from the oxidation process [44] and regains its state when exposed to air. In the presence of reducing gas such as NH_3, resistance of p-type PANI, rGO and PANI-rGO composite increased [48].

4.5 CONCLUDING REMARK

Directly or indirectly, industrial pollutant gaseous by-products, urban by-products affect humans and animals. Continuous monitoring of pollutant gases is a necessary requirement. Both carbon nanotube and graphene-based material are potential candidates for gas sensing applications, however in some cases graphene/rGO showed superior performance than CNTs based sensors. This may be due to large surface area of graphene and rGO. Sensing performance of pristine CNT and graphene is very limited, but their composites with metal oxide semiconductors, polymers and nanoparticles of Au, Ag, Pt, and Pd surely enhance their potential. Sensing properties of metal oxide composites are mainly due to oxygen adsorption and change in barrier height potential in the presence of reducing and oxidizing analytes. Sensing mechanism can be explained with the simplified band diagram of the composites. So far, carbon nanotube and graphene-based sensors are superior to any other sensors from the point of view of sensitivity, quick response and recovery time, repeatability, selectivity, low working temperature, low power consumption, and robustness.

REFERENCES

1 Li, J., Lu, Y., Ye, Q., Cinke, M., Han, J., Meyyappan, M. 2003. Carbon Nanotube Sensors for Gas and Organic Vapor Detection. *Nano Lett. 3*: 929–933.

2 Lim, Y., Lee, Y., Heo, J.-I., Shin, H. 2015. Highly Sensitive Hydrogen Gas Sensor Based on a Suspended Palladium/Carbon Nanowire Fabricated via Batch Microfabrication Processes. *Sensors and Actuators B: Chemical 210*: 218–224.

3 Lee, Y.D., Cho, W.-S., Moon, S.-I., Lee, Y.-H., Kim, J.K., Nahm, S., Ju, B.-K. 2006. Gas Sensing Properties of Printed Multiwalled Carbon Nanotubes Using the Field Emission Effect. *Chemical Physics Letters 433*: 105–109.

4 Sacco, L., Forel, S., Florea, I., Cojocaru, C.-S. 2020. Ultra-Sensitive NO_2 Gas Sensors Based on Single-Wall Carbon Nanotube Field Effect Transistors: Monitoring from Ppm to Ppb Level. *Carbon 157*: 631–639.

5 Kumar, D., Chaturvedi, P., Saho, P., Jha, P., Chouksey, A., Lal, M., Rawat, J.S.B.S., Tandon, R.P., Chaudhury, P.K. 2017. Effect of Single Wall Carbon Nanotube Networks on Gas Sensor Response and Detection Limit. *Sensors and Actuators B: Chemical 240*: 1134–1140.

6 Zandi, A., Gilani, A., Ghafoori fard, H., Koohsorkhi, J. 2019. An Optimized Resistive CNT-Based Gas Sensor with a Novel Configuration by Top Electrical Contact. *Diamond and Related Materials 93*: 224–232.

7 Khamis, S.M., Jones, R.A., Johnson, A.T.C. 2011. Optimized Photolithographic Fabrication Process for Carbon Nanotube Devices. *AIP Advances 1*: 022106.

8 Cao, J., Wang, Q., Dai, H. 2005. Electron Transport in Very Clean, as-Grown Suspended Carbon Nanotubes. *Nature Mater 4*: 745–749.

9 Muoth, M., Helbling, T., Durrer, L., Lee, S.-W., Roman, C., Hierold, C. 2010. Hysteresis-Free Operation of Suspended Carbon Nanotube Transistors. *Nature Nanotech 5*: 589–592.
10 Martin-Fernandez, I., Gabriel, G., Rius, G., Villa, R., Perez-Murano, F., Lora-Tamayo, E., Godignon, P. 2009. Vertically Aligned Multi-Walled Carbon Nanotube Growth on Platinum Electrodes for Bio-Impedance Applications. *Microelectronic Engineering 86*: 806–808.
11 Moser, J., Eichler, A., Güttinger, J., Dykman, M.I., Bachtold, A. 2014. Nanotube Mechanical Resonators with Quality Factors of up to 5 Million. *Nature Nanotech 9*: 1007–1011.
12 Eichler, A., Moser, J., Chaste, J., Zdrojek, M., Wilson-Rae, I., Bachtold, A. 2011. Nonlinear Damping in Mechanical Resonators Made from Carbon Nanotubes and Graphene. *Nature Nanotech 6*: 339–342.
13 Sharma, B., Kim, J.-S. 2018. MEMS Based Highly Sensitive Dual FET Gas Sensor Using Graphene Decorated Pd-Ag Alloy Nanoparticles for H2 Detection. *Sci Rep 8*: 5902.
14 Hong, S., Shin, J., Hong, Y., Wu, M., Jang, D., Jeong, Y., Jung, G., Bae, J.-H., Jang, H.W., Lee, J.-H. 2018. Observation of Physisorption in a High-Performance FET-Type Oxygen Gas Sensor Operating at Room Temperature. *Nanoscale 10*: 18019–18027.
15 Jeong, Y., Shin, J., Hong, Y., Wu, M., Hong, S., Kwon, K.C., Choi, S., Lee, T., Jang, H.W., Lee, J.-H. 2019. Gas Sensing Characteristics of the FET-Type Gas Sensor Having Inkjet-Printed WS2 Sensing Layer. *Solid-State Electronics 153*: 27–32.
16 Zhou, S., Xiao, M., Liu, F., He, J., Lin, Y., Zhang, Z. 2021. Sub-10 Parts per Billion Detection of Hydrogen with Floating Gate Transistors Built on Semiconducting Carbon Nanotube Film. *Carbon 180*: 41–47.
17 Kim, Y., Kang, S.-K., Oh, N.-C., Lee, H.-D., Lee, S.-M., Park, J., Kim, H. 2019. Improved Sensitivity in Schottky Contacted Two-Dimensional MoS_2 Gas Sensor. *ACS Appl. Mater. Interfaces 11*: 38902–38909.
18 Minh Triet, N., Thai Duy, L., Hwang, B.-U., Hanif, A., Siddiqui, S., Park, K.-H., Cho, C.-Y., Lee, N.-E. 2017. High-Performance Schottky Diode Gas Sensor Based on the Heterojunction of Three-Dimensional Nanohybrids of Reduced Graphene Oxide–Vertical ZnO Nanorods on an AlGaN/GaN Layer. *ACS Appl. Mater. Interfaces 9*: 30722–30732.
19 Abdullah, Q.N., Ahmed, A.R., Ali, A.M., Yam, F.K., Hassan, Z., Bououdina, M., Almessiere, M.A. 2018. Growth and Characterization of GaN Nanostructures under Various Ammoniating Time with Fabricated Schottky Gas Sensor Based on Si Substrate. *Superlattices and Microstructures 117*: 92–104.
20 Punetha, D., Pandey, S.K. 2019. Ultrasensitive NH_3 Gas Sensor Based on Au/ZnO/n-Si Heterojunction Schottky Diode. *IEEE Trans. Electron Devices 66*: 3560–3567.
21 Lee, Y., Kwon, H., Yoon, J.-S., Kim, J.K. 2019. Overcoming Ineffective Resistance Modulation in P-Type NiO Gas Sensor by Nanoscale Schottky Contacts. *Nanotechnology 30*: 115501.
22 Gong, B., Shi, T., Zhu, W., Liao, G., Li, X., Huang, J., Zhou, T., Tang, Z. 2017. UV Irradiation-Assisted Ethanol Detection Operated by the Gas Sensor Based on ZnO Nanowires/Optical Fiber Hybrid Structure. *Sensors and Actuators B: Chemical 245*: 821–827.
23 Xie, T., Sullivan, N., Steffens, K., Wen, B., Liu, G., Debnath, R., Davydov, A., Gomez, R., Motayed, A. 2015. UV-Assisted Room-Temperature Chemiresistive NO2 Sensor Based on TiO2 Thin Film. *Journal of Alloys and Compounds 653*: 255–259.
24 Joshi, N., da Silva, L.F., Shimizu, F.M., Mastelaro, V.R., M'Peko, J.-C., Lin, L., Oliveira, O.N. 2019. UV-Assisted Chemiresistors Made with Gold-Modified ZnO Nanorods to Detect Ozone Gas at Room Temperature. *Microchim Acta 186*: 418.
25 Xie, T., Rani, A., Wen, B., Castillo, A., Thomson, B., Debnath, R., Murphy, T.E., Gomez, R.D., Motayed, A. 2016. The Effects of Surface Conditions of TiO2 Thin Film on the UV Assisted Sensing Response at Room Temperature. *Thin Solid Films 620*: 76–81.
26 Qi, P., Vermesh, O., Grecu, M., Javey, A., Wang, Q., Dai, H., Peng, S., Cho, K.J. 2003. Toward Large Arrays of Multiplex Functionalized Carbon Nanotube Sensors for Highly Sensitive and Selective Molecular Detection. *Nano Lett. 3*: 347–351.
27 Bezdek, M.J., Luo, S.-X.L., Ku, K.H., Swager, T.M. 2021. A Chemiresistive Methane Sensor. *Proc. Natl. Acad. Sci. U.S.A. 118*: e2022515118.

28 Hua, C., Shang, Y., Wang, Y., Xu, J., Zhang, Y., Li, X., Cao, A. 2017. A Flexible Gas Sensor Based on Single-Walled Carbon Nanotube-Fe_2O_3 Composite Film. *Applied Surface Science 405*: 405–411.

29 Navazani, S., Hassanisadi, M., Eskandari, M.M., Talaei, Z. 2020. Design and Evaluation of SnO2-Pt/MWCNTs Hybrid System as Room Temperature-Methane Sensor. *Synthetic Metals 260*: 116267.

30 Le, X.V., Luu, T.L.A., Nguyen, H.L., Nguyen, C.T. 2019. Synergistic Enhancement of Ammonia Gas-Sensing Properties at Low Temperature by Compositing Carbon Nanotubes with Tungsten Oxide Nanobricks. *Vacuum 168*: 108861.

31 Xiao, M., Liang, S., Han, J., Zhong, D., Liu, J., Zhang, Z., Peng, L. 2018. Batch Fabrication of Ultrasensitive Carbon Nanotube Hydrogen Sensors with Sub-Ppm Detection Limit. *ACS Sens. 3*: 749–756.

32 Schedin, F., Geim, A.K., Morozov, S.V., Hill, E.W., Blake, P., Katsnelson, M.I., Novoselov, K.S. 2007. Detection of Individual Gas Molecules Adsorbed on Graphene. *Nature Mater 6*: 652–655.

33 Kim, C.-H., Yoo, S.-W., Nam, D.-W., Seo, S., Lee, J.-H. 2012. Effect of Temperature and Humidity on NO_2 and NH_3 Gas Sensitivity of Bottom-Gate Graphene FETs Prepared by ICP-CVD. *IEEE Electron Device Letters 33*: 1084–1086.

34 Li, H., Wu, J., Qi, X., He, Q., Liusman, C., Lu, G., Zhou, X., Zhang, H. 2013. Graphene Oxide Scrolls on Hydrophobic Substrates Fabricated by Molecular Combing and Their Application in Gas Sensing. *Small 9*: 382–386.

35 Wu, J., Li, H., Qi, X., He, Q., Xu, B., Zhang, H. 2014. Graphene Oxide Architectures Prepared by Molecular Combing on Hydrophilic-Hydrophobic Micropatterns. *Small 10*: 2239–2244.

36 Ren, Y., Zhu, C., Cai, W., Li, H., Ji, H., Kholmanov, I., Wu, Y., Piner, R.D., Ruoff, R.S. 2012. Detection of Sulfur Dioxide Gas with Graphene Field Effect Transistor. *Appl. Phys. Lett. 100*: 163114.

37 Nomani, Md. W.K., Shishir, R., Qazi, M., Diwan, D., Shields, V.B., Spencer, M.G., Tompa, G.S., Sbrockey, N.M., Koley, G. 2010. Highly Sensitive and Selective Detection of NO2 Using Epitaxial Graphene on 6H-SiC. *Sensors and Actuators B: Chemical 150*: 301–307.

38 Cheng, C.-C., Wu, C.-L., Liao, Y.-M., Chen, Y.-F. 2016. Ultrafast and Ultrasensitive Gas Sensors Derived from a Large Fermi-Level Shift in the Schottky Junction with Sieve-Layer Modulation. *ACS Appl. Mater. Interfaces 8*: 17382–17388.

39 Singh, S., Sharma, S., Singh, R.C., Sharma, S. 2020. Hydrothermally Synthesized MoS_2-Multi-Walled Carbon Nanotube Composite as a Novel Room-Temperature Ammonia Sensing Platform. *Applied Surface Science 532*: 147373.

40 Jeevitha, G., Abhinayaa, R., Mangalaraj, D., Ponpandian, N., Meena, P., Mounasamy, V., Madanagurusamy, S. 2019. Porous Reduced Graphene Oxide (RGO)/WO_3 Nanocomposites for the Enhanced Detection of NH_3 at Room Temperature. *Nanoscale Adv. 1*: 1799–1811.

41 Atkinson, Richard. W., Butland, Barbara. K., Anderson, H. Ross., Maynard, Robert. L. 2018. Long-Term Concentrations of Nitrogen Dioxide and Mortality: A Meta-Analysis of Cohort Studies. *Epidemiology 29*: 460–472.

42 Faustini, A., Rapp, R., Forastiere, F. 2014. Nitrogen Dioxide and Mortality: Review and Meta-Analysis of Long-Term Studies. *European Respiratory Journal 44*: 744–753.

43 Ambient (outdoor) air pollution www.who.int/news-room/fact-sheets/detail/ambient-(outdoor)-air-quality-and-health (accessed Mar 14, 2022).

44 Liu, B., Liu, X., Yuan, Z., Jiang, Y., Su, Y., Ma, J., Tai, H. 2019. A Flexible NO_2 Gas Sensor Based on Polypyrrole/Nitrogen-Doped Multiwall Carbon Nanotube Operating at Room Temperature. *Sensors and Actuators B: Chemical 295*: 86–92.

45 Liu, S., Wang, Z., Zhang, Y., Zhang, C., Zhang, T. 2015. High Performance Room Temperature NO2 Sensors Based on Reduced Graphene Oxide-Multiwalled Carbon Nanotubes-Tin Oxide Nanoparticles Hybrids. *Sensors and Actuators B: Chemical 211*: 318–324.

46 Zhou, C., Zhao, J., Ye, J., Tange, M., Zhang, X., Xu, W., Zhang, K., Okazaki, T., Cui, Z. 2016. Printed Thin-Film Transistors and NO2 Gas Sensors Based on Sorted Semiconducting Carbon Nanotubes by Isoindigo-Based Copolymer. *Carbon 108*: 372–380.

47 Mattmann, M., Roman, C., Helbling, T., Bechstein, D., Durrer, L., Pohle, R., Fleischer, M., Hierold, C. 2010. Pulsed Gate Sweep Strategies for Hysteresis Reduction in Carbon Nanotube Transistors for Low Concentration NO_2 Gas Detection. *Nanotechnology 21*: 185501.

48 Helbling, T., Hierold, C., Durrer, L., Roman, C., Pohle, R., Fleischer, M. 2008. Suspended and Non-suspended Carbon Nanotube Transistors for NO2 Sensing—A Qualitative Comparison. *physica status solidi (b) 245*: 2326–2330.

49 You, R., Han, D.-D., Liu, F., Zhang, Y.-L., Lu, G. 2018. Fabrication of Flexible Room-Temperature NO_2 Sensors by Direct Laser Writing of In2O3 and Graphene Oxide Composites. *Sensors and Actuators B: Chemical 277*: 114–120.

50 Van Cat, V., Dinh, N.X., Ngoc Phan, V., Le, A.T., Nam, M.H., Dinh Lam, V., Dang, T.V., Quy, N.V. 2020. Realization of Graphene Oxide Nanosheets as a Potential Mass-Type Gas Sensor for Detecting NO_2, SO_2, CO, and NH_3. *Materials Today Communications 25*: 101682.

51 Bo, Z., Wei, X., Guo, X., Yang, H., Mao, S., Yan, J., Cen, K. 2020. SnO_2 Nanoparticles Incorporated CuO Nanopetals on Graphene for High-Performance Room-Temperature NO_2 Sensor. *Chemical Physics Letters 750*: 137485.

52 Wu, J., Wu, Z., Ding, H., Wei, Y., Huang, W., Yang, X., Li, Z., Qiu, L., Wang, X. 2020. Flexible, 3D SnS_2/Reduced Graphene Oxide Heterostructured NO2 Sensor. *Sensors and Actuators B: Chemical 305*: 127445.

53 Hong, H.S., Phuong, N.H., Huong, N.T., Nam, N.H., Hue, N.T. 2019. Highly Sensitive and Low Detection Limit of Resistive NO_2 Gas Sensor Based on a MoS_2/Graphene Two-Dimensional Heterostructures. *Applied Surface Science 492*: 449–454.

54 Zhang, Z., Gao, Z., Fang, R., Li, H., He, W., Du, C. 2020. UV-Assisted Room Temperature NO_2 Sensor Using Monolayer Graphene Decorated with SnO_2 Nanoparticles. *Ceramics International 46*: 2255–2260.

55 Shaik, M., Rao, V.K., Gupta, M., Murthy, K.S.R.C., Jain, R. 2015. Chemiresistive Gas Sensor for the Sensitive Detection of Nitrogen Dioxide Based on Nitrogen Doped Graphene Nanosheets. *RSC Adv. 6*: 1527–1534.

56 Ma, J., Zhang, M., Dong, L., Sun, Y., Su, Y., Xue, Z., Di, Z. 2019. Gas Sensor Based on Defective Graphene/Pristine Graphene Hybrid towards High Sensitivity Detection of NO_2. *AIP Advances 9*: 075207.

57 He, L., Lv, H., Ma, L., Li, W., Si, J., Ikram, M., Ullah, M., Wu, H., Wang, R., Shi, K. 2020. Controllable Synthesis of Intercalated γ-Bi_2MoO_6/Graphene Nanosheet Composites for High Performance NO_2 Gas Sensor at Room Temperature. *Carbon 157*: 22–32.

58 Yuan, W., Liu, A., Huang, L., Li, C., Shi, G. 2013. High-Performance NO_2 Sensors Based on Chemically Modified Graphene. *Advanced Materials 25*: 766–771.

59 Koenig, J.Q. 2000. Health Effects of Sulfur Oxides: Sulfur Dioxide and Sulfuric Acid. In *Health Effects of Ambient Air Pollution: How safe is the air we breathe?* Ed. Koenig, J.Q., 99–114. Boston, MA: Springer US, 99–114.

60 Septiani, N.L.W., Saputro, A.G., Kaneti, Y.V., Maulana, A.L., Fathurrahman, F., Lim, H., Yuliarto, B., Nugraha, Dipojono, H.K., Golberg, D., et al. 2020. Hollow Zinc Oxide Microsphere–Multiwalled Carbon Nanotube Composites for Selective Detection of Sulfur Dioxide. *ACS Appl. Nano Mater. 3*: 8982–8996.

61 Tyagi, P., Sharma, A., Tomar, M., Gupta, V. 2017. A Comparative Study of RGO-SnO_2 and MWCNT-SnO_2 Nanocomposites Based SO_2 Gas Sensors. *Sensors and Actuators B: Chemical 248*: 980–986.

62 Su, P.-G., Zheng, Y.-L. 2021. Room-Temperature Ppb-Level SO_2 Gas Sensors Based on RGO/WO_3 and MWCNTs/WO_3 Nanocomposites. *Anal. Methods 13*: 782–788.

63 Pandey, S.K., Kim, K.-H., Tang, K.-T. 2012. A Review of Sensor-Based Methods for Monitoring Hydrogen Sulfide. *TrAC Trends in Analytical Chemistry 32*: 87–99.

64 Szabo, C. 2018. A Timeline of Hydrogen Sulfide (H2S) Research: From Environmental Toxin to Biological Mediator. *Biochemical Pharmacology 149*: 5–19.

65 Kim, K.-H., Choi, Y., Jeon, E., Sunwoo, Y. 2005. Characterization of Malodorous Sulfur Compounds in Landfill Gas. *Atmospheric Environment 39*: 1103–1112.

66 Doujaiji, B., Al-Tawfiq, J.A. 2010. Hydrogen Sulfide Exposure in an Adult Male. *Ann Saudi Med 30*: 76–80.

67 Asad, M., Sheikhi, M.H., Pourfath, M., Moradi, M. 2015. High Sensitive and Selective Flexible H2S Gas Sensors Based on Cu Nanoparticle Decorated SWCNTs. *Sensors and Actuators B: Chemical 210*: 1–8.

68 Mubeen, S., Zhang, T., Chartuprayoon, N., Rheem, Y., Mulchandani, A., Myung, N.V., Deshusses, M.A. 2010. Sensitive Detection of H2S Using Gold Nanoparticle Decorated Single-Walled Carbon Nanotubes. *Anal. Chem. 82*: 250–257.

69 Jung, H.Y., Kim, Y.L., Park, S., Datar, A., Lee, H.-J., Huang, J., Somu, S., Busnaina, A., Jung, Y.J., Kwon, Y.-K. 2013. High-Performance H2S Detection by Redox Reactions in Semiconducting Carbon Nanotube-Based Devices. *Analyst 138*: 7206–7211.

70 Moon, S., Vuong, N.M., Lee, D., Kim, D., Lee, H., Kim, D., Hong, S.-K., Yoon, S.-G. 2016. Co_3O_4–SWCNT Composites for H_2S Gas Sensor Application. *Sensors and Actuators B: Chemical 222*: 166–172.

71 Yang, M., Zhang, X., Cheng, X., Xu, Y., Gao, S., Zhao, H., Huo, L. 2017. Hierarchical NiO Cube/Nitrogen-Doped Reduced Graphene Oxide Composite with Enhanced H_2S Sensing Properties at Low Temperature. *ACS Appl Mater Interfaces 9*: 26293–26303.

72 Song, Z., Wei, Z., Wang, B., Luo, Z., Xu, S., Zhang, W., Yu, H., Li, M., Huang, Z., Zang, J., et al. 2016. Sensitive Room-Temperature H2S Gas Sensors Employing SnO_2 Quantum Wire/Reduced Graphene Oxide Nanocomposites. *Chem. Mater. 28*: 1205–1212.

73 Shao, S., Chen, X., Chen, Y., Zhang, L., Kim, H.W., Kim, S.S. 2020. ZnO Nanosheets Modified with Graphene Quantum Dots and SnO_2 Quantum Nanoparticles for Room-Temperature H_2S Sensing. *ACS Appl. Nano Mater. 3*: 5220–5230.

74 Schimmelmann, A., Sauer, P.E. 2018. Hydrogen Isotopes. In *Encyclopedia of Geochemistry: A Comprehensive Reference Source on the Chemistry of the Earth*, Ed. White, W.M., 696–701. Cham: Springer International Publishing, 696–701.

75 Korotcenkov, G., Han, S.D., Stetter, J.R. 2009. Review of Electrochemical Hydrogen Sensors. *Chem. Rev. 109*: 1402–1433.

76 Hübert, T., Boon-Brett, L., Black, G., Banach, U. 2011. Hydrogen Sensors—A Review. *Sensors and Actuators B: Chemical 157*: 329–352.

77 Li, X., Le Thai, M., Dutta, R.K., Qiao, S., Chandran, G.T., Penner, R.M. 2017. Sub-6 nm Palladium Nanoparticles for Faster, More Sensitive H_2 Detection Using Carbon Nanotube Ropes. *ACS Sens. 2*: 282–289.

78 Zhang, M., Brooks, L.L., Chartuprayoon, N., Bosze, W., Choa, Y., Myung, N.V. 2014. Palladium/Single-Walled Carbon Nanotube Back-to-Back Schottky Contact-Based Hydrogen Sensors and Their Sensing Mechanism. *ACS Appl. Mater. Interfaces 6*: 319–326.

79 Choi, B., Lee, D., Ahn, J.-H., Yoon, J., Lee, J., Jeon, M., Kim, D.M., Kim, D.H., Park, I., Choi, Y.-K., et al. 2015. Investigation of Optimal Hydrogen Sensing Performance in Semiconducting Carbon Nanotube Network Transistors with Palladium Electrodes. *Appl. Phys. Lett. 107*: 193108.

80 Lin, T.-C., Huang, B.-R. 2012. Palladium Nanoparticles Modified Carbon Nanotube/Nickel Composite Rods (Pd/CNT/Ni) for Hydrogen Sensing. *Sensors and Actuators B: Chemical 162*: 108–113.

81 Randeniya, L.K., Martin, P.J., Bendavid, A. 2012. Detection of Hydrogen Using Multi-Walled Carbon-Nanotube Yarns Coated with Nanocrystalline Pd and Pd/Pt Layered Structures. *Carbon 50*: 1786–1792.

82 Alamri, M.A., Liu, B., Walsh, M., Doolin, J.L., Berrie, C.L., Wu, J.Z. 2021. Enhanced H_2 Sensitivity in Ultraviolet-Activated Pt Nanoparticle/SWCNT/Graphene Nanohybrids. *IEEE Sensors Journal 21*: 19762–19770.

83 Dhall, S., Kumar, M., Bhatnagar, M., Mehta, B.R. 2018. Dual Gas Sensing Properties of Graphene-Pd/SnO_2 Composites for H2 and Ethanol: Role of Nanoparticles-Graphene Interface. *International Journal of Hydrogen Energy 43*: 17921–17927.

84 Soleimanpour, A.M., Jayatissa, A.H., Sumanasekera, G. 2013. Surface and Gas Sensing Properties of Nanocrystalline Nickel Oxide Thin Films. *Applied Surface Science 276*: 291–297.

85 Soleimanpour, A.M., Khare, S.V., Jayatissa, A.H. 2012. Enhancement of Hydrogen Gas Sensing of Nanocrystalline Nickel Oxide by Pulsed-Laser Irradiation. *ACS Appl. Mater. Interfaces 4*: 4651–4657.

86 Kamal, T. 2017. High Performance NiO Decorated Graphene as a Potential H2 Gas Sensor. *Journal of Alloys and Compounds 729*: 1058–1063.
87 Das, S., Roy, S., Bhattacharya, T.S., Sarkar, C.K. 2021. Efficient Room Temperature Hydrogen Gas Sensor Using ZnO Nanoparticles-Reduced Graphene Oxide Nanohybrid. *IEEE Sensors Journal 21*: 1264–1272.
88 Sun, K., Tao, L., Miller, D.J., Pan, D., Golston, L.M., Zondlo, M.A., Griffin, R.J., Wallace, H.W., Leong, Y.J., Yang, M.M., et al. 2017. Vehicle Emissions as an Important Urban Ammonia Source in the United States and China. *Environ. Sci. Technol. 51*: 2472–2481.
89 Walters, W.W., Song, L., Chai, J., Fang, Y., Colombi, N., Hastings, M.G. 2020. Characterizing the Spatiotemporal Nitrogen Stable Isotopic Composition of Ammonia in Vehicle Plumes. *Atmospheric Chemistry and Physics 20*: 11551–11567.
90 Li, Y., Thompson, T.M., Van Damme, M., Chen, X., Benedict, K.B., Shao, Y., Day, D., Boris, A., Sullivan, A.P., Ham, J., et al. 2017. Temporal and Spatial Variability of Ammonia in Urban and Agricultural Regions of Northern Colorado, United States. *Atmospheric Chemistry and Physics 17*: 6197–6213.
91 Xue, L., Wang, W., Guo, Y., Liu, G., Wan, P. 2017. Flexible Polyaniline/Carbon Nanotube Nanocomposite Film-Based Electronic Gas Sensors. *Sensors and Actuators B: Chemical 244*: 47–53.
92 Du, W.X., Lee, H.-J., Byeon, J.-H., Kim, J.-S., Cho, K.-S., Kang, S., Takada, M., Kim, J.-Y. 2020. Highly Sensitive Single-Walled Carbon Nanotube/Polypyrrole/Phenylalanine Core–Shell Nanorods for Ammonia Gas Sensing. *J. Mater. Chem. C 8*: 15609–15615.
93 Seekaew, Y., Pon-On, W., Wongchoosuk, C. 2019. Ultrahigh Selective Room-Temperature Ammonia Gas Sensor Based on Tin–Titanium Dioxide/Reduced Graphene/Carbon Nanotube Nanocomposites by the Solvothermal Method. *ACS Omega 4*: 16916–16924.
94 Gao, J., Qin, J., Chang, J., Liu, H., Wu, Z.-S., Feng, L. 2020. NH_3 Sensor Based on 2D Wormlike Polypyrrole/Graphene Heterostructures for a Self-Powered Integrated System. *ACS Appl. Mater. Interfaces 12*: 38674–38681.
95 Hung, C.M., Dat, D.Q., Van Duy, N., Van Quang, V., Van Toan, N., Van Hieu, N., Hoa, N.D. 2020. Facile Synthesis of Ultrafine RGO/WO_3 Nanowire Nanocomposites for Highly Sensitive Toxic NH3 Gas Sensors. *Materials Research Bulletin 125*: 110810.
96 Su, P.-G., Yang, L.-Y. 2016. NH3 Gas Sensor Based on Pd/SnO_2/RGO Ternary Composite Operated at Room-Temperature. *Sensors and Actuators B: Chemical 223*: 202–208.
97 Pang, Z., Yang, Z., Chen, Y., Zhang, J., Wang, Q., Huang, F., Wei, Q. 2016. A Room Temperature Ammonia Gas Sensor Based on Cellulose/TiO_2/PANI Composite Nanofibers. *Colloids and Surfaces A: Physicochemical and Engineering Aspects 494*: 248–255.
98 Pang, Z., Fu, J., Luo, L., Huang, F., Wei, Q. 2014. Fabrication of PA6/TiO_2/PANI Composite Nanofibers by Electrospinning–Electrospraying for Ammonia Sensor. *Colloids and Surfaces A: Physicochemical and Engineering Aspects 461*: 113–118.
99 Seif, A.M., Nikfarjam, A., Hajghassem, H. 2019. UV Enhanced Ammonia Gas Sensing Properties of PANI/TiO_2 Core-Shell Nanofibers. *Sensors and Actuators B: Chemical 298*: 126906.
100 Kukla, A.L., Shirshov, Yu. M., Piletsky, S.A. 1996. Ammonia Sensors Based on Sensitive Polyaniline Films. *Sensors and Actuators B: Chemical 37*: 135–140.
101 Chabukswar, V.V., Pethkar, S., Athawale, A.A. 2001. Acrylic Acid Doped Polyaniline as an Ammonia Sensor. *Sensors and Actuators B: Chemical 77*: 657–663.
102 Hadano, F.S., Gavim, A.E.X., Stefanelo, J.C., Gusso, S.L., Macedo, A.G., Rodrigues, P.C., Mohd Yusoff, A.R. bin, Schneider, F.K., Deus, J.F. de, José da Silva, W. 2021. NH_3 Sensor Based on RGO-PANI Composite with Improved Sensitivity. *Sensors 21*: 4947.
103 Turner, A.J., Frankenberg, C., Kort, E.A. 2019. Interpreting Contemporary Trends in Atmospheric Methane. *Proceedings of the National Academy of Sciences 116*: 2805–2813.
104 Aldalbahi, A., Feng, P., Alhokbany, N., Al-Farraj, E., Alshehri, S.M., Ahamad, T. 2017. Synthesis and Characterization of Hybrid Nanocomposites as Highly-Efficient Conducting CH_4 Gas Sensor. *Spectrochimica Acta Part A: Molecular and Biomolecular Spectroscopy 173*: 502–509.

105 Humayun, M.T., Divan, R., Stan, L., Rosenmann, D., Gosztola, D., Gundel, L., Solomon, P.A., Paprotny, I. 2016. Ubiquitous Low-Cost Functionalized Multi-Walled Carbon Nanotube Sensors for Distributed Methane Leak Detection. *IEEE Sensors Journal 16*: 8692–8699.

106 Humayun, M.T., Divan, R., Stan, L., Gupta, A., Rosenmann, D., Gundel, L., Solomon, P.A., Paprotny, I. 2015. ZnO Functionalization of Multiwalled Carbon Nanotubes for Methane Sensing at Single Parts per Million Concentration Levels. *Journal of Vacuum Science & Technology B 33*: 06FF01.

107 Humayun, M.T., Sainato, M., Divan, R., Rosenberg, R.A., Sahagun, A., Gundel, L., Solomon, P.A., Paprotny, I. 2017. Effects of O_2 Plasma and UV-O_3 Assisted Surface Activation on High Sensitivity Metal Oxide Functionalized Multiwalled Carbon Nanotube CH_4 Sensors. *Journal of Vacuum Science & Technology A 35*: 061402.

108 Yan, J., Wei, T., Shao, B., Fan, Z., Qian, W., Zhang, M., Wei, F. 2010. Preparation of a Graphene Nanosheet/Polyaniline Composite with High Specific Capacitance. *Carbon 48*: 487–493.

109 Wu, Z., Chen, X., Zhu, S., Zhou, Z., Yao, Y., Quan, W., Liu, B. 2013. Room Temperature Methane Sensor Based on Graphene Nanosheets/Polyaniline Nanocomposite Thin Film. *IEEE Sensors Journal 13*: 777–782.

110 Zhang, D., Sun, Y., Jiang, C., Zhang, Y. 2017. Room Temperature Hydrogen Gas Sensor Based on Palladium Decorated Tin Oxide/Molybdenum Disulfide Ternary Hybrid via Hydrothermal Route. *Sensors and Actuators B: Chemical 242*: 15–24.

111 Li, Y., Deng, D., Xing, X., Chen, N., Liu, X., Xiao, X., Wang, Y. 2016. A High Performance Methanol Gas Sensor Based on Palladium-Platinum-In_2O_3 Composited Nanocrystalline SnO_2. *Sensors and Actuators B: Chemical 237*: 133–141.

112 Absalan, S., Nasresfahani, Sh., Sheikhi, M.H. 2019. High-Performance Carbon Monoxide Gas Sensor Based on Palladium/Tin Oxide/Porous Graphitic Carbon Nitride Nanocomposite. *Journal of Alloys and Compounds 795*: 79–90.

113 Nasresfahani, Sh., Sheikhi, M.H., Tohidi, M., Zarifkar, A. 2017. Methane Gas Sensing Properties of Pd-Doped SnO_2/Reduced Graphene Oxide Synthesized by a Facile Hydrothermal Route. *Materials Research Bulletin 89*: 161–169.

114 Niu, F., Liu, J.-M., Tao, L.-M., Wang, W., Song, W.-G. 2013. Nitrogen and Silica Co-Doped Graphene Nanosheets for NO_2 Gas Sensing. *J. Mater. Chem. A 1*: 6130–6133.

115 Roshan, H., Sheikhi, M.H., Faramarzi Haghighi, M.K., Padidar, P. 2020. High-Performance Room Temperature Methane Gas Sensor Based on Lead Sulfide/Reduced Graphene Oxide Nanocomposite. *IEEE Sensors Journal 20*: 2526–2532.

116 Li, D., Tang, Y., Ao, D., Xiang, X., Wang, S., Zu, X. 2019. Ultra-Highly Sensitive and Selective H_2S Gas Sensor Based on CuO with Sub-Ppb Detection Limit. *International Journal of Hydrogen Energy 44*: 3985–3992.

117 Poloju, M., Jayababu, N., Ramana Reddy, M.V. 2018. Improved Gas Sensing Performance of Al Doped ZnO/CuO Nanocomposite Based Ammonia Gas Sensor. *Materials Science and Engineering: B 227*: 61–67.

118 Rzaij, J.M., Abass, A.M. 2020. Review on: TiO_2 Thin Film as a Metal Oxide Gas Sensor. *Journal of Chemical Reviews 2*: 114–121.

119 Murali, G., Reddeppa, M., Seshendra Reddy, Ch., Park, S., Chandrakalavathi, T., Kim, M.-D., In, I. 2020. Enhancing the Charge Carrier Separation and Transport via Nitrogen-Doped Graphene Quantum Dot-TiO_2 Nanoplate Hybrid Structure for an Efficient NO Gas Sensor. *ACS Appl. Mater. Interfaces 12*: 13428–13436.

120 Umar, A., Ibrahim, A.A., Kumar, R., Albargi, H., Zeng, W., Alhmami, M.A.M., Alsaiari, M.A., Baskoutas, S. 2021. Gas Sensor Device for High-Performance Ethanol Sensing Using α-MnO_2 Nanoparticles. *Materials Letters 286*: 129232.

121 Bigiani, L., Zappa, D., Maccato, C., Comini, E., Barreca, D., Gasparotto, A. 2020. Quasi-1D MnO_2 Nanocomposites as Gas Sensors for Hazardous Chemicals. *Applied Surface Science 512*: 145667.

122 Liu, C., Navale, S.T., Yang, Z.B., Galluzzi, M., Patil, V.B., Cao, P.J. , Mane, R.S., Stadler, F.J. 2017. Ethanol Gas Sensing Properties of Hydrothermally Grown α-MnO_2 Nanorods. *Journal of Alloys and Compounds 727*: 362–369.

123 Gui, Y., Tian, K., Liu, J., Yang, L., Zhang, H., Wang, Y. 2019. Superior Triethylamine Detection at Room Temperature by {-112} Faceted WO_3 Gas Sensor. *Journal of Hazardous Materials 380*: 120876.

124 Li, S., Xie, L., He, M., Hu, X., Luo, G., Chen, C., Zhu, Z. 2020. Metal-Organic Frameworks-Derived Bamboo-like CuO/In_2O_3 Heterostructure for High-Performance H_2S Gas Sensor with Low Operating Temperature. *Sensors and Actuators B: Chemical 310*: 127828.

125 Ma, J., Fan, H., Zhang, W., Sui, J., Wang, C., Zhang, M., Zhao, N., Kumar Yadav, A., Wang, W., Dong, W., et al. 2020. High Sensitivity and Ultra-Low Detection Limit of Chlorine Gas Sensor Based on In2O3 Nanosheets by a Simple Template Method. *Sensors and Actuators B: Chemical 305*: 127456.

126 Han, D., Zhai, L., Gu, F., Wang, Z. 2018. Highly Sensitive NO_2 Gas Sensor of Ppb-Level Detection Based on In2O3 Nanobricks at Low Temperature. *Sensors and Actuators B: Chemical 262*: 655–663.

127 Nakate, U.T., Ahmad, R., Patil, P., Yu, Y.T., Hahn, Y.-B. 2020. Ultra Thin NiO Nanosheets for High Performance Hydrogen Gas Sensor Device. *Applied Surface Science 506*: 144971.

128 Zhou, Q., Zeng, W., Chen, W., Xu, L., Kumar, R., Umar, A. 2019. High Sensitive and Low-Concentration Sulfur Dioxide (SO_2) Gas Sensor Application of Heterostructure NiO-ZnO Nanodisks. *Sensors and Actuators B: Chemical 298*: 126870.

129 Park, J., Mun, J., Shin, J.-S., Kang, S.-W. 2018. Highly Sensitive Two-Dimensional MoS_2 Gas Sensor Decorated with Pt Nanoparticles. *R Soc Open Sci 5*: 181462.

130 Baek, D.-H., Kim, J. 2017. MoS_2 Gas Sensor Functionalized by Pd for the Detection of Hydrogen. *Sensors and Actuators B: Chemical 250*: 686–691.

131 Franke, M.E., Koplin, T.J., Simon, U. 2006. Metal and Metal Oxide Nanoparticles in Chemiresistors: Does the Nanoscale Matter? *Small 2*: 36–50.

132 Hong, J., Lee, S., Seo, J., Pyo, S., Kim, J., Lee, T. 2015. A Highly Sensitive Hydrogen Sensor with Gas Selectivity Using a PMMA Membrane-Coated Pd Nanoparticle/Single-Layer Graphene Hybrid. *ACS Appl. Mater. Interfaces 7*: 3554–3561.

133 Zhang, S., Park, J.G., Nguyen, N., Jolowsky, C., Hao, A., Liang, R. 2017. Ultra-High Conductivity and Metallic Conduction Mechanism of Scale-up Continuous Carbon Nanotube Sheets by Mechanical Stretching and Stable Chemical Doping. *Carbon 125*: 649–658.

134 Ressler, T., Walter, A., Scholz, J., Tessonnier, J.-P., Su, D.S. 2010. Structure and Properties of a Mo Oxide Catalyst Supported on Hollow Carbon Nanofibers in Selective Propene Oxidation. *Journal of Catalysis 271*: 305–314.

135 Marichy, C., Russo, P.A., Latino, M., Tessonnier, J.-P., Willinger, M.-G., Donato, N., Neri, G., Pinna, N. 2013. Tin Dioxide–Carbon Heterostructures Applied to Gas Sensing: Structure-Dependent Properties and General Sensing Mechanism. *J. Phys. Chem. C 117*: 19729–19739.

136 Van Duy, N., Van Hieu, N., Huy, P.T., Chien, N.D., Thamilselvan, M., Yi, J. 2008. Mixed SnO_2/TiO_2 Included with Carbon Nanotubes for Gas-Sensing Application. *Physica E: Low-dimensional Systems and Nanostructures 41*: 258–263.

137 Wu, T.-C., De Luca, A., Zhong, Q., Zhu, X., Ogbeide, O., Um, D.-S., Hu, G., Albrow-Owen, T., Udrea, F., Hasan, T. 2019. Inkjet-Printed CMOS-Integrated Graphene–Metal Oxide Sensors for Breath Analysis. *npj 2D Mater Appl 3*: 1–10.

138 Amiri, V., Roshan, H., Mirzaei, A., Neri, G., Ayesh, A.I. 2020. Nanostructured Metal Oxide-Based Acetone Gas Sensors: A Review. *Sensors 20*: 3096.

5 Carbon Nanotube and Graphene Oxide Reinforced Composites for Lightning Strike Protection of Composite Structures

Vipin Kumar, Sukanta Das, Xianhui Zhao, and Yu Zhou

CONTENTS

5.1 INTRODUCTION

Carbon fiber reinforced plastics (CFRP) or composites are becoming a prevalent structural material in applications where strength and lightweight are desired [1]. Aerospace, automotive, energy, sports, and marine sectors are predominantly adopting CFRP composites [2]. Composites are also known for their corrosion resistance and high specific stiffness and strength compared to traditional metallic materials. However, composites are also known to have poor electrical conductivities. Carbon fibers are inherently electrically conductive (10^3 S/cm), but when they are embedded in an insulating resin matrix (10^{-14} S/cm), the overall electrical conductivity of the CFRP is reduced drastically (10^{-1} S/cm) [3].

The low electrical conductivity of CFRP is a significant concern in the event of a lightning strike on CFRP structures [4]. CFRPs are the main structural components in advanced airplanes, and their usage in wind turbine blades is also increasing. Unfortunately, airplanes and wind turbine blades are often struck by lightning [5]. When CFRP gets hit by a lightning strike, a massive amount of electric current passes through the CFRP structure, generating a vast amount of Joule's heat or resistive heating [6]. The temperature around the lightning arc can reach up to 30,000 degrees Celsius [7]. The extreme amount of heat accompanied by the shock waves and other

DOI: 10.1201/9781003231943-5

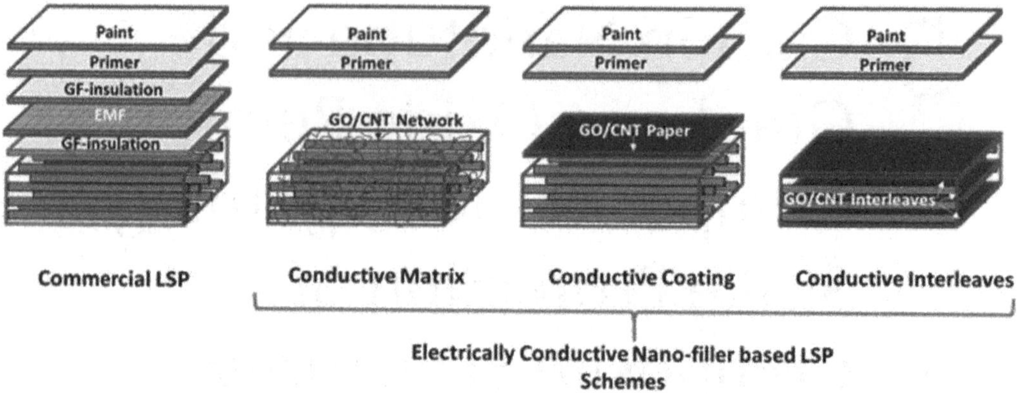

FIGURE 5.1 Commercially used expanded metal foil (EMF)- LSP scheme is compared with nano-filler-based LSP fabrication approaches (i.e., conductive matrix, conductive coating, and interleaved).

mechanical loads generated from lightning can cause catastrophic damage to the CFRP structure [8]. Damages to the CFRP can be in the form of fiber-burnout, matrix evaporation, degradation, ply-delamination, charring, matrix cracking, etc. [9–11]. Therefore, it is paramount to utilize lightning protection systems for these composite structures to protect them from catastrophic damages.

The most common methodology to make this CFRP safe against lightning strikes is to provide a highly conductive coating on top of CFRP, which acts as a Faraday cage and restricts the flow of current to the exterior of the CFRP structure [12]. Commercially available solutions for providing lightning protection to CFRP structures are based on metal-based expanded foils/films [13]. However, metal-based LSP raises serious concerns, such as unwanted additional weight, galvanic corrosion, and complicated integration methods [14]. Due to the issues mentioned earlier with the current LSP system, researchers have been looking for better alternative LSP solutions. Therefore, an essential requirement for the future LSP system is to make the CFRP electrically conductive itself or provide a high electrically conductive coating on the top surface of the CFRP structures [8, 15].

Various studies to improve the electrical conductivity of CFRP have been conducted in the past. Most research methodologies have revolved around improving the electrical conductivity of the CFRP structures by adding conductive fillers into the insulating matrix. In the realm of the material world, the discovery of highly electrically conductive carbon-based nano-fillers has opened a new research paradigm, especially CNTs and graphene. Since the first report of CNT synthesis by Ijima et al. in 1991 [16] and graphene by Novoselov and Geim et al. in 2004 [17], they have always been the focus for researchers for various applications. In this chapter, CNT and GO-based conductive composites/LSP system developments are discussed. Three fabrication approaches are discussed to develop a CNT/GO-based lightning strike(LS) resistive composite/LSP system, i.e. (a) CNT/GO-filled resin, (b) CNT/GO coating or buckypaper as a surface coating layer (c) CNT/GO sheet as interleaves between CFRP laminates as shown in Figure 5.1. Finally, the future scope of this development is covered in the conclusion section.

5.2 FABRICATION APPROACH FOR CNT-BASED COMPOSITES

CNTs with electrical conductivity and mechanical properties close to metals prompted many researchers to investigate their performance as a lightning strike protection material. High aspect ratio and low density make CNTs ideal candidates for achieving unprecedented high electrical conductivity and mechanical properties in composites.

5.2.1 CNT-Based Conductive Resin

One of the most studied ways to import electrical conductivity in an insulating polymer is to create a conductive network using conductive fillers inside the insulating matrix. In this approach, conductive fillers are mixed with insulating polymers (i.e., epoxy) to make the electrically conductive matrix. Further, the conductive matrix is used to fabricate the CFRP composites. The amount of conductive filler should reach a percolation threshold to create conductive pathways inside the insulating matrix. CNTs, due to their extremely high aspect ratio, can reach a percolation at a small loading. However, homogeneous dispersion of CNT in the polymer is a big challenge. Therefore, the highest potential of CNT-based composite can only be achieved by achieving a good dispersion. The quality of CNT dispersion can directly affect the resultant electrical and mechanical properties of the composites. Poor dispersion of CNTs or agglomerated CNTs can lead to deficient mechanical properties due to the stress concentration zones inside the polymer composite. However, higher loading of the CNTs is required to achieve the percolation threshold to get a better electrical conductivity. Higher loading of the CNT leads to an increment in viscosity of the matrix system, which is also detrimental to the processability of the composite manufacturing.

Many researchers took this challenge and studied CNT-filled composites as LSP materials. Chakravarthi et al. worked on developing novel nanocomposites for LSP application [18]. They used nickel-coated single-walled carbon nanotubes (Ni-SWNTs) as conductive fillers inside the high-temperature bismaleimide (BMI) resin. The authors first created Ni-SWCNTs dispersants and then sprayed the Ni-SWNTs directly on the carbon fabrics. Ni-SWNT coated carbon fabric was then impregnated with BMI resin to address the fabrication challenges. Next, they tested a baseline and Ni-SWCNTs-BMI composite against simulated lightning strike according to MIL-STD 1757A standard (Zone 2A lightning strike). They reported that adding 4 wt% Ni-SWCNTs into the BMI composite reduced the electrical resistivity by ten orders of magnitude. In addition, the lightning strike test showed that the central-fiber damage areas indicated by the carbon fiber pull-out for composites filled with and without Ni-SWNTs were 5.17 cm^2 and 32.85 cm^2, respectively. Thus, a significant reduction in lightning damage areas was reported. The two-step approach used by authors helped in creating a homogenous CNT network inside the composite. Results from the study are shown in Figure 5.2.

E. Logakis et al. followed a similar approach and combined multiwall carbon nanotubes (MWNTs) with aerospace-grade epoxy (HexFlow RTM6, Hexcel) resin [19]. Their approach was novel in their resin preparation method; a standard aerospace resin was used. First, they prepared epoxy/CNTs (0.3 wt%) masterbatch, followed by high shear mixing to achieve good CNT dispersion and dilution to 0.1 wt%. The obtained epoxy/CNTs mixture was used to impregnate carbon preform using the vacuum-assisted resin transfer molding (VARTM) process. The authors reported that 0.1 wt% addition of CNTs did not increase the viscosity of the epoxy system, and no problem in resin infusion was observed. In this work, lightning strike tests were performed on four CFRP plates of 320 × 320 mm size as described below:

1. Unprotected carbon fiber/epoxy laminate (control).
2. Unprotected carbon fiber/epoxy/0.1 wt% CNTs laminate (CNT).
3. Mesh-protected carbon fiber/epoxy laminate (mesh).
4. Mesh-protected carbon fiber/epoxy /0.1 wt% CNTs laminate (mesh-CNT).

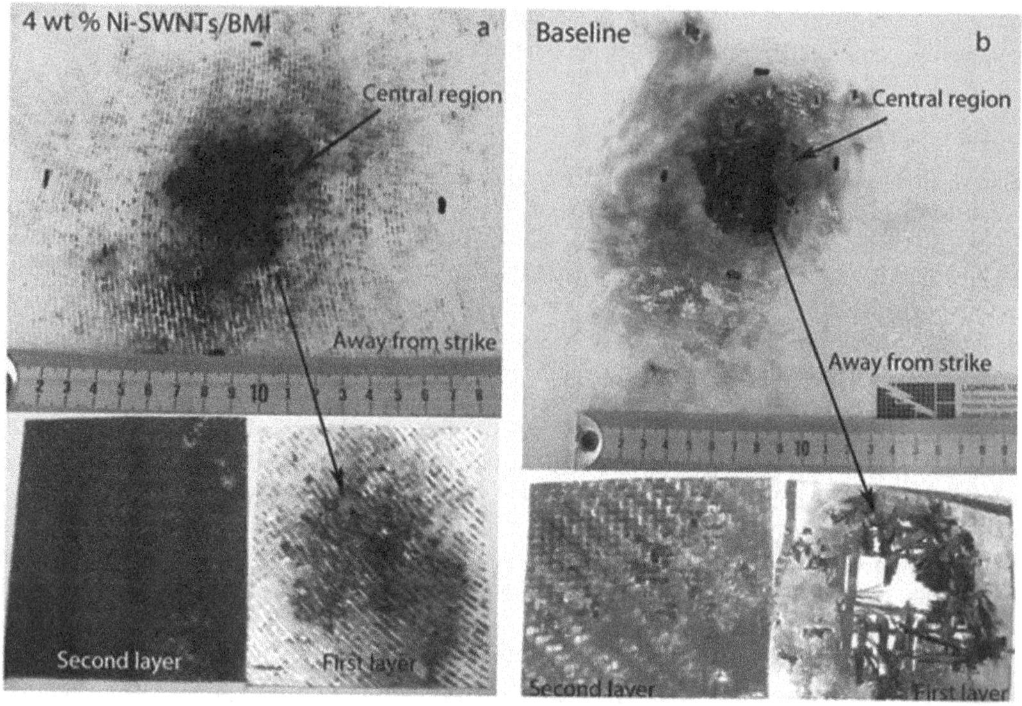

FIGURE 5.2 Photographs of BMI composite samples hit by simulated lightning showing a) minimal damage to the top layer of the Ni-SWCNTs filled composite and b) carbon fiber pull-out in the baseline composite damaging the top layer [18]. Reprinted with permission from [18]. Copyright {2011} John Wiley and Sons.

In this study, simulated lightning strikes, comprising three current components: (i) D (100 kA, ≤ 500 μs); (ii) B (2 kA, ≤ 5 ms); and (iii) C (200—800 A, 0.25—1 s) was applied. It is summarized that the extent of surface damage decreased drastically with the addition of CNTs when compared to unprotected CFRP and metal-mesh protected CFRP. Analysis showed that with the addition of CNTs, the damage was reduced by about 40% in the case of carbon fiber/epoxy/0.1 wt% CNTs laminate and about 60% for the mesh-protected carbon fiber/epoxy /0.1 wt% CNTs laminate.

Many other studies are available in the literature, where researchers used CNTs as conductive filler into the insulation epoxy as a potential lightning strike protection material. However, none of them performed or could not perform an acceptable experimental lightning strike test to confirm their hypothesis. Lampkin et al. also utilized a similar approach and tried to enhance the electrical conductivity of the epoxy system and further tested it against lightning strikes [20]. However, they used hand layup followed by a vacuum bagging fabrication approach. Therefore, they mixed the CNTs directly with the resin and then impregnated them to carbon fibers. They reported a about 31% reduction in electrical resistance of CNT-filled epoxy compared to neat epoxy. A peak current of 100 kA was also used, and the grounding of the CFRP specimen was found not correct to conclude any result from this study. However, they emphasized that the CNT-filled composites can potentially replace the metal-mesh-based LSP system in the aerospace industry; unfortunately, the study was inconclusive.

Similarly, X. Ma et al. and C. Wu et al. did not perform any lightning test but theoretically demonstrated that CNT/CF/epoxy composite can be optimized for fuselage panel case against lightning strike [21] [22]. They demonstrated that adding 2% volume fraction of CNTs in matrix resin improved the through-thickness electrical conductivity of the composite by 81%. However, they also did not perform an actual lightning strike test. The addition of CNTs into epoxy was the most tried method, but none of the researchers could confirm its superiority compared to the commercially available LSP systems. Poor dispersion, poor processability, and difficulty in achieving high electrical conductivity from this method limit its applicability as a lightning strike protection solution.

To counter processability and dispersion, researchers utilized CNTs in bulk to create thin papers or coatings and employed them as LSP materials. The conductive coating/layer fabrication approach also helps retain all the substrate CFRP material properties and fabrication processes, which is highly desirable to incorporate into the aerospace industry. However, CFRP manufacturing processes have already matured, and modifying them is not desirable.

5.2.2 CNT-Based Conductive Coating

In this approach, conductive coating or a conductive layer spray/attached to the top surface of the CFRP composites. This approach has several advantages like ease of installation, easy fabrication steps, cost-effectiveness, and retaining the CFRP composites' mechanical properties. This approach is also closest to the commercially available LSP systems and offers minimal changes in the whole composite manufacturing process. Several researchers worked on CNT-based LSP coating/layer, where a high conductive coating/layer is fabricated on the CFRP composites to guide the lightning current through the top coating/layer and mitigate the damage. This approach is also known as the Faraday Cage approach, as lightning current never penetrates inside the substrate CFRP structure and dissipates via the exterior conductive coating. For Example, Q. Xia et al. developed highly conductive silver modified carbon nanotube film (SMCNF) to protect CFRP structures and components of the aircraft [23]. The authors used the electrophoretic deposition (EPD) method to apply the SMCNF coating on the CFRP substrate. The authors claimed that the LSP system could reduce the weight by 27.4% compared to commercial copper-mesh LSP material. The electrical conductivity of the conductive film was achieved at 5091.65 S/cm with its areal density of 53.3 g/m^2. After the lightning strike test, the LSP film retained the 91.05% residual compressive strength of CFRP composites. The authors struck the panels with a 100-kA lightning waveform A. Although SMCNF film seems to perform extremely well as an LSP material, silver material in the LSP is the most significant disadvantage. Silver is a highly hazardous material; therefore, making LSP with silver is not acceptable for the industry.

J. Han et al. also introduce a high conductive layer of buckypaper (BP) on CFRP composites along with a thick insulating adhesive layer [24]. The thin BP guide dissipates the lightning current, where a thick insulating adhesive layer hinders the transfer of the lightning current through the thickness direction to CFRP laminates. The conductivity of the BP was reported as 5.7 x 10^3 S/m with a thickness of ~ 70 μm. The authors claimed that the proposed LSP system can achieve a weight reduction of up to 30% compared to the commercial Cu-LSP coating and could sustain the lightning strike with a peak current up to 100 kA. This study mimics the commercially available LSP system, where conductive metal mesh helps dissipate the incident, and an underneath glass fabric help to retain the current at the exterior. One challenge with this approach is that the effect of the glass layer on the LSP performance is not fully understood, as many studies show that the insulating layer between the conductive LSP layer and the CFRP substrate can even cause severe damage the CFRP [8]. This behavior can be attributed to the dielectric breakdown voltage of the insulating layer. If the current does not breach the insulating layer, it acts as a barrier and helps reduce lightning damage. However, once the current breaks this barrier, it causes severe damage to the CFRP substrate due to a large amount of current lingering at one place for a

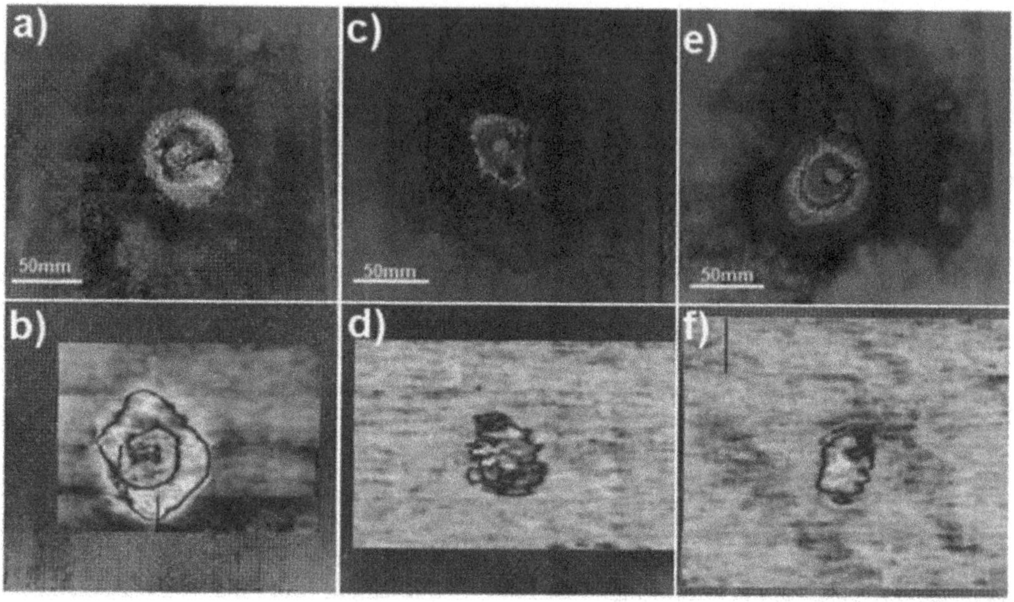

FIGURE 5.3 Images of composites flat panels after a lightning strike (upper row) and ultrasonic C-Scan visualization (lower row) for: (a,b) reference Panel, (c,d) SWCNT-Tuball LSP and (e,f) ECF-based LSP [25].

more extended period. Therefore, the proposed methodology by Han et al. needs to be completed and reconfirmed at various current and voltage levels.

In one of the most recent studies, K. Dydake et al. used single-walled carbon nanotubes (SWCNTs)-Tuball paper to develop an LSP system for CFRP composites [25]. Two different loadings (i.e., 75 and 90 wt%) of tuball paper were installed on the top of the CFRP to create an LSP system. It is reported that adding tuball paper improved the surface and volume conductivity of the CFRP by 8800% and 300%, respectively. Specimens were tested according to the SAE standard with a peak current of 100 kA. For comparison, a commercially used copper-mesh LSP system was also tested as a reference. The results are presented in Figure 5.3. The authors claimed that the proposed LSP system could be a better alternative to the current metal-based LSP system due to significant mass savings. However, it is worth noticing that the authors used the SWCNTs (up to \$2000/kg) [26], which is a few orders costly than MWCNTs (approx. \$200/kg) [27] and many orders than counterpart 400 GSM Cu-Mesh (approx. \$125/m^2) [28]. Furthermore, they also utilized a much higher CNT loading, i.e., 75% and 90%, which further increased the cost of the proposed LSP system. However, with a further increase in the production capacity of SWCNTs manufacturers, the SWCNT may be available more economically to users in the future, which makes this material a strong contender for LSP technology.

S. Mall et al. also studied various combinations of CNT BP and Nickel-coated nano strands (NiNS) carbon fibers as LSP coating [29]. They used two types of CNT BP, randomly aligned CNT mat and aligned CNT mat with understanding the LSP performance. For the performance evaluation, the authors considered the compressive strength of the composites after the lightning test. In their study, five different LSP systems were evaluated, and the random buckypaper (RBP) based LSP consisting of randomly oriented SWCNT performed best. RBP established that LSP retained more than 70% of the compressive strength after a simulated lightning test with a peak current of 100 kA, while NiNS based LSP residual compressive strength was just 25%.

These comparisons show that the most effective lightning strike protection system, in addition to the nickel-coated carbon woven fabric (NiCCF), was the RBP, while the least effective was the NiNS. Although an apparent reason for why RBP performed better than other LSP coating was not provided in the paper, it can be assumed that RBP dissipated the lightning current faster in all the directions, while aligned BP forced the current to take only one primary direction leading to delay the dissipation time and hence more damage occurred to the CFRP. From this study, CNT BP once again proved to be a good candidate as an LSP material.

5.2.3 CNT-Interleaved Composites

Since Kumar et al. reported that the through-thickness electrical conductivity of the CFRPs is the primary factor affecting its performance against a lightning strike [30], much research has been conducted to improve the through-thickness electrical conductivity of the CFRPs. A practical and easy way to impart electrical conductivity to the CFRP laminate is to replace the resin insulating layers between the conductive CF layers with conductive interleaves comprised of CNT BP. CNT BP can be infused with resin; therefore, it can act as a conductive bridge between CF layers in a CFRP laminate, allowing current flow in the through-thickness direction.

V. Kumar et al. also studied the interleaving approach for LSP of CFRPs [31]. They introduced a straightforward approach to fabricating the interleave composites. Therefore, they made BP from multi-wall carbon nanotube (MWCNT) and placed it in between the CF/epoxy prepreg plies before consolidating using a carver hot-press at optimum curing temperature and pressure. Different numbers of BP were studied, and their effect on through-thickness electrical conductivity and eventually their performance against 40 kA lightning strike current was investigated. It is summarized from the test that interleaved BP can reduce the damage compared with specimens without BP. The mechanism for this approach is to provide multiple conductive pathways to dissipate the current as quickly as possible. Interleaved BP helped dissipate the current in the in-plane direction and transferred from top CF laminate to bottom CF laminate, creating very efficient current discharge channels. However, the authors confirmed that as the number of BP increases, the mechanical properties of the composites decrease. Therefore, further study is needed to improve through-thickness electrical conductivity simultaneously without degrading the mechanical properties.

Similarly, X. Zhang et al. and J. Zhang et al. introduced CNT-based conductive films between the plies of CF/epoxy in their respective works to increase the through-thickness electrical conductivity of the composites. X. Zhang added CNTs in phenolphthalein modified polyether ketone (PEK-C) to prepare carbon nanotube films [32, 33]. They also prepared different samples with different interleaved CNT films before testing them for their lighting damage performance. These studies also employed a peak current of 100 kA of waveform A. In conclusion, the authors stated that CNT films could provide adequate lightning strike protection for CFRP compared with the reference laminate. Damage in the laminate with CNT films decreased by 77.6% and 68.0% in area and depth, respectively, compared to reference CFRP laminate. Furthermore, this study summarized that; carbon nanotubes could reduce the in-plane damage by consuming lightning strike energy in the depth direction. Finally, they recommended increasing the number of interleaves further to minimize the lightning damage to the CFRP. However, this study also did not provide any answer to the reduced mechanical properties of CFRPs due to the integration on interleaved BP.

5.3 FABRICATION APPROACH FOR GO-BASED COMPOSITES

Although it has been almost two decades since the first report of graphene's discovery, scientific reports on GO-based LSPs are limited compared to CNT-based LSPs. Although interestingly,

GO-based LSP is commercially available, Haydale (Ammanford, UK) has launched a range of graphene-enhanced prepreg materials for lightning strike protection recently [34]. Therefore, it is worth dividing the GO-LSP literature in the same way as we did for the CNT-LSP, i.e., (1) GO-filled resin, (2) GO coating, and (3) GO-paper interleaving.

5.3.1 GO-Based Conductive Resin

W. Lin et al. studied the effects of premixing highly conductive graphene nanoplatelets (GnP) on CFRP's lightning strike damage response [35]. Effects of GnP on the overall electrical resistance of the composite and its response to lightning damage were investigated. The authors prepared four panels, two baseline panels without GnP, two other panels with 0.5 and 1.0 wt% graphene, respectively. A simulated lightning strike of 100 kA intensity was utilized to assess the performance of composite materials under extreme environments. They claimed that adding GnP significantly decreases the surface electrical resistance of the composite laminate by 56.5% and 51.7% for the 0.5 wt% and 1.0 wt% samples, respectively. Improved electrical conductivity due to 0.5 wt % graphene reduces the severity of the lightning strike damage. The delamination area for each panel was 872 mm^2, 498 mm^2, and 1395 mm^2 for the baseline, 0.5 wt%, and 1.0 wt% panels, respectively. The authors argued that the 1.0 wt% panel exhibited much more significant damage/delamination due to poor GnP dispersion. These results suggest that increasing the weight content of GnP is redundant until a better dispersion method is applied.

In a similar attempt, M. Raimondo et al. prepared a new multi-functional GO/polyhedral oligomeric silsesquioxane (POSS) filled conductive epoxy [36]. They claimed that this modified epoxy is designed for aerospace applications, such as lightning strike protection, fire resistance, and good electrical and mechanical properties. No experimental lightning strike tests were done to support their claim, but they used Tunneling Atomic Force Microscopy (TUNA) and atomic force microscope (AFM) to successfully demonstrate the prepared composite's quantitative electrical characterization. In addition, they identified multiple conductive pathways due to the presence of graphene and, therefore, argued that this material can be a potential LSP material.

Overall, it can be concluded that GO-filled epoxy has not been studied extensively to support the hypothesis that it can mitigate the lightning strike damages significantly. However, a commercial product in the market confirms that it has an excellent potential to be a good LSP material. Nevertheless, a more scientifically validated result is still yet to be published.

5.3.2 GO-Based Conductive Coating

On the other hand, one of the most detailed studies on the performance of GO coating on CFRP substrate as LSP material was conducted by B. Wang et al. [37]. They used the percolating-assisted resin film infusion method to create an extremely thin (thickness varying between 25–50 μm) coating of reduced graphene oxide (RGO) on top of the CFRP substrate as shown in Figure 5.4. They confirmed that incorporating GO-layer enhanced the surface electrical conductivity to 440 S/cm compared to the 16 S/cm of pristine CFRP. In addition, RGO coating significantly improved the lightning damage resistance as compared to pristine CFRP. Finally, the author performed a 3-point bending test on a sample after a simulated lightning strike to quantify the results. The authors reported a 23% reduction in residual bending strength in the GO-coated sample, while a 66% reduction was observed in the sample without GO coating.

This work is critical to understand that GO-LSP can show promising LSP results if designed and manufactured correctly. S. Kumar et al. used the BP approach to create highly conductive graphene thin films (GTF) and used them on top of CFRP as the scarifying coatings to reduce

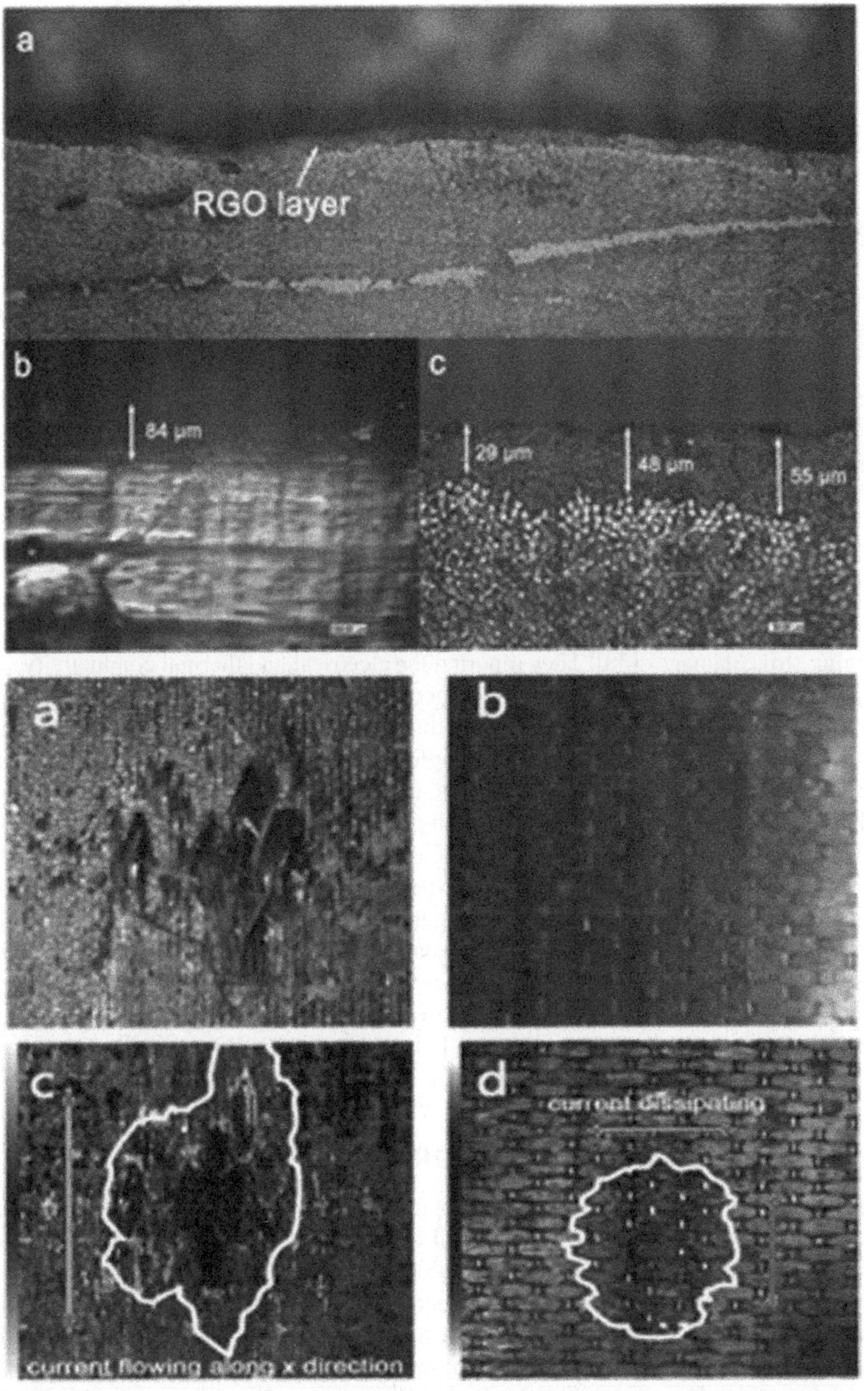

FIGURE 5.4 (i) Cross-section view of the prepared sample using side-light transmission photomicrographs in the y-direction (a and c) and the x-direction (b). (ii) images of the surface damage of (a, c) CFRP and (b, d) 0.05 g-RGO/CFRP samples observed using a digital camera (a and b) and using C-scope inspection (c and d) [37]. Reprinted from [37], Copyright (2018), with permission from Elsevier.

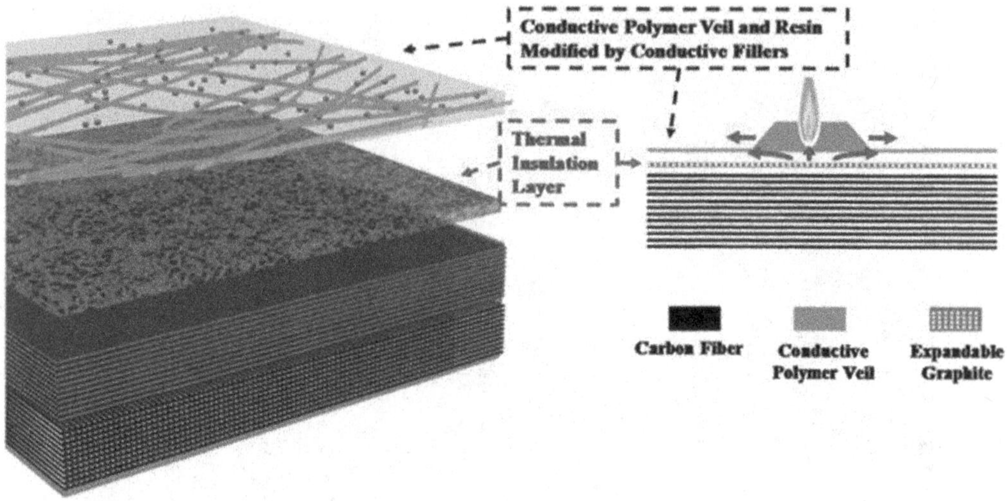

FIGURE 5.5 The structure of NM-LSP film [39]. Reprinted from [39], Copyright (2020), with permission from Elsevier.

the lightning strike damages [38]. They reported the electrical and thermal conductivity values of GTF papers as 1,800 S/cm and 425 W/m.K, respectively. However, this study reported poor adhesion between the GTF layer and the CFRP and did not perform any experimental lightning strike test to verify their hypothesis. Therefore, a complete study to validate their hypothesis is needed.

After lessons learned from the above studies, a 5-step complicated GO coating was prepared by Z. Zhao et al. and tested against a simulated lightning strike of 100 kA [39]. The LSP film was composed of two layers, where the first graphene-doped conductive veil was used to dissipate the incident lightning current, and the second expanded graphite film was used to deflect the heat shock from the lightning strike as depicted in Figure 5.5. Their approach seems to be working, and they confirmed that this new LSP was able to reduce the lighting damage by 79% and the residual compression strength by 21%. Furthermore, they compared the results of the new LSP with the expanded metal foil (EMF)-based LSP and showed that newly developed film is as effective as commercially available LSP, while it was 37% lighter than the EMF-based LSP. These results are very promising to create future LSP solutions for emerging markets such as UAM.

5.4 CONCLUSION AND FUTURE WORKS

These studies highlighted in this chapter explained some promising approaches to reducing lightning strike damage by improving electrical conductivity (conductive resin and interleaves) or guiding the current through the conductive coating using CNT and GO nanoparticles. All approaches are discussed concerning their advantages and limitations. It is concluded that a good design and manufacturing process, CNT/GO-LSP can provide adequate resistance to lighting damages while saving weight compared to traditional LSP systems. However, wrongly designed LSP can be catastrophic to the CFRPs. Reduced mechanical properties of CFRP and extra complicated integration steps are some of the questions that need to be answered to reach the commercialization of nano-filler-based LSPs. Surprisingly, there are no reports on the GO-paper interleaving to improve the through-thickness electrical conductivity and its performance as an LSP material; therefore, readers are encouraged to investigate this area. Also, GO-LSP research is not mature as CNT-LSP research; therefore, it is believed that much more work is needed to gain more understanding of the GO as an LSP material. From this chapter, we have confirmed that the higher electrical conductivity of CFRP can be a viable solution for LSP development. However, a well-known approach to improve the electrical conductivity, i.e., using CNT/GO as sizing for CF fibers to increase the electrical conductivity of CFRP, has not been studied for LSP application. In future works, it would be worth investigating the effect of nano-filler sizing on the LSP performance of the CFRPs.

ACKNOWLEDGMENTS

This research was supported by the DOE Office of Energy Efficiency and Renewable Energy (EERE), Advanced Manufacturing Office, and used resources at the Manufacturing Demonstration Facility (MDF), a DOE-EERE User Facility at Oak Ridge National Laboratory.

REFERENCES

1. Zhang X, Chen Y, Hu J. Recent advances in the development of aerospace materials. Prog Aerosp Sci 2018;97:22–34. doi:10.1016/j.paerosci.2018.01.001.
2. Rajak DK, Pagar DD, Kumar R, Pruncu CI. Recent progress of reinforcement materials: A comprehensive overview of composite materials. J Mater Res Technol 2019;8:6354–74. doi:10.1016/j.jmrt.2019.09.068
3. Yokozeki T, Goto T, Takahashi T, Qian D, Itou S, Hirano Y, et al. Development and characterization of CFRP using a polyaniline-based conductive thermoset matrix. Compos Sci Technol 2015;117:277–81. doi:10.1016/j.compscitech.2015.06.016
4. Manomaisantiphap S, Kumar V, Okada T, Yokozeki T. Electrically conductive carbon fiber layers as lightning strike protection for non-conductive epoxy-based CFRP substrate. J Compos Mater 2020:002199832093594. doi:10.1177/0021998320935946
5. Gagné M, Therriault D. Lightning strike protection of composites. Prog Aerosp Sci 2014;64:1–16. doi:10.1016/j.paerosci.2013.07.002
6. Lee J, Lacy TE, Pittman CU, Mazzola MS. Comparison of lightning protection performance of carbon/epoxy laminates with a non-metallic outer layer. J Reinf Plast Compos 2019;38:301–13. doi:10.1177/0731684418817144
7. Rakov VA. The Physics of Lightning. Surv Geophys 2013;34:701–29. doi:10.1007/S10712–013–9230–6/FIGURES/19
8. Kumar V, Yokozeki T, Karch C, Hassen AA, Hershey CJ, Kim S, et al. Factors affecting direct lightning strike damage to fiber reinforced composites: A review. Compos Part B Eng 2020;183:107688. doi:10.1016/j.compositesb.2019.107688
9. Zhou Y, Raghu SNV, Kumar V, Okada T, Yokozeki T. Simulated lightning strike investigation of CFRP comprising a novel polyaniline/phenol based electrically conductive resin matrix. Compos Sci Technol 2021;214:108971. doi:10.1016/j.compscitech.2021.108971
10. Das S, Yokozeki T. A brief review of modified conductive carbon/glass fibre reinforced composites for structural applications: Lightning strike protection, electromagnetic shielding, and strain sensing. Compos Part C Open Access 2021;5:100162. doi:10.1016/j.jcomc.2021.100162
11. Kumar V, Yeole PS, Hiremath N, Spencer R, Masum Billah KM, Vaidya U, et al. Internal Arcing and Lightning Strike Damage in Short Carbon Fiber Reinforced Thermoplastic Composites. Compos Sci Technol 2020:108525. doi:10.1016/j.compscitech.2020.108525
12. Kumar V, Spencer R, Smith T, Condon JC, Yeole PS, Hassen AA, et al. Replacing metal-based lightning strike protection layer of cfrps by 3d printed electrically conductive polymer layer. Compos. Adv. Mater. Expo (CAMX 2019)—Anaheim, California, USA, 2019.
13. Millen SLJ, Murphy A, Abdelal G, Catalanotti G. A lightning plasma and composite specimen damage simulation framework for SAE test waveform B. n.d.

14. Karch C, Metzner C. Lightning protection of carbon fibre reinforced plastics—An Overview. 2016–33rd Int. Conf. Light. Prot. ICLP 2016, Institute of Electrical and Electronics Engineers Inc.; 2016. doi:10.1109/ICLP.2016.7791441
15. Pati S, Kumar V, Goto T, Takahashi T, Yokozeki T. Synthesis and characterization of PANI/P-2M conductive composites: Thermal, rheological, mechanical, and electrical properties. Polym Compos 2019;40:4321–8. doi:10.1002/pc.25293
16. Iijima S. Helical microtubules of graphitic carbon. Nat 1991–3546348 1991;354:56–8. doi:10.1038/354056a0.
17. Novoselov KS, Geim AK, Morozov S V., Jiang D, Zhang Y, Dubonos S V., et al. Electric field in atomically thin carbon films. Science (80-) 2004;306:666–9. doi:10.1126/SCIENCE.1102896.
18. Chakravarthi DK, Khabashesku VN, Vaidyanathan R, Blaine J, Yarlagadda S, Roseman D, et al. Carbon fiber-bismaleimide composites filled with nickel-coated single-walled carbon nanotubes for lightning-strike protection. Adv Funct Mater 2011;21:2527–33. doi:10.1002/adfm.201002442.
19. Logakis E, Skordos AA. Lightning strike performance of carbon nanotube loaded aerospace composites. ECCM 2012—Compos Venice, Proc 15th Eur Conf Compos Mater 2012:24–8.
20. Lampkin S, Lin W, Rostaghi-Chalaki M, Yousefpour K, Wang Y, Kluss J. Epoxy resin with carbon nanotube additives for lightning strike damage mitigation of carbon fiber composite laminates. Proc Am Soc Compos—34th Tech Conf ASC 2019–2019. doi:10.12783/asc34/31338.
21. Ma X, Scarpa F, Peng HX, Allegri G, Yuan J, Ciobanu R. Design of a hybrid carbon fibre/carbon nanotube composite for enhanced lightning strike resistance. Aerosp Sci Technol 2015;47:367–77. doi:10.1016/j.ast.2015.10.002.
22. Wu C, Lu H, Liu Y, Leng J. Study of carbon nanotubes/short carbon fiber nanocomposites for lightning strike protection. Behav Mech Multifunct Mater Compos 2010–2010;7644:76441H. doi:10.1117/12.847491.
23. Xia Q, Mei H, Zhang Z, Liu Y, Liu Y, Leng J. Fabrication of the silver modified carbon nanotube film/carbon fiber reinforced polymer composite for the lightning strike protection application. Compos Part B Eng 2020;180. `doi:10.1016/j.compositesb.2019.107563.
24. Han JH, Zhang H, Chen MJ, Wang D, Liu Q, Wu QL, et al. The combination of carbon nanotube buckypaper and insulating adhesive for lightning strike protection of the carbon fiber/epoxy laminates. Carbon N Y 2015;94:101–13. doi:10.1016/j.carbon.2015.06.026.
25. Dydek, Kamil, Anna Boczkowska, Rafał Kozera, Paweł Durałek, Łukasz Sarniak, Małgorzata Wilk and WŁ. Effect of SWCNT-Tuball Paper on the Lightning Strike Protection of CFRPs and Their Selected Mechanical Properties. Materials (Basel) 2021;14.
26. Carbon Nanotubes—Price—Chinese Academy of Sciences, Chengdu Organic Chemistry Co., Ltd. n.d. www.timesnano.com/en/article.php?prt=4,31,105 (accessed December 21, 2021).
27. Multi Walled Carbon Nanotubes Products n.d. www.cheaptubes.com/product-category/multi-walled-carbon-nanotubes/ (accessed December 21, 2021).
28. Astroseal Product List by Astrostrike Code—Astrostrike® Premier Lightning Strike Protection by Astroseal Products Manufacturing Corp. n.d. www.astrosealproducts.com/product_codes.html (accessed December 21, 2021).
29. Mall S, Ouper BL, Fielding JC. Compression strength degradation of nanocomposites after lightning strike. J Compos Mater 2009;43:2987–3001. doi:10.1177/0021998309345337.
30. Kumar V, Yokozeki T, Okada T, Hirano Y, Goto T, Takahashi T, et al. Effect of through-thickness electrical conductivity of CFRPs on lightning strike damages. Compos Part A Appl Sci Manuf 2018;114. doi:10.1016/j.compositesa.2018.09.007.
31. Kumar V, Sharma S, Pathak A, Singh BP, Dhakate SR, Yokozeki T, et al. Interleaved MWCNT buckypaper between CFRP laminates to improve through-thickness electrical conductivity and reducing lightning strike damage. Compos Struct 2019;210:581–9. doi:10.1016/j.compstruct.2018.11.088.
32. Zhang X, Zhang J, Cheng X, Huang W. Carbon nanotube protected composite laminate subjected to lightning strike: Interlaminar film distribution investigation. Chinese J Aeronaut 2021;34:620–8. doi:10.1016/j.cja.2020.04.033.

33. Zhang J, Zhang X, Cheng X, Hei Y, Xing L, Li Z. Lightning strike damage on the composite laminates with carbon nanotube films: Protection effect and damage mechanism. Compos Part B Eng 2019;168:342–52. doi:10.1016/j.compositesb.2019.03.054.
34. Morris EW. Huntsman Advanced Materials and Haydale Composite Solutions deliver significant step change in the market for nanocomposites. Reinf Plast 2016;60:214–7. doi:10.1016/j.repl.2016.05.006.
35. Lin W, Jony B, Yousefpour K, Wang Y, Park C, Roy S. Effects of graphene nanoplatelets on the lightning strike damage response of carbon fiber epoxy composite laminates. Proc Am Soc Compos—35th Tech Conf ASC 2020–2020:539–53. doi:10.12783/asc35/34878.
36. Raimondo M, Guadagno L, Speranza V, Bonnaud L, Dubois P, Lafdi K. Multifunctional graphene/POSS epoxy resin tailored for aircraft lightning strike protection. Compos Part B Eng 2018;140:44–56. doi:10.1016/j.compositesb.2017.12.015.
37. Wang B, Duan Y, Xin Z, Yao X, Abliz D, Ziegmann G. Fabrication of an enriched graphene surface protection of carbon fiber/epoxy composites for lightning strike via a percolating-assisted resin film infusion method. Compos Sci Technol 2018;158:51–60. doi:10.1016/j.compscitech.2018.01.047.
38. Kumar SSA, Uddin MN, Rahman MM, Asmatulu R. Introducing graphene thin films into carbon fiber composite structures for lightning strike protection. Polym Compos 2019;40 :E517–25. doi:10.1002/pc.24850.
39. Zhao Z, Ma Y, Yang Z, Yu J, Wang J, Tong J, et al. Light weight non-metallic lightning strike protection film for CFRP. Mater Today Commun 2020;25:101502. doi:10.1016/j.mtcomm.2020.101502.

6 Carbon Nanotubes and Graphene for Conducting Wires

Pallvi Dariyal, Manoj Sehrawat, Sanjay R. Dhakate, and Bhanu Pratap Singh

CONTENTS

DOI: 10.1201/9781003231943-6

6.1 INTRODUCTION

The evolution of lightweight conductors with the ability to carry equivalent current as copper is expected to be an economy transformation for weight critical applications, especially for electrical motors, power lines, data transmission cables and transformer wirings etc. Traditional wires used till today such as copper (Cu), aluminum (Al), silver (Ag) has high density, wherein Cu wire is the most commercially used for conducting cables in daily life as in refrigerators, vehicle engine's motors, induction plates etc. owing to its cheaper availability. On the contrary, Ag is used specifically in niche applications i.e., satellites, as it has higher conductivity along with less density than Cu. Though Ag and Au cables have superior electrical properties, yet due to emergence need of ultralight conductors for progressive and compact nano electronics field, new conducting materials such as carbon-based manufactured allotropes i.e., carbon nanotube (CNT) and graphene, have got extensive attention among researchers as these nanomaterials are superior quantum conductors displaying magnificent electrical performance. Though these nanomaterials have ultrahigh conductivity, their dimensions limit their applications. In search of obtaining a "single domain" and continuous CNT and graphene fiber, their macro counterparts with improved properties have been explored, yet have been limited by scalable and effective production methods. Indeed, for having efficient conductors, industries are also pushing their barriers toward these new advanced macro assemblies that is expected to bring revolutionary technologies as they have great commercial and academic value because of their extraordinary properties. Their assemblies, i.e., fiber, yarn, sheet and ribbons, have been synthesized via various techniques, as will be discussed further. These conducting macro assemblies can be used in broad application areas, such as smart textiles and electronic fibers, and in the aerospace industry. In aircrafts and satellites, the wiring of engines imparts most weight, for example narrow-body, wide-body and double-decker aircrafts, contain up to 40 miles, 150 miles and 320 miles of cables, respectively. By using lightweight conductors such as CNT and graphene fibers, weight trimming in these spaceflights can be performed which can minimize the launching costs and also allow auxiliary engineering equipments and more fuel for lifetime extension of the spaceflights, aircrafts, and automobiles.

In this chapter, we will mainly summarize the demonstrated as well as futuristic applications of carbon nanotube, graphene, and their hybrid fibers. Firstly, the conduction mechanism of CNTs and graphene has been described. After that, various synthesis technique of these macro assemblies and numerous applications of CNT and graphene fibers, used for electrical applications, have been discussed.

6.2 CONDUCTION MECHANISM OF CNTS, GRAPHENE AND THEIR MACRO ASSEMBLIES

The electronic properties of carbon allotropes more or less overlap in terms of conductivity due to presence of free electron, however, differs at macroscale. Both CNT and graphene show ballistic conduction of charge carriers, the band gap of both materials can be engineered by controlling their structural configurations. The electronic properties of CNTs mainly depends upon the diameter and chirality of the nanotube whereas number of layers defines the same in graphene. This section describes the conductivity mechanism at atomic scale higher assemblies.

6.2.1 Carbon Nanotubes as Futuristic Conducting Material

CNTs are one-dimensional quantum wires due to ballistic transportation of electrons where the electrons experience quantum confinement along their circumference. These nanotubes possess higher electrical conductivity than any other metal because the electrons in CNTs can travel large distance (~1 μm) without any collision. For better understanding of electron transport in 1D CNTs, two basic models are explained below:

6.2.1.1 Ballistic Conductivity in Individual CNT

Based on electron scattering, conduction can be ballistic or diffusive. CNTs have been reported as ballistic or quasi-ballistic conductors without dissipation of heat as the mean free path is greater than an electron's wavelength. In other words, there is negligible scattering of electrons in conducting channel. These 1-D wires show ballistic transportation and are termed as quantum wire [1] (negligible electrical resistivity). In quantum wires, the electrons experience quantum confinement along the circumference of wire which results in discrete quantum electrical conductance (G_o) [2]. The conductance in 1D quantum wire, can be explained via model (shown in Figure 6.1).

$$\mathrm{I} = \int_{\mu 1}^{\mu 2} ev\left(2\frac{1}{2}g_1\mathrm{D}(E)dE\right)$$

Where, $v, \frac{1}{2}g_1\mathrm{D}(E)$ are velocity and density of states for 1-D system, respectively and μ_1 and μ_2 are the chemical potentials on either side of 1-D conductor.

$$\mathrm{I} = \int_{\mu 1}^{\mu 2}\left(\frac{2}{hvE}\right)dE$$

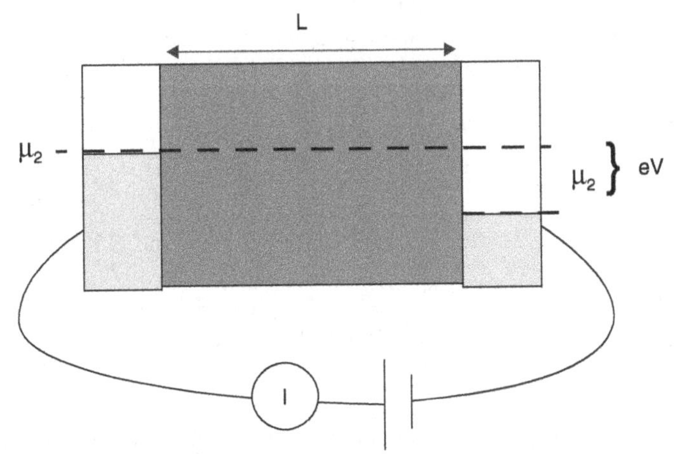

FIGURE 6.1 Schematic of a ballistic conductor with length L is connected to two electrodes with Fermi energies μ_1 and μ_2 respectively.

$$I = \frac{2e}{h}(\mu_2 - \mu_1)$$

$$\text{Conductivity (G)} = I/V = 2e^2 / h = 7.74\times10^{-5}\,\text{S}$$

Where e is charge of an electron and h is Plank's constant. The integer 2 is spin degeneracy of electron. Generally, for metallic single walled carbon nanotubes (SWCNTs) with ideal contacts, the conductivity is $2G_o$ due to the presence to 2 sub bands and if contacts are not ideal, then ballistic transport of electrons may occurr in 1 channel (π^* channel) because the other channel (π channel) may turn off due to charge transfer between CNTs and contacts. Further the conductance is modified by the Landauer formula given below:

$$G = {}^{2e}\!/_{h} \times N \times T$$

Where N and T are the number of sub bands playing role and the transmission probability of electrons respectively. The reported value of electrical conductance and resistivity for SWCNT are 0.15 mS and ~10^{-6} Ωcm^{-1}, respectively[3].

6.2.1.2 Quantum Tunneling Between CNTs in CNT Bundles

As reported, CNTs are generally found in bundles, connected via weak van der Waal forces. Therefore, electrons can hop from one nanotube to other without any thermal activation. The only reason of electron hopping is tunneling mechanism, which plays an important role not only in intertube coupling, but also for intershell coupling where intershell spacing is very small (~0.34 nm).

Due to the tube deformation, the intershell conductance get enhanced and the estimated value of intershell conductance, g, is calculated by

$$g = G_{atom}\frac{2\pi r}{S}$$

Where S, r and G_{atom} are the surface occupied by one atom, shell radius, and tunneling conductance through the π-orbital overlap between two atoms of nearby shell, respectively. Assuming elastic tunneling, G_{atom} is described as

$$G_{atom} = \frac{4\Pi e^2}{\hbar}N^2E^2$$

where E is the binding energy due to electronic delocalization, and N is the density of state per atom [4], which is given by:

$$N = \frac{2n}{hv}\frac{S}{2\pi r}$$

Where v and n are the Fermi velocity, and number of modes due to doping, respectively.

Moreover, using a resistive transmission line model, the intershell conductance is ~ $(10\ k\Omega)^{-1}$ /μm, which resembles with the estimated value based on electrons tunneling through the atomic orbitals of neighboring shells while considering the conservation of energy but not momentum.

The intershell interaction in double walled carbon nanotubes (DWCNTs)

$$\beta = kS$$

Where β and S are resonance integral and overlap integral respectively.

In MWCNTs, the shells are incommensurate and non-periodic. In this case, the transfer of the electronic wave packet is followed by a non-ballistic propagation.

$$y\,L(t) \sim At^{\eta}$$

The coefficient η is found to decrease from ~ 1 to ~ ½ by increasing the number of coupled incommensurate shells. Moreover, the conductivity of CNT decreases by enhancing its diameter as the bandgap (E_{gap}) between conductance and valance band is inversely proportional to the diameter of CNT [5].

$$Egap = \frac{2\gamma_o l}{d}$$

Where γ_o = c-c tight binding overlap energy (~2.5–3 eV).

Till now, laser ablation, arc discharge and chemical vapor deposition (CVD) are well-known methods to synthesize powdered CNTs. Though these nanotubes have tremendous properties, yet it is next to impossible to weave them for commercialization. Therefore, the only route to have macroscopic CNT fibers, is an interlocked chain of CNTs.

Due to their astonishing electrical properties, researchers are attracting toward CNT-based electrical wires where these CNTs can be used as CNT tape, CNT film as outer conductors and CNT fibers as inner conductors for economic saving. For having pristine CNT cables, thousands of CNTs were interconnected via van der Waal forces where the electrical properties of these cables depend upon the as-synthesized CNTs nature and their packaging manner. After understanding the possible electron flow mechanism in individual CNTs or in CNT bundles, the transportation mechanism in CNT macro assemblies is necessary to be discussed where the current will flow through CNT bundles as well as across junctions made between neighboring bundles.

6.2.1.3 Resistance at Junction Between Nanotube Bundles

Till now, the conduction between macro assembly's junctions is not clear, but the basic phenomenon is quite similar as described before, termed as tunneling. The only difference between transportation occurring in CNT to CNT and transportation occurring in bundle to bundle is the tunneling distance. Previous reports says that the tunneling junction shows power law dependency, explained by Luttinger Liquid theory. The conductance, due to tunneling can be explained as:

$$G = T^{\alpha}$$

When electrons are added to CNT, electrons spread in one direction from one CNT to neighboring CNT and if electron flows in both direction [6]

$$\alpha_{end} = \frac{g^{-1} - 1}{4} \qquad \alpha_{bulk} = \frac{g^{-1} + g - 2}{8}$$

Where g is the Luttinger parameter.

6.2.2 Graphene as Conducting Material

The other most-researched carbon allotrope, graphene nano sheets, show exceptional physical, electrical and thermal properties owing to its high quantum battle in its structure. It consists of flat hexagonal units laid side by side in a seamless manner where each sp^2 hybridized carbon make bonds with its three neighboring mates and has an unaffected p-orbital with one electron leading to the formation of half-filled π-band as shown in Figure 6.2(a). The surprising electronic properties displayed by graphene are generated due to the introduction of inversion symmetry by the presence of same atoms at all sites in its planar honeycomb lattice. One of the four available electrons on carbon, the naïve π-electron plays a vital role in demonstrating low energy electronic properties. The π-electron energy bands are explained via tight binding approximation, i.e., assuming that an electron is affected by a single ion only [7].

In reciprocal lattice, the first brillouin zone is found to be a hexagon. The corners of the BZ can be distinguished into two sets of K-points as others can be connected via reciprocal lattice vectors as shown in Figure 6.2(b). The tight binding Hamiltonian based on Huckel model for the nearest neighbor interaction is:

$$\hat{H} = -t\sum_{|\vec{R}|}\left(\left|\vec{R}\right\rangle\left\langle\vec{R}+\vec{\tau}\right| + \left|\vec{R}\right\rangle\left\langle\vec{R}-\vec{a}_1+\vec{\tau}\right| + \left|\vec{R}\right\rangle\left\langle\vec{R}-\vec{a}_2+\vec{\tau}\right| + h.c.\right)$$

Where, t is Hopping integral (hopping energy between atoms of different states) [8].

Since there are two interpenetrating triangular sublattice, two sets of bloch orbitals are needed, each having energy $\varepsilon(\vec{k}) = \pm\left|e(\vec{k})\right|$, where positive and negative energy depicts conduction and valence bands respectively as depicted in Figure 6.2. The energy vanishes at band contact point termed as Dirac point. Pure graphene is often considered as a semiconductor with zero band gap because its density of states vanishes at Fermi level (E=0) as can be seen by the expression

$$D(E) = \frac{|E|}{2\pi\hbar^2 v^2}$$

Referring in the same context, a graphene can also be regarded as a metal since its conductivity seems to be independent of Fermi energy level. However, these are quantum confinement effects which govern its electronic behavior. All the Dirac points in graphene are degenerate in due to presence of C_3, center of inversion and time reversal symmetries which prevents opening of band gap by level repulsions. In absence of degeneracy, graphene would be an insulator. Schrodinger equation when considered for a point close to Dirac point takes the form:

$$H_K\psi_K = -i\hbar v_F\begin{pmatrix} 0 & \partial_x - i\partial_y \\ \partial_x + i\partial_y & 0 \end{pmatrix}\begin{pmatrix} \psi_{KA} \\ \psi_{KB} \end{pmatrix} = \varepsilon\begin{pmatrix} \psi_{KA} \\ \psi_{KB} \end{pmatrix}$$

This equation can also be written in a concise manner as:

$$H_K = \hbar v_F \vec{\sigma}.\vec{p}$$

The above equation is of the same form as Dirac-Weyl equation for 2-D structures. This equation is generally applied for relativistic particles (having zero rest mass) with half integral spin and moving with velocity of light. On comparing, it can be inferred that electrons in graphene undergo

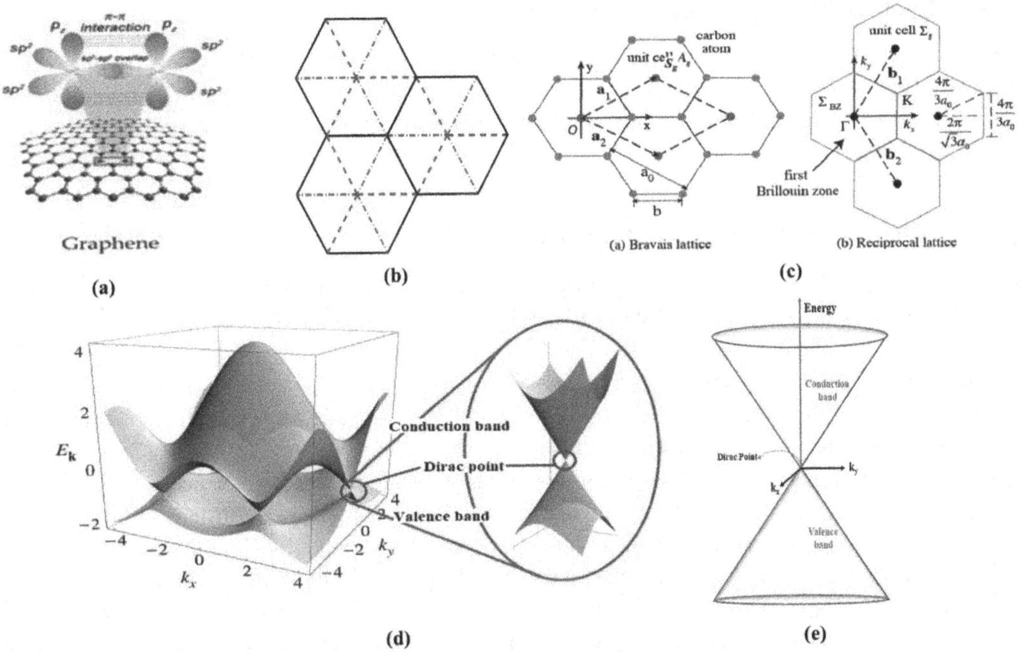

FIGURE 6.2 Electronic properties of graphene. (a) Orbital structure of graphene (reprinted with permission from Creative Common License 3.0 Ref. [9]. Copyright (2017)) (b) sketch showing two distinct corner in each brillouin zone (reprinted with permission from Ref. [7]. Copyright (2009) American Physical Society) (c) Bravais and reciprocal lattice reprinted with permission from Creative Common License 3.0 Ref. [10]. Copyright (2014)) (d) valence and conduction bands (reprinted with permission from Ref. [7]. Copyright (2009) American Physical Society) (e) Dirac cone structure of VB and CB of graphene.

low energy excitations, behaving as massless Dirac fermions having Fermi velocity around 300 times smaller than that of light. The energy of relativistic particles can be expressed according to special theory of relativity as:

$$\varepsilon=\sqrt{m^2c^4+c^2\boldsymbol{p}^2}$$

Since the rest mass m for graphene electrons is zero, the dispersion takes a linear form, where c is replaced by v. This linear dispersion of charge carriers along with confined thickness and high surface to volume ratio offers exceptionally high carrier mobility and transport properties thus declaring the excellent electrical properties of graphene.

It can be seen in a neutrino wave function that on rotating the motion of particle by 2π, its phase changes by $\pm\pi$ along with its sign. This sign difference among the amplitudes of rotations thus nullifies each other. As opposed to local electrons, which are prone to localization effects, Dirac fermions in graphene can propagate up to several micrometer distances without backscattering. This phenomenon is termed as ballistic transport mechanism [11]. However, the presence of dominant scatterers such as charged impurities leads to non-zero density of states at E_f=0, because scattering potential is increased due to poor screening by electrons. Topological defects like dislocations and grain boundaries also disturb the smooth delocalization of sp^2 charge carriers. It has been reported that the carrier concentration can be influenced by external gate voltage and therefore conductivity. The band gap of graphene can also be engineered by choosing an appropriate substrate to introduce asymmetry.

6.3 SYNTHESIS PROCESS FOR FIBERS

As previously mentioned, CNTs and graphene fibers possess extraordinary electrical properties i.e. large current carrying capability, high specific conductivity. Hence, for their commercial usage, various well-established synthesis techniques for fibers have been explored which are discussed in this upcoming section.

6.3.1 Methods for Synthesis of CNT Fiber

There are two synthesis techniques to produce CNT fibers: wet spinning (Figure 6.3(a)), and dry spinning i.e., forest spinning (Figure 6.3(b)) and direct spinning (Figure 6.3(c)).

6.3.1.1 Wet Spinning Technique

For CNT fiber production via wet spinning [12], pre-synthesized CNTs are dispersed in either surfactant-assisted solvents or in super acids (Chlorosulfonic acid, sulfuric acid). These solvents promote their protonation, resulting in homogeneously dispersed solution termed as CNT dough. The as-made dough is injected in storage cylinder whereby applying a constant pressure, CNT fibers of different diameters are extruded via a disc consisting holes of different sizes as per one's requirement. After that, these fibers are deprotonated by passing them through coagulation bath tank having DI water, acetone, or diethyl ether etc. The morphologies and properties of CNT fibers can be controlled by controlling the spinneret hole size. Though the technique has great potential to be scaled up for commercialization, still it is difficult to have pristine CNT fibers with this method as the complete removal of ions is quite difficult. On the contrary, for having hybrid CNT fibers, the technique is very beneficial.

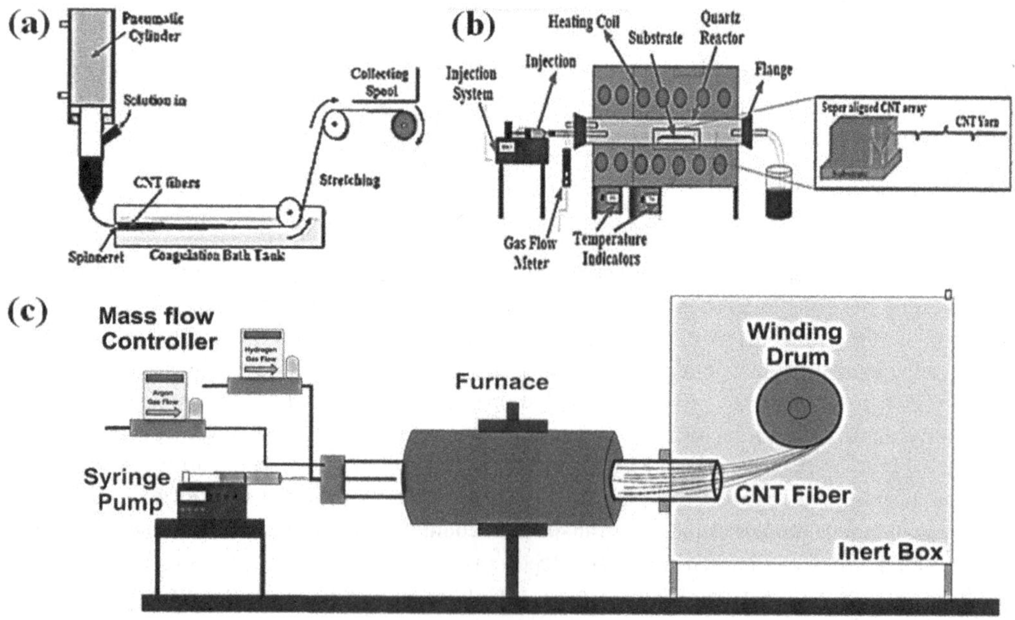

FIGURE 6.3 Schematic of (a) wet spinning, (b) forest spinning, (c) Direct spinning technique (reprinted with permission from RightsLink: Ref. [13]. Copyright (2021) Springer).

6.3.1.2 Dry Spinning Technique

6.3.1.2.1 Forest Spinning Technique

CNT fibers can be synthesized via forest spinning technique which consists of two-step process. Initially, vertically aligned CNT array known as CNT forest is grown by CVD technique on a substrate, and subsequently CNT fiber can be drawn out using a rotating probe. In this technique, while pulling the CNTs, present at the edge of a CNT forest, the adjacent CNTs are consecutively pulled out (due to van der Waal attraction), resulting in a long CNT fiber formation [14]. Though the fibers spun with this method show higher electrical conductivity than the fibers spun via other methods, the major concern is to grow equal length CNT arrays otherwise it would be difficult to draw a continuous fiber. However, this technique is the most preferred on lab-scale as the fiber synthesized via the same, has high conductivity but it is difficult to have limitless fiber with this technique.

6.3.1.2.2 Direct Spinning

It is a single step and most preferred technique in which CNT fiber is synthesized from vapor phase of the reactor where porous cluster, known as CNT aerogel [15], is formed which is collected to produce a continuous filament. In this technique, the feedstock solution containing carbon source, a catalyst and a promoter, is introduced from the one side of reactor and fiber is directly drawn out as CNT organogel [16] which is formed due to thermophoresis phenomena from the other end of reactor.

6.3.2 Methods for Synthesis of Graphene Fiber

Alike CNT fiber synthesis, the same synthesis techniques i.e., wet and dry spinning techniques are used for graphene fibers where wet spinning is the most preferred. However, the pristine two-dimensional graphene plates do not hold enough interactions among its solid flakes and have poor solubility in common solvents thus functionalized graphene oxide (GO) sheets dissolved in aqueous medium served as liquid crystal assembly for spinning of GO fibers which are subsequently reduced into neat graphene fibers. As mentioned before, in the first step, functionalization of graphene oxide is performed to have oxygen functionalities for its easy dispersibility in water and other polar solvents. Hummer's and modified Hummer's method are the two common and convenient pathways for GO synthesis at large scale. In both of the synthesis techniques, hydrogen peroxide, potassium permanganate and sulfuric acid induce delamination and subsequently oxidize graphite sheets into GO powder.

6.3.2.1 Wet Spinning Method

The GO fibers came into existence in 2011 by the efforts of Xu and Gao, who successfully spun meters of GO fibers from liquid crystals of GO via wet spinning technique [17]. It is the most widely adopted method for GO fibers synthesis, due to its easy processability and no special prerequisites. In this technique, a sufficiently concentrated solution of pre-synthesized GO in desired solvent is prepared and is ejected out from a spinneret through a coagulating bath onto a rotating collection unit as depicted in Figure 6.4(a). The as-spun GO fibers hold poor electrical properties, therefore after spinning, GO fibers are reduced subsequently to restore its continuous sp^2 structure and the electron delocalization. Furthermore, an air gap can be introduced as shown in Figure 6.4 between the injection tip and the coagulant bath which decreases the velocity gradient and induces the gravity resulting in high alignment of GO fibers [18]. In another work, a continuous handedness twisting procedure for fabrication of graphene fiber has been done. Initially, GO belts are collected via wet spinning of GO liquid crystalline dope in Dimethylformamide.

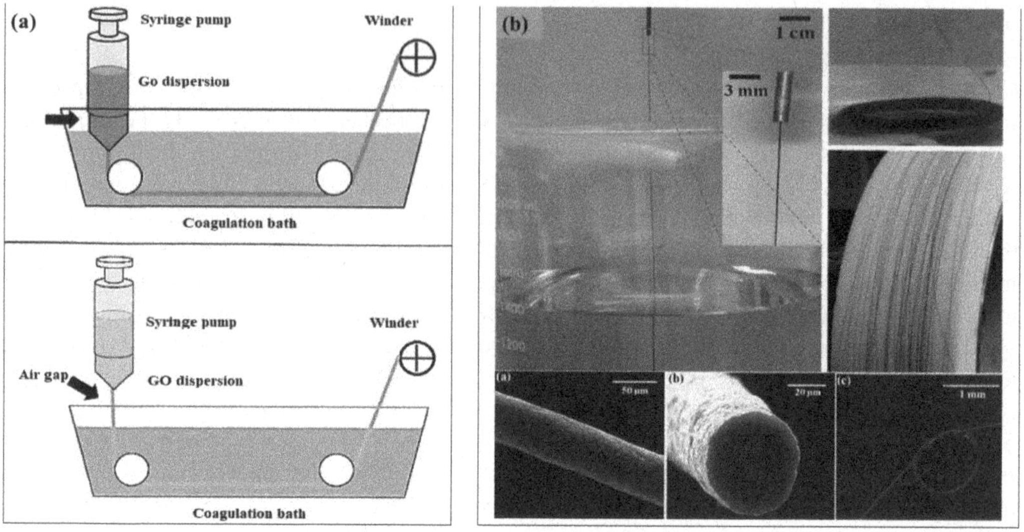

FIGURE 6.4 (a) Schematic of wet spinning and dry jet wet spinning of GO fiber (reprinted with permission from (reprinted with permission from Creative Common License 4.0 Ref. [23]. Copyright (2018)) (b) dry jet wet spinning setup and cross-sectional images of fiber (reprinted with permission from (reprinted with permission from Ref. [18]. Copyright (2013) American Chemical Society).

These GO belts are then drawn and twisted simultaneously via a roller revolving out of plane and thus twisting GO belts axially into GOF having large tensile strain and twists [19].

6.3.2.2 Dry Spinning

Unlike wet spinning, purification is not necessary in dry spinning which somewhat adds to complicacy and also causes potential pollution [20]. In this process, either a polypropylene (PPy) pipe or a hollow glass/ceramic tube is used as autoclave where the length and diameters of GO fiber depends upon the tube's dimensions. The first step of making GO dispersion is quite similar as in wet spinning method. After that, the GO gel is heated in a sealed hydrothermal autoclave and then dried in air which also leads to spontaneous alignment and close packing of graphene sheets because of water loss. Further for in-situ reduction of GO, Li et al. used vitamin C in a specific proportion with GO gel. In this consent, after heating, the GO and vitamin C suspension in the PPy pipe shaped reactor, the reduced GO fibers were extracted and dried in air. In this technique, the length and diameter of GF can be controlled by the dimensions of the PPy tube shaped reactor used such as 15 cm long having 4.5mm diameter porous GO fibers is obtained via heating in 18 cm long PPy tube where the shrinkage is because of water loss [21, 22].

6.3.2.3 Directly Assembled GFs via CVD

Similar with CNT fibers, graphene fibers can also be spun directly from graphene sheets synthesized via CVD technique. Though by using this technique, acidic damages made in wet spinning technique can be avoided, but still, it is a complex method having four step processing. Initially, graphene sheets were synthesized on a substrate. Subsequently, substrate is etched out to have free-standing as-grown graphene sheets. After substrate removal, the graphene assemblies have to be passed though different organic solvents such as ethanol and acetone due to which the assembly's edges immediately scroll up such that graphene is unable to attain its planar structure.

As a result, the fiber structured graphene is drawn out by tweezers and dried naturally. As above-mentioned, the process is very complex, therefore in order to reduce the processing complexity, graphene nano sheets are grown on wire-based substrate instead of film substrate. By doing so, the shrinkage step of passing through organic solvents can be skipped. In 2015, hollow tubes of graphene sheets have been synthesized on Cu wire as substrate, using methane under H_2 and Ar environment, via CVD method. After that, the substrate is removed via $FeCl_3$, and passed through DI water in order to remove $FeCl_3$, $CuCl_2$ and HCl. After repeated washing, the fiber was either extracted directly as a porous structure or spun using a rotating probe to get twisted graphene fiber [24]. The resulting fiber possess electrical conductivity and is stable toward bending as demonstrated by tests performed, however it is difficult to prepare uniform graphene fibers of desired length on large scale through this method.

6.3.2.4 Chemical Unzipping of CNT Sheets

Other than aforementioned techniques, GO fibers can be prepared from MWCNT sheet via a novel conversion technique of chemical unzipping. Carretero et al. used a MWCNT sheet, prepared from CNT forest and deposited it on polytetrafluorethylene (PTFE) framework and chemically oxidized to unzip the aligned nanotubes. The obtained GO nanoribbon gel phase when taken out from the liquid phase turns into a flexible gelated fiber. The GO nano yarn (GONY) gets reduced when exposed to N_2H_4 vapors for 10–15 minutes at 95°C. The conductivity of GONY could be increased by annealing yarns at high temperature under Argon and Diluted Hydrogen gaseous environment [25].

6.4 APPLICATIONS OF CNT AND GRAPHENE FIBERS

A lot of electrical applications of CNT as well as graphene fibers are demonstrated, explained in this section.

6.4.1 Applications of CNT Fibers

Due to high conductivity, CNT fibers are the most promising candidate for numerous technological applications.

6.4.1.1 CNT Fibers as Winding Material in Transformer

For electric power applications, transformers are widely used in electronic and electrical based power applications i.e., generating stations, transmission, distribution, and utilization of alternating current. In conventional transformers, Cu wire is commonly used as windings, but these windings have less efficiency at high frequencies due to skin and proximity effect [26, 27]. On the contrary, CNTs possess negligible skin effects at high-frequency AC conditions because the current flows only along its circumference unlike metallic wires. Moreover, their light density decreases the energy losses such as core losses, heat losses and eddy current losses. Therefore, researchers are exploring macroscopic CNT wires for transformer's winding. For an instance, Kurzepa et al. (2012) has demonstrated CNT wires as core-type transformer's winding [28]. The transformer studied, has 21 turns of Cu wire in primary winding and 12 turns of CNT wire in secondary winding or vice-versa. The as-made transformer was tested at various input voltages (10 mV and 150 mV). The voltage to turn ratio of as-made transformer has shown results in agreement with conventional transformer. However, at different electric loads, CNT windings show lesser response at 400 KHz than Cu winding due to higher resistance of CNT coil.

Further, to achieve hypothetical estimations, Jaspal et al. (2016) has analyzed three-phase transformer model with Cu, Al and CNT wires via Finite Element Method technique and reported that resistive losses (I^2R losses) are decreased by 45% and 67% when Cu (937W) and Al (1564W) individually are supplanted by CNWs (519W) as winding material [29]. Due to less value of I^2R losses in CNT wire-based transformer, the emissivity also decreased. Additionally, the weight of the transformer having CNT wire has also reduced. Moreover, the working efficiency of transformer increased due to less eddy current losses in CNT winding. In the same year, they have simulated E-core transformer model via COMSOL Multiphysics software and concluded that CNT winding shows lesser resistive losses than Cu winding. These theoretical results suggested that CNT fibers can replace traditional wires for having more efficient transformers.

6.4.1.2 CNT Fibers in Electric Motors

Keeping an eye on the progression of electrical machine industries, it is a matter of concern that traditional Litz wires are attaining their thresholds and so a revolution in wirings is needed. Till now, Cu coils are used to make stator of electric machines, but for today's advanced technologies, lightweight electric motors, having less eddy current losses and other power losses, are required. Moreover, the operating temperature of conventional motors is quite high. At such high temperature, electrical conductivity of Cu wire decrease. Hence Cu coils must be replaced by more compatible material and CNT wires may be the best substitute of Cu wires as they have less losses even at high temperature. Therefore, for the first time, researchers from Lappeenranta University of Technology (LUT), Finland have designed a prototype electric motor replacing Cu wires with CNT wire [30]. The motor works at 15,000 rpm at 40 W and shows 70% efficiency. In 2015, Pyrhonen et al. have demonstrated a very low voltage tooth-coil permanent magnet machine having three stator teeth and two rotor poles with rotational speed (15000 rpm) experimentally as well as theoretically [31]. In as-demonstrated motor, conductors are made of ten parallel and 1.2 m long CNT yarns. For theoretical studies, CEDRAT Flux 2D software has been used. By analyzing the results, it has been concluded that the total losses of machine having CNT yarn and Cu fiber-based conductors are 13.7 and 7.7 W, respectively. Though the simulated results evidenced the poor efficiency (0.69) of motor having CNT conductors, yet it suggests that CNT technology will lead to a revolution in cabling. Furthermore, the experimental studies show that CNT fiber conductors show very small positive temperature coefficient (PTC) which is 40% of Cu wire's PTC. The other explained experiment is no-load measurements where the induced voltage got raised linearly with speed until 10000 rpm. On the other hand, at higher speeds, the eddy currents decreased the voltage slightly. In another study, researchers from the University of Kentucky studied a multidisc coreless axial flux permanent magnet (AFPM) machine consisting of 4 rotor and 3 stator disks with 16 poles and 12 coils on each rotor and stator stack, respectively [32]. ANSYS Maxwell 17.2 software has been used on the machine having different metallic cable coils and reported that the efficiency of AFPM machines has enhanced by the use of low-density CNT windings in place of Cu and Al coils. The as-designed machine's performance is studied at different rotational speed: 3600,7000 and 10000 rpm and it is reported that by increasing the rotational speed, the power per unit mass of CNT-based machine improves in comparison with Cu-based machine as CNT wires possesses negligible skin effect.

6.4.1.3 CNT Fibers as Data Transmission Cables

In telecommunication industries, progressed CNT-based ethernet cables are needed for high-rate data transmission or propagation [33] between two electronic gadgets as they show exceptionally less skin impact at high frequencies. Commonly, coaxial cables consist of center conductor covered with an insulating layer and an outer conductor. Jarosz et al. (2012) has synthesized coaxial cable using CNT macro assemblies as inner as well as outer conductor [34]. These CNT-based coaxial cables have been post-treated with $KAuBr_4$ and these densified CNT fibers are used to demonstrate USB cable [35]. The as-made cable transmitted data from a computer to external devices with several hundred Mbps without any errors. Moreover, the outer CNT assemblies possess high EMI shielding and less signal attenuation. Furthermore, Harvey et al. reported coaxial cables having either macroscopic CNTs as shielding material (CNT shield) or CNT assemblies as inner as well as outer conductors (all-CNT cable) [36]. The performance comparison in terms of losses for both cables is analyzed with standardize RG-316 cable. The signal attenuation is observed in all-CNT cable that can be attributed to its poor conductivity as well as irregularities in the CNT fiber. Recently, White et al. (2020) reported the uplink and downlink speed of a CNT ethernet cable (Figure 6.5(d)) and concluded that MWCNT wires were capable of data transfer rates of at least 99 Mbps [37]. In another reported work, Raus et al. (2013) demonstrated CNT fibers as ethernet cables and able to transfer data with the speed of 10 Mbps successfully [38]. In the same year, an electronic stuffs manufacturing company named as TE connectivity in USA has designed CNT prototype space bus cable and further demonstrated it as MIL-STD-1553B. In this context, a twisted-pair cable of CNT wires (26AWG wire equivalent diameter), covered with ethylene tetrafluoroethylene (ETFE) insulation is made [36]. The voltage drops across as-designed CNT data bus is 13.067 V which is only 1% higher than the Cu data bus. In 2016, Mirri et al. synthesized lightweight and flexible cables which has Cu and CNT assemblies as inner and outer conductor, respectively (shown in Figure 6.5(b)) [39]. The as-made cable was attached to female Sub Miniature version A connector and AC electrical properties of cables with varying thickness, were measured over 50–300GHz with a broadband, multiline-thru-reflect technique using an open-short-load-through corrected vector network analyzer. Figure 6.5(c) depicted the thickness effect of CNT coating w.r.t. insertion losses. The CNT cable having diameter 90 ± 14 μm, met the reference frequency (1GHz) for military specifications as it shows negligible effect of fatigue. The cable showed 30% higher signal attenuation at 400 MHz and approximately two-times higher below 100 MHz. Though the losses are higher in CNT-based cable, the weight of coaxial cable reduce up to 50%. Moreover, the enhancement in thickness of CNT layer, decreased the attenuation constant because the CNT layer decreases the cable resistance leading

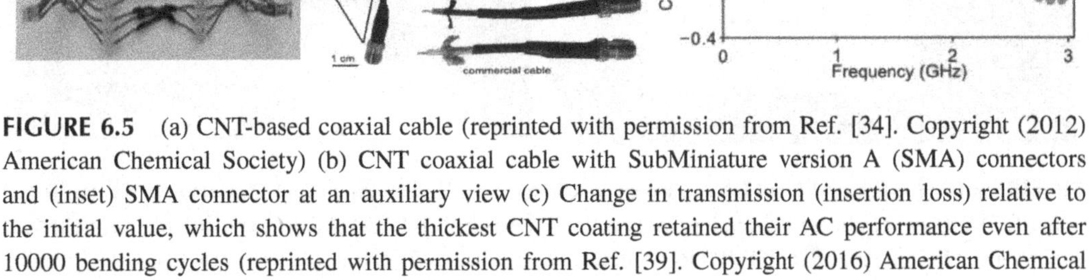

FIGURE 6.5 (a) CNT-based coaxial cable (reprinted with permission from Ref. [34]. Copyright (2012) American Chemical Society) (b) CNT coaxial cable with SubMiniature version A (SMA) connectors and (inset) SMA connector at an auxiliary view (c) Change in transmission (insertion loss) relative to the initial value, which shows that the thickest CNT coating retained their AC performance even after 10000 bending cycles (reprinted with permission from Ref. [39]. Copyright (2016) American Chemical Society), (d) CNT ethernet cable made using polystyrene-toluene feedstock (reprinted with permission from Creative Common License 4.0 Ref. [37]. Copyright (2021)).

to the lower attenuation values and improved cable quality at low frequencies. Further, 1m long CNT cable, weighing 7 g possess 1.5 db signal attenuation. In another research work, an ethernet cable had fabricated via CNT fibers which can transmit data at 100 Mbps and did not show any error up to 1 Gbps, certified via LanTek II cable test equipment. The cable shows performance resembling with conventional twisted cable for computer network (Category 5 cable).

The advanced aircraft technologies and rapid growth in space reconnaissance has encouraged the researchers to find smart and lightweight materials [40]. Hence, various space defense and aeronautics agencies in United States such as NASA, AFRL and DARPA are investing into various projects focusing on for mass saving in satellites launching. Consequently, an exigency for advanced wiring associated with least losses that can enhance the electrical stability of spaceships is paramount. In this way, CNT fibers are the most promising material for weight reductions in satellites, unmanned aerial vehicles (UAVs), military aircrafts [41]. In this frame, S. Harvey demonstrated CNT cable electric boards in replacement of Cu cable boards in high-altitude long-endurance unnamed aerial vehicle (HALE UAV) [36]. When CNTFs replaced the shielding metal, then the weight can be cut down up to 300 pounds. Moreover, if the cables contain CNTs as outer as well as inner conductor, then the total weight reduction can be 400 pounds. Recently, in 2019, Abbe et al. performed first in-situ space material experiment of CNTs entitled as CiREX (Carbon Nanotubes—Resistance Experiment) to investigate the properties variation of CNT fibers under solar ultraviolet light (UV) and vacuum ultraviolet light (VUV) [42]. Figure 6.6(a) presents the CiREX, FIPEX and TEG integrated in SOMP2. In their work, the CNT-EXT (Figure 6.6(b)), connected with electrical circuit board (CNT-ELEK) via flat cable, have been exposed to UV and VUV. By analyzing the ground validation test under solar light, it is indicated that the molecular interaction of CNTs and oxygen causes an increase in resistance during the evacuation process.

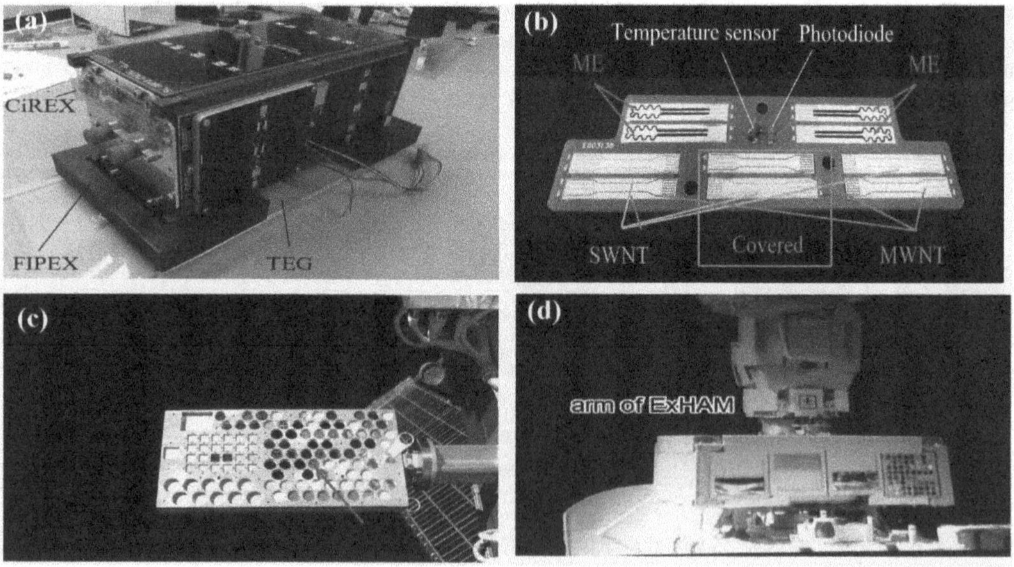

FIGURE 6.6 (a) Ground level CNT yarn exposure to space environment on robot arm 'Kibo' (b) RAM side of CNT yarns (indicated via red arrow) on ISS experiment- 8 ORMaTE III exposure plate (c) CiREX, FIPEX and TEG integrated in SOMP2 Reprinted with permission from Ref. [42]. Copyright (2019) Elsevier, (d) CNT-EXT without cover plate (reprinted with permission from Ref. [45]. Copyright (2019) Elsevier).

Further, the oxygen ion gets desorbed under solar light and no notable structural change is observed before and after irradiation. Similarly, Hopkins et al. (2016) exposed CNT fibers in high-fluence atomic oxygen (AO) natural space environment in (low earth orbit) LEO for 2.14 years (May 20, 2011 to July 9, 2013) on the exterior of the International Space Station (ISS) as part of a Materials International Space Station Experiment (MISSE-8), Payload and Optical Reflector Materials Experiment III [43]. They have placed CNT fibers in ram and wake direction of satellite (Figure 6.6(c)) where CNT fibers in ram direction receives higher intensity of AO than the wake direction. From structural microanalysis, it is observed that CNT yarns in both directions remains undamaged. However, on nanoscale, an outer layer of ashes can be observed on CNT fiber for both ram and wake exposures. Though, higher amorphization or thicker amorphous ash layer resulted in ram direction still none of the CNTs in fiber has damaged. Raman analysis of exposed CNT fiber in ram orientation (higher I_D/I_G) is also in agreement with structural micrographs. Further, the higher erosion yield was calculated for ram direction. The CNT fibers showed increase in resistance in ram (26.1%) and wake (28.5%) orientation as compared to reference fiber. They have also showed a decrease in mechanical strength of CNT fibers in ram side while the wake side showed little change. In another research, Kemnitz et al. (2019) have studied the effect of space environment i.e., AO and ultraviolet C (UVC) exposure, on the properties of acid-treated CNT fibers so that they can be incorporated in future spacecrafts [44]. For this purpose, AO beam was radiated at 2 Hz for 1, 50,000 pulses which is equivalent to about one year of AO exposure in LEO. They reported that the conductivity of CNT fibers decreased up to 22% during AO exposure. Moreover, erosion is also observed in SEM images of CNT fibers after AO exposure, resulting in the reduced effective cross-sectional area for electron transport. Further, for UVC exposure, UVC lamps, having peak intensity at 254 nm is used which is about 3.77 years of equivalent constant LEO UVC exposure. Under UVC light, the acid of fiber gets diffused and goes on increasing the alignment of CNTs in fiber which increases the electrical conductivity. Ishikawa et al. (2019) reported the experimental exposure of CNT fiber on lab-level as well as in space [45]. Firstly, the fibers were exposed in AO, EB and UV radiation on lab-scale and observed that there is negligible effect after EB and UV exposure. Though the conductivity should be decreased due to damaging of fibers but still no impact is notable. Furthermore, the CNT fibers were exposed on ISS at LEO via attaching them on Exposed Experiment Handrail Attachment Mechanism (see Figure 6.6(d)) for 395 and 780 days at wake orientation and 484 days at ram orientation. The conductivity remains same in wake-facing side for both thick and thin CNT fibers. On the contrary, the conductivity of thin fiber showed lesser conductivity in ram-facing side, on which the reason is debatable.

6.4.1.4 CNT Fibers as Antenna

The future forefront antennas in RF applications will require the integration of fundamental research concerning the development of lightweight wires. The major problem with traditional wire is their higher resistivity at high frequencies. Therefore, for weight critical devices, multiband and ultrawide band operation and bandwidth enhancement of futuristic telecommunication applications, CNTs-based antennas have been researched as they possess high-frequency electrical properties. In these terms, various experimental and theoretical studies on CNT bundles as antenna have been done and the path has been paved for industrial scale. On the basis of these results, CNT macro assemblies such as CNT films and fibers are demonstrated and studied as RF patch antennas and dipole antenna [46]. The CNT fibers, having different diameters (12.5, 37.5 and 112.5 μm named as 1ply, 3ply and 3*3ply, respectively) and different lengths (4inch and 3mm), are simulated using Hallèn's integral equation and compared with 30 Gauge Cu wire (shown in Figure 6.7(c)). The calculated radiation efficiency of 4-inch-long dipole antenna (resonant frequency=1.475 GHz) having 1ply, 3ply and 3*3ply CNT fiber are 28db, 19.4db and

11.6db lower than Cu wire. Furthermore, for 3mm long antenna (resonant frequency =50GHz), the calculated radiation efficiencies for 1ply, 3ply and 3*3ply CNT fibers are 4.9db, 1.8db and 0.9db, respectively. In addition, a half-wavelength dipole antenna, synthesized via dimethyl siloxane densified CNT fiber (spun via Forest spinning), was demonstrated by Keller et al. Figure 6.7(a) and Figure 6.7(b) show the front and back view of 3-ply CNT thread dipole antenna prototype. The H-plane radiation pattern of as-made 3-ply CNT thread dipole, having resonant frequency ~2.45 GHz, is reported and the realized gain −10.5 dBi is in agreement with theoretical values. Puchades et al. (2016) designed dipole antenna resonating at 1.5–2 GHz (Figure 6.7(d)) and compared the efficiencies of SWCNT and MWCNT based dipole antenna [47]. Additionally, the effect of oxidation temperature is also studied. The numeric value of reflection coefficient (S_{11}) of chlorosulfonic acid doped SWCNT based dipole antenna is calculated as -21.7, -19.9, -15.2 and -11.6 dB at ambient temperature, 50, 100 and 250°C (shown in Figure 6.7(e)), respectively. In 2017, Bengio et al. demonstrated a quarter wavelength monopole antenna (see Figure 6.7(f)) with varying diameter and twisting angle of CNT fibers (spun via wet spinning). They reported that as-made antennas showed 20 folds higher efficiency than Cu wire-based antenna (shown in Figure 6.7(g)). Furthermore, Bengio et al. (2019) demonstrated CNT-based patch antenna for wireless communication (shown in Figure 6.7(h)) and reported 94% efficiency at 10GHz and 14 GHz [48]. By analyzing their results, they concluded that though the sheet resistance drops with the increase in film thickness, but the radiation efficiency of as-made antenna increased. The evaluated resonance frequencies (Hammersatd formula) of as-made patch antennas intended for 5 GHz, 10 GHz and 14 GHz is very close with a relative error of 4%, 10% and 28%, respectively. Figure 6.7(i) presents the relative radiation efficiencies plot for various kinds of wires such as Cu wire, as-grown CNTs and post-treated CNT wires.

6.4.2 Application of Graphene Fibers

Upcoming era of portable and transparent electronics demands devices that are flexible, easy cut and inexpensive electronic textiles. With heaps of expectations associated to astonishing properties of graphene, tremendous proportion of research is underway to achieve new horizons with this futuristic material. Moving toward its microstructure, graphene-based fibers have and is being continuously explored for daily life applications because of their remarkable electrical and mechanical properties. In this section, use of graphene-based fibers and yarns in current carrying applications has been described thoroughly.

6.4.2.1 Graphene Fiber as Electric Cable

For electricity transmission to distant places, low-density transmission electric cables of high power and good mechanical strength are desired. In this frame, various research groups are working on graphene fibers as they can replace and even overshadow the conventional electric wires for daily use applications. In 2015, Xu et al. engineered specifically controlled parameters to fabricate chemically reduced graphene fibers with least defects with additional stretching during collection. The as-spun fiber highly efficient GF has been used in candescent bulb as an alternative of tungsten filament and achieved higher irradiance intensity for longer periods of time (500 hours) [50]. In another significant work, Though GFs have proved to be good power transmission cables, yet hybrid GFs are explored for having improved strength, excellent durability and fire-resistant property. In 2015, for having ultra-strong graphene-based fibers, graphene nanosheets (GNS) was grown on Ni films and were scrolled up and shrunk into solutions in the form of monolithic fiber using a roller. Further, Cu is electroplated to have graphene-based composite fibers achieving 10 and 100 folds higher current density than Cu wire and pristine GFs, respectively.

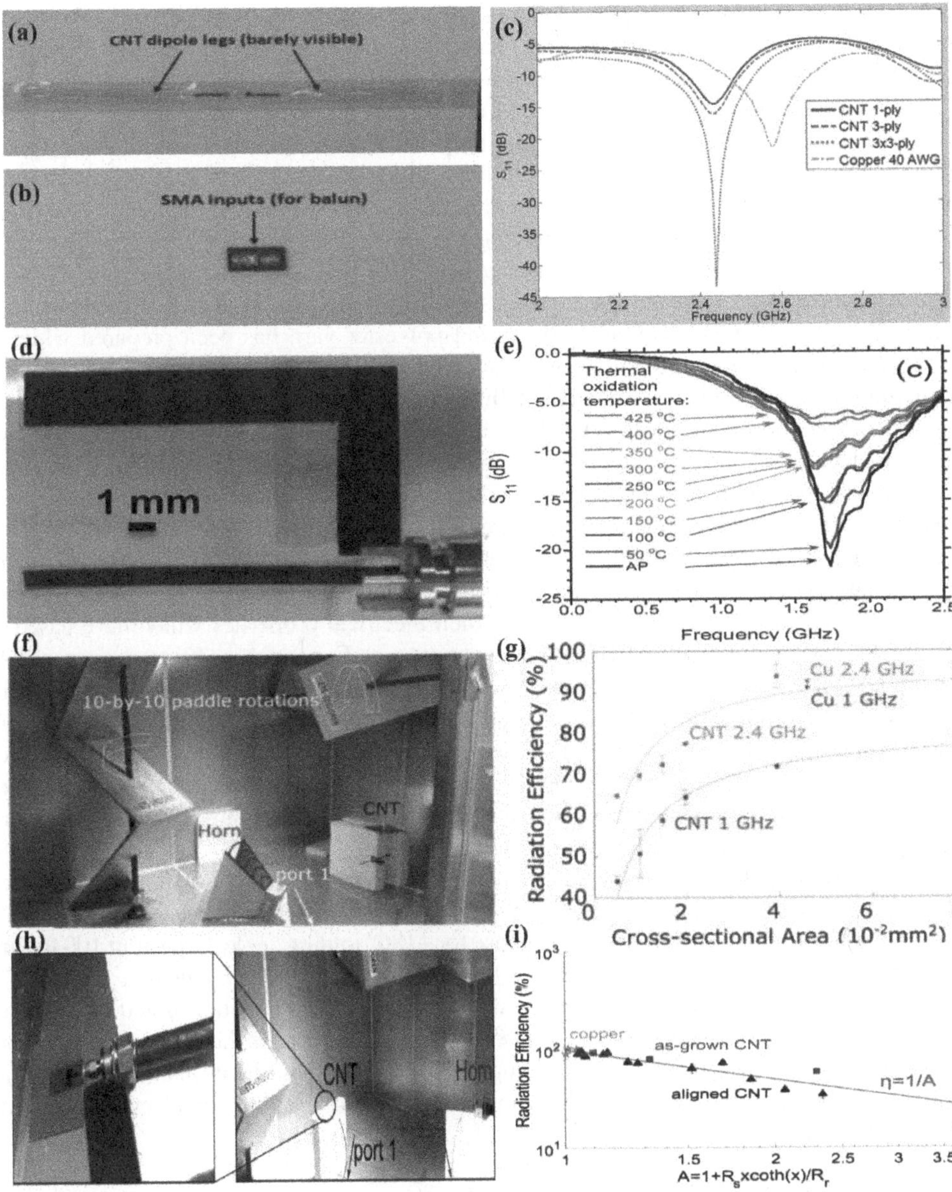

FIGURE 6.7 3-ply CNT thread dipole antenna prototype (a) front view, (b) back view, (c) measured reflection coefficient of Cu and CNT thread dipole antenna prototype (reprinted with permission from Creative Common License Ref. [46]. Copyright (2014)), (d) image of diploe antenna fabricated with NCTI MWCNT sheet material on a soda-lime glass substrate, covered with Kapton tape and showing the SMA connector, (e) plot summarizing measured progression of S11 of a dipole SWCNT thin-film antenna after thermal oxidation at different temperatures (reprinted with permission from Ref. [47]. Copyright (2016) American Chemical Society), (f) measurement setup in reverberation chamber, (g) summary of radiation efficiencies on CNT thread and Cu control antenna (reprinted from ref. [49] with the permission of AIP Publishing), (h) image of CNT patch antenna, (i) radiation efficiencies plot for various kind of wires (reprinted from Ref. [48] with the permission of AIP Publishing).

Further, these hybrid fibers are used for lighting a 9 V LED bulb (shown in Figure 6.8) [51]. In another work, Fang et al. successfully spun graphene-based composite fibers from montmorillonite nanoplatelets, and liquid crystal GO followed by high temperature treatment which reduces GO to graphene. Along with good electric and mechanical properties, these hybrid fibers offer fire-resistant properties for high temperature applications. A well-performed circuit working well circuit has been made by compacting dozens of fibers and is exposed to combustion environment through an alcohol lamp (Figure 6.8(d)). The circuit perform well even when combusted part has turned glowing red as indicated by glowing LED bulb [52]. For exploiting the advantage of hybrid GFs-based LED lightning wire, Yun et al. (2015) have synthesized wearable NO_2 gas sensor that turns on the LED light while the gas exposure reaches a certain concentration [53]. For this, reduced GO (rGO) decorated cotton and polyester yarn has been prepared which is integrated into commercial fabric along with an electric circuit. One of the studies reveals the excellent durability of gold/graphene wire which has been demonstrated by illuminating an LED lamp. Furthermore, due to high ampacity, the GFs can also be used to supply power to a circuit having high electric load i.e., lamp [54]. In this context, Liu et al. synthesized highly conducting GF-Br_2 filament for a desk lamp (220 V, 9 W). The lamp showed stable irradiation intensity of 1060 lux matching well with benchmark Cu wire [55].

Furthermore, due to high flexibility of graphene nanosheets, GFs-based conductors are highly eligible for flexible electronics as the high flexibility allows graphene fibers to bend, knit and fold with different structures while maintaining their electrical properties which have eased its way past many conventional and newly explored conductors. For instance, Xu et al. prepared Ag doped graphene nanowires, via wet spinning of giant GO. These doped fibers show conductivity of 9.3×10^6 S/m with high current carrying capacity and appreciable mechanical strength and flexibility. These doped fibers have then been used for constructing a stretchable electric circuit which kept on working even after 50 cycles of stretching and relaxing as indicated by glowing LED. Though bending shows repeated results for LED lightning but the surface developed crease at folded points which remains even after unfolding[57].

In another study, graphene circuit has been prepared on a paper by vacuum filtration and selective transfer printing through a membrane filter. These graphene circuits show minor decrement in conductance with maximum 6.7% decrease for $+180^0$ folding angle. Even after 100 folding cycles, nearly 83% of conductivity is retained by the circuit and the decrement is attributed to fractures induced with each folding angle. The effectiveness has been demonstrated by attaching an LED chip onto a paper substrate through silver paste. The substrate is then subjected to various twist and turns including -180^0 & $+180^0$ folding angles, 3-D circuit boards and before and after crumpling of paper, however the LED glowed in every, describing the good electrical performance of graphene material[56].

6.4.2.2 Graphene Fiber as Transistor Electrodes

For transparent, flexible, compact and wearable electronic circuits, i.e., morph phones, human health monitoring devices televisions etc., flexible transistor electrodes are in great demand. In 2015, Yoon and his team have synthesized Ag doped GFs having electrical conductivity of 1.58×10^6 S/m and utilized the same for production of fiber type transistors. These hybrid fibers have been used as source, drain and gate electrodes on flexible substrates in the prepared ion gel gated transistor which showed similar electrical characteristics as conventional transistor electrodes. The schematic diagram of prepared fiber type transistor electrode is shown in Figure 6.8(e–f). The performance of the device has been found stable even after 1000 cycles of bending and rolling [58]. In another work, Xu et al. have employed synergetically engineered ultrafine GFs to connect homemade microchips. The performance of these economical and lightweight GFs

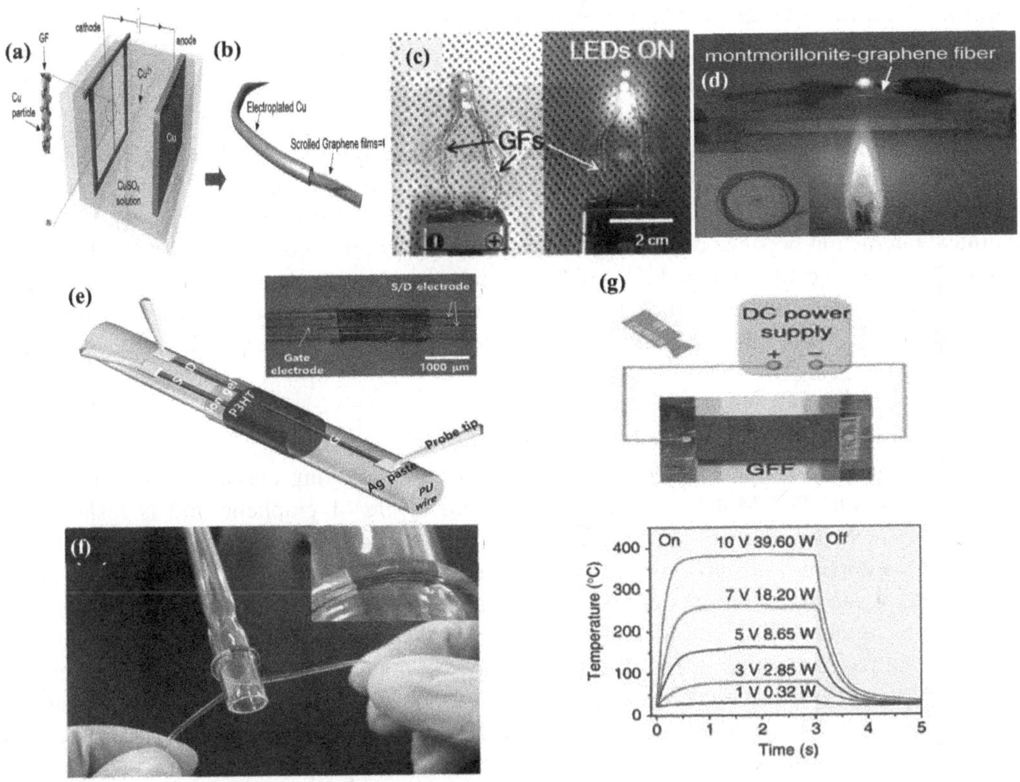

FIGURE 6.8 (a) Electroplating copper on graphene fiber (b) graphene wire with electroplated copper (c) working LED demonstration using Cu shell graphene fibers (reprinted with permission from Ref. [51]. Copyright (2018) American Chemical Society) (d) working electric circuit when the wire is heated to glowing red (reprinted with permission from Ref. [52]. Copyright (2015) American Chemical Society) (e) schematic showing fiber type transistor (f) and digital image of flexible fiber type transistor on graphene/Ag hybrid electrode (reprinted with permission from Creative Common License 4.0 Ref. [58]. Copyright (2015)) (g) graphene fiber fabric based electrothermal heater and its temperature profile at different voltages (reprinted with permission from Creative Common License Ref. [59]. Copyright (2016)).

was same as that of expensive gold fibers as shown by voltammetry curves which proved that GFs have great potential in order to replace traditional metallic wires in future carbon-based electronics [50].

6.4.2.3 Graphene Fibers in Electric Motor

Alike CNTs, the unmatched electrical characteristics of graphene have opened up possibilities for their use in metal free electric motors. Xu et al. have demonstrated the working of an electric motor by replacing the Cu coil (4.4 g) with lightweight graphene filaments (0.55 g). For this, the as-obtained GO fibers were graphitized at high temperature to eliminate any atomic defects resulting in high stiffness (282 GPa) and conductivity of 10^5 S/m. The designed motor can undergo 350 rpm at 8 V using DC current supply. Though this work has highlighted the potential use of weight economical graphene fibers-based devices in drones and robotics, yet a lot of work has to be done in this field [50].

6.4.2.4 Graphene Fibers as Data Transferring Cable

Alike power transmission cables, GFs can also be exemplified as data transmission cables. For instance, Liu et al. have doped their pure graphene filaments with a donor potassium (p-type) and acceptor type Br_2 and $FeCl_3$ molecules (n-type) via two zone vapor transport method. These doped GFs show high electrical conductivity (10^7 S/m) outperforming the metals in terms of specific conductivity and easy scalability. However, only Br_2 doped GF have conferred appreciable stability in air. On the contrary, GF-K and GF-$FeCl_3$ show degraded electrical behavior due to the chemical interaction between dopants and water. Inspired by long term stability of GF-Br_2 fiber, a multiply graphene fiber-based USB cable was fabricated via replacing conventional copper wire. The developed data transmission device was tested at usual reading and writing speeds of commercial USB cables, hence show good signs for use in high-frequency signal transmission function [55].

6.4.2.5 Graphene Fibers for Aerospace Engineering

Authors synthesized graphene fiber doped with $MoCl_5$ employing classical two zone vapor transport reaction. The $MoCl_5$ gets intercalated into layers of graphene and is responsible for increase in carrier density and mobility resulting in record conductivity of 1.73×10^7 S/m. Since this dopant is a stable and efficient intercalating compound, the GF-$MoCl_5$ is quite stable under realistic conditions and shows much effective EMI shielding than pristine GF [60]. Due to graphene's shielding capacity helps to produce paints that can reduce the radar footprint and paved the path for stealth technology. Moreover, graphene can be used for de-icing systems integrated with aero vehicle wings as graphene sheets can be used for designing of flexible heaters [61]. Hence, researchers are focusing on graphene-based electrothermal thermal. For this instance, authors have produced non-woven graphene fiber fabric (GFF) via fusing short length graphene fibers into a whole fabric using wet fusing assembly and then annealing them at 3000^{0}C to enhance their mechanical and electrical properties (4.5×10^4 S/m). The synthesized GFF is employed as electrothermal heater in large area (4×2 cm^2) at lower voltages. Figure 6.8(g) shows the temperature variation of the fabricated heater at different operating voltages. The temperature achieved and ultrafast thermal response has been found better than other carbon papers and is due to their high specific conductivity. These synthesized GFF heaters showed uniform temperature distribution in both flat and bending state thus can be explored in large area flexible heaters [59].

6.4.2.6 Graphene Fibers as Antennas

With constantly evolving communication strategies, materials that can serve as higher data transmission rates with broad operating spectra are desired. Practical application requiring the transport of significant current such as NFC antennas in power electronics require however sheet resistance of $\ll$1 ohm/cm^2. Highly conducting and flexible graphene paper has been employed for the fabrication of flexible antennas for near field communication device. The graphene paper is patterned into antennas via mechanical cutting and laser ablation. The fabricated antennas show remarkable stability in resonance and inductance upon bending and is found better than commercialized metal antennas. This antenna is also found working in exchanging data with a smartphone via NFC app and in other daily purpose applications as well as described in Figure 6.9(a–g) [62]. In case of metals, scaling down the antenna size causes high attenuation losses due to poor electron mobilities at higher frequencies. Graphene, however, supports the propagation of surface plasmon polariton waves and can radiate electromagnetic waves even in THz band. George et al. have designed a microstrip patch antenna using graphene radiating patch to radiate at 2.6 THz, furthermore it can also be made to resonate at dual frequencies by increasing substrate height

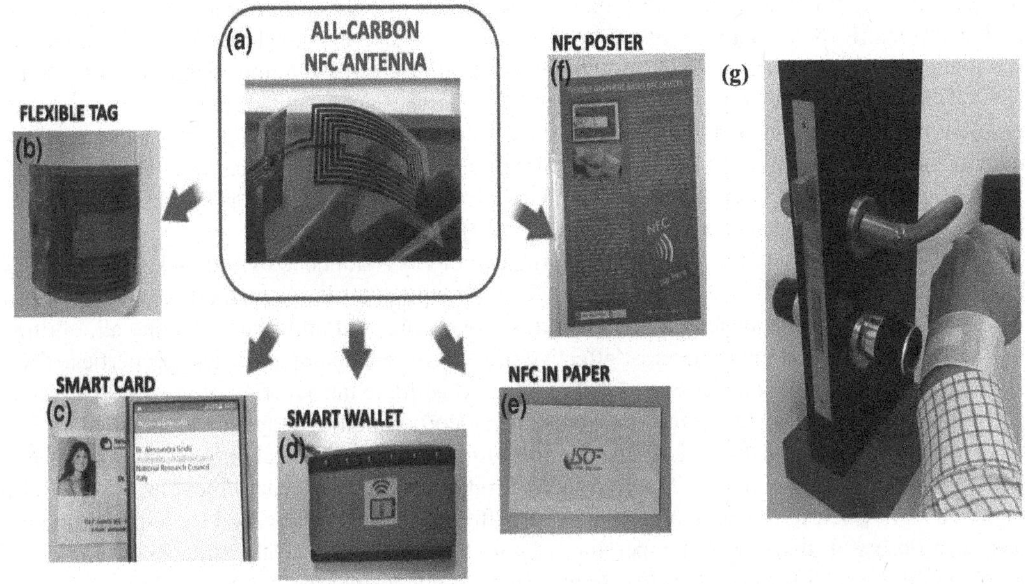

FIGURE 6.9 (a) GF-based antennas used in different NFC based applications (b) flexible tag (c) smart card (d) smart wallet (e) NFC in paper (f) digital poster (g) antenna-based silk band for smart lock (reprinted with permission from Ref. [62]. Copyright (2018) Elsevier).

with 87.3% efficiency [63]. In another study by Hafizah et al., the authors have proposed fabrication of single and array graphene antennas for 5G applications. The antennas developed have been tested at higher frequencies (15 GHz) to generate large bandwidths and higher speeds. Both single and array graphene show higher efficiencies of 67.44% and 72.98% respectively, however array element produced higher gain and beam scanning capability of up to 39.05^0. Thus, they infer that graphene material is suitable for 5G materials, however more conducting graphene can be employed for further improvement in parameters [64].

6.5 CNT GRAPHENE HYBRID

As mentioned in previous sections, both graphene and nanotubes show marvelous property individually, which persuaded researchers to hybridize both materials to get some interesting features. The idea is to use the two-dimensional planar structure of graphene in bridging on dimensional nanotubes in a fiber to complement the electrical properties.

Zhu et al. have used a novel method that covalently attached high quality graphene with less wall CNT carpets. They initially grow graphene on the copper foil, followed by deposition of catalyst and alumina particles in series using electron beam evaporation. In whole, the catalyst layer has been sandwiched between the graphene and alumina layer to maintain the diameter of CNT while bond formation with graphene takes place. This growth strategy resulted in CNT forest that grew out of graphene layer forming covalent bonds [65]. It has been argued that covalent bond formation during the functionalization of carbon nanostructures hampers the facile electron transport, thereby reducing the conductivity. CNT-graphene hybrid material offers self-functionalization due to their geometry difference thus no additional material is needed for dispersion. Oh et al. have fabricated CNT/graphene/PDMS nanocomposite via solution mixing technique. Simultaneous presence of CNT and graphene inhibits agglomeration and restacking and enhances interconnection among them which leads to reduction in contact resistance and increased electrical conductivity [66]. Foroughi et al. have described the fabrication of conductive CNT-graphene yarn via deposition of graphene through electrospinning technique. In whole, MWCNT sheets drawn from vertically aligned CNT arrays are attached with a DC motor, and graphene flakes are deposited within and on the surface of sheet. Resulting graphene coated CNT hybrid sheet has been twisted in clockwise direction to get CNT/graphene yarn. The developed yarn showed several fold increase in conductivity as compared to pristine CNT and graphene yarn due to low contact resistance among the bundles [67].

Lepak-Kuc et al. study the electrical performance of CNT-graphene hybrid and doped CNT-G hybrid networks. Results have been analyzed from a range of available methods of hybridization including rubbing of graphene with CNT sheet, spray coating and infiltration. Among all, infiltration technique is found to be the most effective method for enhancing electrical properties. They also find that the doping efficiency of CNTs increased with the introduction of graphene. Current carrying capacity is also found to be enhanced by incorporation of graphene as justified by the formation of additional pathways for the flow of current [68]. Zhong et al. have synthesized CNT thick stick like assembly in a 1600 mm long horizontal quartz CVD reactor kept at 1150^{0}C. The obtained CNT stick has been analyzed at three different regions differenced by temperatures. It has been observed that in high temperature region, CNTs are extensively wrapped around with graphene like a cauliflower as described by the author, suggesting that GNS growth is favored at high temperature in a floating catalyst CVD reaction. They have argued that growth of graphene take place on CNT bundles as no free graphene was isolated under TEM observations. Two probe conductivity measurements of prepared CNT-GNS multiple hybrid yarn show value around 10^5 S/m and high tensile strength of 300 MPa. Real time potential use of these fibers has been demonstrated in replacing conventional wire of a lamp with it and operating it at 24 V DC supply [69].

With the launching of electric vehicles, the usage load over Cu wires enhanced with increase in wiring network requirements. To cope with increased energy demands, lightweight conductors are desired to cut down the usage for cutting extra weight. Park et al. have synthesized CNT-Cu-Gr wire by electrodepositing copper on prepared CNTFs followed by growth of graphene over the surface of deposited copper. These wires have been reported to have approximate 60% reduction in density with increased thermal and electrical conductivities. The specific ampacity (4.42×10^4 Acm/g) is also found significantly greater than Cu wire (2.8 time) indicating higher attainable power [70]. From all these studies, it is clear that graphene increases the conductance and current in the CNT materials therefore can pave a way toward realization of maximum electrical performance of CNT fibers.

6.6 CONCLUSION

In this chapter we have dealt with two pioneers of carbon nanostructures having miraculous electrical properties and potential to greatly change the current scenario of electrical world. Both CNTs and graphene show ballistic conduction of electron through their structures leading to superior electrical performances. However, their scalability are greatly limited by their synthesis and processability. Direct spinning of CNT's is a profitable technique that can meet the huge market requirement of conducting wires however lacks in terms of cost economy as it is an expensive process. Production of graphene fibers is economical however not scalable to industrial scale level, but this field is rapidly evolving, and much positive results are expected in upcoming time. Both these fibers have been demonstrated in various current carrying and conducting applications in several forms. Researchers are also exploiting their hybrids to address the issue even better and are getting good results. Although, the current scenario does not hold past the conventional metal counterpart specification, yet the upcoming era of electronics and electric cables will be circuited by these carbon structures. Both Authors Pallvi Dariyal and Manoj Sehrawat contributed equally.

REFERENCES

1. Tans, S.J., et al., *Individual single-wall carbon nanotubes as quantum wires.* Nature, 1997. **386**(6624): p. 474.
2. Datta, S., *Electronic transport in mesoscopic systems* 1997: Cambridge university press.
3. Dai, H., *Carbon Nanotubes: Synthesis, Structure, Properties, and Applications*, in *Topics in Applied* Physics 2001, Springer.
4. Bourlon, B., et al., *Determination of the intershell conductance in multiwalled carbon nanotubes.* Physical review letters, 2004. **93**(17): p. 176806.
5. Gelao, G., R. Marani, and A. Perri, *A formula to determine energy band gap in semiconducting carbon nanotubes.* ECS Journal of Solid State Science and Technology, 2019. **8**(2): p. M19.
6. Kane, C., L. Balents, and M.P. Fisher, *Coulomb interactions and mesoscopic effects in carbon nanotubes.* Physical review letters, 1997. **79**(25): p. 5086.
7. Neto, A.C., et al., *The electronic properties of graphene.* Reviews of modern physics, 2009. **81**(1): p. 109.
8. Andrei, E.Y., G. Li, and X. Du, *Electronic properties of graphene: a perspective from scanning tunneling microscopy and magnetotransport.* Reports on Progress in Physics, 2012. **75**(5): p. 056501.
9. Suvarnaphaet, P. and S. Pechprasarn, *Graphene-based materials for biosensors: a review.* Sensors, 2017. **17**(10): p. 2161.
10. Maffucci, A. and G. Miano, *Electrical properties of graphene for interconnect applications.* Applied Sciences, 2014. **4**(2): p. 305–317.
11. Ando, T., *The electronic properties of graphene and carbon nanotubes.* NPG asia materials, 2009. **1**(1): p. 17–21.
12. Headrick, R.J., et al., *Structure–property relations in carbon nanotube fibers by downscaling solution processing.* Advanced Materials, 2018. **30**(9): p. 1704482.
13. Dariyal, P., et al., *A review on conducting carbon nanotube fibers spun via direct spinning technique.* Journal of Materials Science, 2021. **56**(2): p. 1087–1115.
14. Kuznetsov, A.A., et al., *Structural model for dry-drawing of sheets and yarns from carbon nanotube forests.* Acs Nano, 2011. **5**(2): p. 985–993.
15. Motta, M., et al., *Mechanical properties of continuously spun fibers of carbon nanotubes.* Nano letters, 2005. **5**(8): p. 1529–1533.
16. Dariyal, P., et al., *Synthesis of carbon nanotube fiber via direct spinning for conducting wires.* 2020.
17. Xu, Z. and C. Gao, *Graphene chiral liquid crystals and macroscopic assembled fibres.* Nature communications, 2011. **2**(1): p. 1–9.
18. Xiang, C., et al., *Graphene nanoribbons as an advanced precursor for making carbon fiber.* Acs Nano, 2013. **7**(2): p. 1628–1637.
19. Fang, B., et al., *Handedness-controlled and solvent-driven actuators with twisted fibers.* Materials Horizons, 2019. **6**(6): p. 1207–1214.
20. Feng, L., et al., *Dry Spin Graphene Oxide Fibers: Mechanical/Electrical Properties and Microstructure Evolution.* Scientific reports, 2018. **8**(1): p. 1–7.
21. Dong, Z., et al., *Facile fabrication of light, flexible and multifunctional graphene fibers.* Advanced Materials, 2012. **24**(14): p. 1856–1861.
22. Li, J., et al., *Flexible graphene fibers prepared by chemical reduction-induced self-assembly.* Journal of Materials Chemistry A, 2014. **2**(18): p. 6359–6362.
23. DeFrates, K.G., et al., *Protein-based fiber materials in medicine: A review.* Nanomaterials, 2018. **8**(7): p. 457.
24. Chen, T. and L. Dai, *Macroscopic Graphene Fibers Directly Assembled from CVD-Grown Fiber-Shaped Hollow Graphene Tubes.* Angewandte Chemie, 2015. **127**(49): p. 15160–15163.

25. Carretero–González, J., et al., *Oriented graphene nanoribbon yarn and sheet from aligned multi-walled carbon nanotube sheets*. Advanced Materials, 2012. **24**(42): p. 5695–5701.
26. Lekawa-Raus, A., et al., *Electrical transport in carbon nanotube fibres*. Scripta Materialia, 2017. **131**: p. 112–118.
27. Xi, N. and C.R. Sullivan. *An improved calculation of proximity-effect loss in high-frequency windings of round conductors*. In *Power Electronics Specialist Conference, 2003. PESC '03. 2003 IEEE 34th Annual*. 2003.
28. Kurzepa, L., et al., *Replacing copper wires with carbon nanotube wires in electrical transformers*. Advanced Functional Materials, 2014. **24**(5): p. 619–624.
29. Jaspal, P. and P. Sharma, *Performance Comparison of Carbon Nanotubes with Copper and Aluminium as Winding Material in Transformer*. Indian Journal of Science and Technology, 2016. **9**(40).
30. Johnson, D., *Carbon nanotube yarns could replace copper windings in electric motors*. IEEE Spectrum: 03–Oct–2014.
31. Pyrhönen, J., et al., *Replacing copper with new carbon nanomaterials in electrical machine windings*. International Review of Electrical Engineering (IREE), 2015.
32. Rallabandi, V., et al., *Coreless multidisc axial flux PM machine with carbon nanotube windings*. IEEE Transactions on Magnetics, 2017. **53**(6): p. 1–4.
33. Kukowski, T. *Lightweight CNT cables for aerospace*. in *NESC NDE technology assessment NASA NDI workshop, Johnson Space Center Houston, TX*. 2012.
34. Jarosz, P.R., et al., *High-performance, lightweight coaxial cable from carbon nanotube conductors*. ACS Applied Materials & Interfaces, 2012. **4**(2): p. 1103–1109.
35. Alvarenga, J., et al., *High conductivity carbon nanotube wires from radial densification and ionic doping*. Applied Physics Letters, 2010. **97**(18): p. 182106.
36. Harvey, S.E. *Carbon as conductor: a pragmatic view*. In *International Wire Cable Connectivity Symposium (IWCS), Providence, RI, Nov*. 2012.
37. Orbaek White, A., et al., *On the use of carbon cables from plastic solvent combinations of polystyrene and toluene in carbon nanotube synthesis*. Nanomaterials, 2021. **12**(1): p. 9.
38. Łękawa-Raus, A.E., *Carbon Nanotube Fibres for Electrical Wiring Applications*, 2013, University of Cambridge.
39. Mirri, F., et al., *Lightweight, flexible, high-performance carbon nanotube cables made by scalable flow coating*. ACS Applied Materials & Interfaces, 2016. **8**(7): p. 4903–4910.
40. Gohardani, O., M.C. Elola, and C. Elizetxea, *Potential and prospective implementation of carbon nanotubes on next generation aircraft and space vehicles: A review of current and expected applications in aerospace sciences*. Progress in Aerospace Sciences, 2014. **70**: p. 42–68.
41. Alvarenga, J., Carbon Nanotube Materials for Aerospace Wiring. 2010. Thesis, Rochester Institute of Technology. https://scholarworks.rit.edu/cgi/viewcontent.cgi?article=3752&context=theses
42. Abbe, E., et al., *A material experiment for small satellites to characterise the behaviour of carbon nanotubes in space–development and ground validation*. Advances in Space Research, 2019. **63**(7): p. 2312–2321.
43. Hopkins, A., et al., *Space survivability of carbon nanotube yarn material in low Earth orbit*. Carbon, 2016. **107**: p. 77–86.
44. Kemnitz, R.A., et al., *Characterization of simulated low earth orbit space environment effects on acid-spun carbon nanotube yarns*. Materials & Design, 2019. **184** : p. 108178.
45. Ishikawa, Y., et al., *Survivability of carbon nanotubes in space*. Acta Astronautica, 2019. **165** : p. 129–138.
46. Keller, S.D., et al., *Electromagnetic simulation and measurement of carbon nanotube thread dipole antennas*. IEEE Transactions on Nanotechnology, 2014. **13**(2): p. 394–403.
47. Puchades, I., et al., *Carbon nanotube thin-film antennas*. ACS applied materials & interfaces, 2016. **8**(32): p. 20986–20992.
48. Amram Bengio, E., et al., *Carbon nanotube thin film patch antennas for wireless communications*. Applied Physics Letters, 2019. **114**(20): p. 203102.

49. Amram Bengio, E., et al., *High efficiency carbon nanotube thread antennas*. Applied Physics Letters, 2017. **111**(16): p. 163109.
50. Xu, Z., et al., *Ultrastiff and strong graphene fibers via full-scale synergetic defect engineering*. Advanced Materials, 2016. **28**(30): p. 6449–6456.
51. Kim, S.J., et al., *Ultrastrong graphene–copper core–shell wires for high-performance electrical cables*. ACS Nano, 2018. **12**(3): p. 2803–2808.
52. Fang, B., et al., *Wet-spinning of continuous montmorillonite-graphene fibers for fire-resistant lightweight conductors*. ACS Nano, 2015. **9**(5): p. 5214–5222.
53. Ju Yun, Y., et al., *Ultrasensitive and highly selective graphene-based single yarn for use in wearable gas sensor*. Scientific Reports, 2015. **5**(1): p. 1–7.
54. Yun, Y.J., et al., Highly conductive and environmentally stable gold/graphene yarns for flexible and wearable electronics. Nanoscale, 2017. **9**(32): p. 11439–11445.
55. Liu, Y., et al., *Superb electrically conductive graphene fibers via doping strategy*. Advanced Materials, 2016. **28**(36): p. 7941–7947.
56. Hyun, W.J., O.O. Park, and B.D. Chin, *Foldable graphene electronic circuits based on paper substrates*. Advanced Materials, 2013. **25**(34): p. 4729–4734.
57. Xu, Z., et al., *Highly electrically conductive Ag-doped graphene fibers as stretchable conductors*. Advanced Materials, 2013. **25**(23): p. 3249–3253.
58. Yoon, S.S., et al., *Highly conductive graphene/Ag hybrid fibers for flexible fiber-type transistors*. Scientific Reports, 2015. **5**(1): p. 1–12.
59. Li, Z., et al., *Multifunctional non-woven fabrics of interfused graphene fibres*. Nature Communications, 2016. **7**(1): p. 1–11.
60. Liu, Y., et al., *Environmentally stable macroscopic graphene films with specific electrical conductivity exceeding metals*. Carbon, 2020. **156** : p. 205–211.
61. Vertuccio, L., et al., *Effective de-icing skin using graphene-based flexible heater*. Composites Part B: Engineering, 2019. **162** : p. 600–610.
62. Scidà, A., et al., *Application of graphene-based flexible antennas in consumer electronic devices*. Materials Today, 2018. **21** (3): p. 223–230.
63. George, J.N. and M.G. Madhan, *Analysis of single band and dual band graphene based patch antenna for terahertz region*. Physica E: Low-dimensional Systems and Nanostructures, 2017. **94**: p. 126–131.
64. Sa'don, S.N.H., et al., *Analysis of graphene antenna properties for 5G applications*. Sensors, 2019. **19**(22): p. 4835.
65. Zhu, Y., et al., *A seamless three-dimensional carbon nanotube graphene hybrid material*. Nature Communications, 2012. **3**(1): p. 1–7.
66. Oh, J.Y., et al., *Enhanced electrical networks of stretchable conductors with small fraction of carbon nanotube/graphene hybrid fillers*. ACS Applied Materials & Interfaces, 2016. **8**(5): p. 3319–3325.
67. Foroughi, J., et al., *Highly conductive carbon nanotube-graphene hybrid yarn*. Advanced Functional Materials, 2014. **24**(37): p. 5859–5865.
68. Lepak-Kuc, S., et al., *Highly conductive doped hybrid carbon nanotube–graphene wires*. ACS Applied Materials & Interfaces, 2019. **11**(36): p. 33207–33220.
69. Zhong, X., et al., *Carbon nanotube and graphene multiple-thread yarns*. Nanoscale, 2013. **5**(3): p. 1183–1187.
70. Park, M., et al., *Performance enhancement of graphene assisted CNT/Cu composites for lightweight electrical cables*. Carbon, 2021. **179**: p. 53–59.

7 Carbon Nanotubes and Graphene in Photovoltaics

Pankaj Kumar

CONTENTS

7.1 INTRODUCTION

Owing to their excellent electronic and optoelectronic properties, along with several electronic applications like light-emitting diodes, transistors, integrated circuits, touch panels, the carbon nanotubes (CNTs) and graphene have attracted great interest in photovoltaic (PV) applications as well. PV devices convert the incident light directly into electricity and are also known as solar cells. PV devices incorporate semiconducting materials which absorb the incident light photons and convert their energy into generation of electrons and holes, the flow of which causes electric current. PV devices have enormous potential to solve the world energy crisis via harvesting the never-ending solar energy, however, the potential of PV devices is not being utilized to its maximum yet due to some limitations associated with them. PV devices have been known since 1883, when Charles Fritts build the first solid state PV device by depositing a thin layer of gold on selenium (Se) semiconductor; that cell had a power conversion efficiency of around 1% only [1]. After that, lots of efforts were made to improve the performance of these PV devices but nothing really appreciable came out until the Bell Laboratory, USA, prepared the first practical PV device in 1954 from silicon (Si) semiconductor that had power conversion efficiency (PCE) of around 6% [2]. This put lots of thrill in the scientific community and several international groups started developing Si-based PV devices and the performance of these devices has improved a lot with PCE over 26% [3]. Silicon is a very stable material and does not lose its properties in ambient atmosphere, therefore it provides quite high stability to PV devices. III-V compound semiconductor like GaAs-based single junction solar cells have shown PCE up to 27.8% whereas the multijunction solar cells with concentrator have shown PCE up to 47.1% [3]. Since the multijunction solar cells are very costly, they are only used for space applications and for terrestrial applications Si solar cells are dominating the PV market with over 90% shares. The other less than 10% of PV market shares are shared by thin film PV technologies like a-Si, CIGS, and CdTe solar cells. Though the cost of Si PV technology is reducing regularly, it is still beyond the reach of the majority of people in both developed and developing countries. One of the promising approaches to reduce the cost of Si solar cells is simpler fabrication. Efforts are being made

DOI: 10.1201/9781003231943-7

globally to reduce the cost of Si solar cells and also develop alternate PV technologies that are more cost effective with simpler fabrication. The quest of alternate PV technologies has given birth to new emerging PV technologies, like organic photovoltaics (OPV), dye-sensitized solar cells (DSSCs) and perovskite solar cells (PSCs). Initial performance of OPV, DSSCs and PSCs were not high enough to be sought for commercialization and efforts were made to improve their performance via incorporating advanced innovative nanomaterials like CNTs and graphene. In this chapter we discuss the different PV technologies and how incorporation of CNTs and graphene has led to improve their performance.

CNTs have shown excellent thermal, mechanical, and electrical properties, which make them suitable for numerous advanced technological applications. Depending on the diameter and chirality (the degree of their twist), the CNTs could behave as semiconducting or metallic [4, 5]. A typical PV device incorporates semiconducting materials to absorb the incident light and a built-in electric field causes separation of the photogenerated holes and electrons followed by their transport and collection at corresponding electrodes. The metallic CNTs do not absorb incident light, but due to high conductivity along the tube length they could be used as charge transport and electrode materials. Depending upon the applications, both the semiconducting and metallic CNTs make their use in PV devices. Thinking of this, Ji Ung Lee, prepared a diode using individual SWCNTs and studied the PV effect in that itself [6]. The diodes were prepared on a heavily doped Si wafer, with a 400 nm of SiO_2 grown via thermal oxidation on it. The SiO_2 layer was lithographically etched to formulate the split gate, source, and drain contacts. SWCNTs were grown using catalytic chemical vapor deposition technique between source and drain electrode such that one end of SWCNT touched the source and the other end touched the drain electrode. To form the ideal p-n junction along the SWCNT, the split gate was applied two biases of opposite polarities VG1 =—VG2 = +10V. The fit of the measured I-V characteristics of the diodes exhibited ideality factor $n = 1.0$. For PV response the diode was exposed to a 1.5 μm (0.8 eV) continuous wave laser diode with variable output power up to 10 mW. The diode showed PV response and the photogenerated current increased linearly with illumination intensity. This PV response was purely from SWCNT and the contribution from the underlying Si wafer was ruled out as the energy of the laser light was less than the band gap of Si. The p-type and n-type regions in SWCNT are formed due to split gate voltages, which also provide built-in electric field in SWCNT. In the middle of SWCNT the photogenerated electron-holes get separated due to built-in electric field. The photogenerated electron-holes transported opposite to the applied electric field. Though the measured PCE of the diode was only 0.2%, this too was significant as the diode absorbed only a small fraction of the incident light. Based on the actual power absorption an estimate suggested the PCE of these diodes to be over 5% [6]. As CNTs exhibit high charge carrier mobility and photo-absorption, they have been incorporated in various PV technologies to improve their performance. Let us discuss the efforts and progress made in different PV technologies via integration of CNTs and graphene.

7.2 APPLICATIONS OF CNTS AND GRAPHENE IN SI SOLAR CELLS

CNTs have shown many advantages and potential for simple, low-cost, high-efficiency solar cells fabrication with Si. CNTs behave like p-type material and form a p-n junction with n-type Si, which eliminates high temperature diffusion of dopants used in conventional Si solar cells. The CNT-Si junction provides sufficient potential for separation of photogenerated charge carriers. In addition to that, a thin CNT film (2D CNT network) is also transparent that allows the incident light to get absorbed in Si and collects charge carriers. Because of transparency and high conductivity, CNT network eliminates the deposition of top metal fingers, which are used in traditional Si solar cells and block some of the incident light. However other networks made of some other

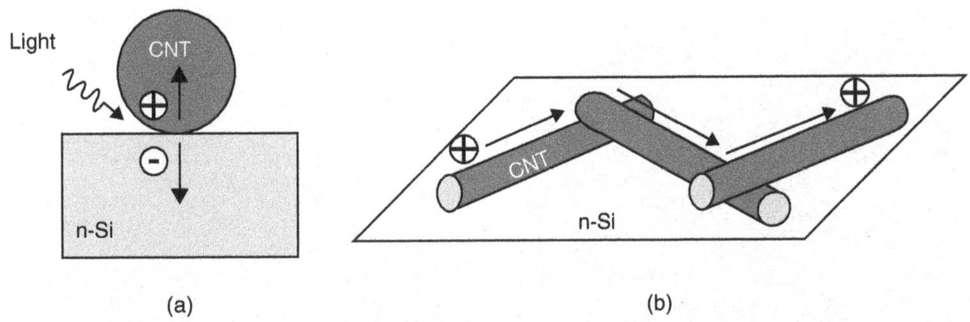

FIGURE 7.1 Schematic representation of (a) photo-generation of charge carriers at CNT-Si interface and (b) transport of holes through percolating pathways of interconnected CNTs.

semiconducting nanowires like InAs, the nanostructures are well separated, and they still require a common electrode on the top [7]. In addition to that, the efficiency of CNT-Si solar cells can be enhanced quite easily via tailoring the properties of CNTs and their functionalization [8]. CNT-Si solar cells make use of distinct properties of both the CNTs (high charge carrier mobility) and Si (diffusion length). Figure 7.1 shows schematically the absorption of light and separation of charge at the CNT-Si interface (p-n junction) and holes transportation in the percolating network of CNTs.

In 2007, Wei et al. reported the fabrication of CNT-Si solar cells via direct deposition of a semitransparent thin film of double-walled carbon nanotubes (DWCNTs) on n-type Si substrate [9]. The nanotubes created high-density p-n heterojunctions between DWCNTs and Si substrate that led to charge separation and extraction of electrons through Si substrate and holes through DWCNTs. Figure 7.2 shows the SEM and HRTEM images of the DWCNTs along with the cross-sectional SEM image of the DWCNTs on the Si substrate and optical transmission of DWCNTs for different thicknesses. The DWCNTs with 2–3 nm diameter has a band gap of ~ 0.5 eV [10], and the work function of about 4.8 eV with Fermi level in the middle of the band gap [11], which develops a built-in voltage of around 0.6 V with moderately doped n-type Si. DWCNTs in DWCNT Si solar cells serve as the sites for light absorption and holes transport. Wei et al. prepared the DWCNTs films of > 100 cm^2 by chemical vapor deposition (CVD) technique and treated them with H_2O_2 and HCl solutions followed by rinsing with distilled water [9]. The DWCNT films were then conformally transferred on n-type Si wafer and dried at room temperature. The adhesion of the DWCNT films on Si substrate was quite firm. Though the DWCNT films were transparent, for the contact purpose Ag fingers were painted on one side of the DWCNT film where it was separated from Si substrate by isinglass insulator. For the back contact a Ti/Pd/Ag layer was sputter coated on the back of Si substrate. The DWCNT films were quite porous and semitransparent but virtually free from catalyst and amorphous carbon. Semitransparency of DWCNT films allowed the incident light to get absorbed in both the DWCNT and Si substrate. The sheet resistance of these DWCNT films with thickness 35–200 nm varied from 0.5–5 Ω/sq, which was even less than that usually observed in ITO. Under illumination the CNT-Si solar cells having DWCNT films of 50–100 nm exhibited V_{oc} in the range 0.35–0.55 V and J_{sc} in the range 5–15 mA/cm^2. Though these solar cells exhibited PCE of about 1.38% but it was proposed that with careful optimization of thickness and density of DWCNT films, one could get higher efficiencies. Later they achieved 7.4% PCE with spider-web like DWCNT film deposited on an oxide free n-type Si wafer [12]. The native SiO_2 layer was etched out from the top of Si wafer to form a window of 7 mm × 7 mm. A 10 nm of Au/Ti film was deposited on the back of Si wafer and the electrical

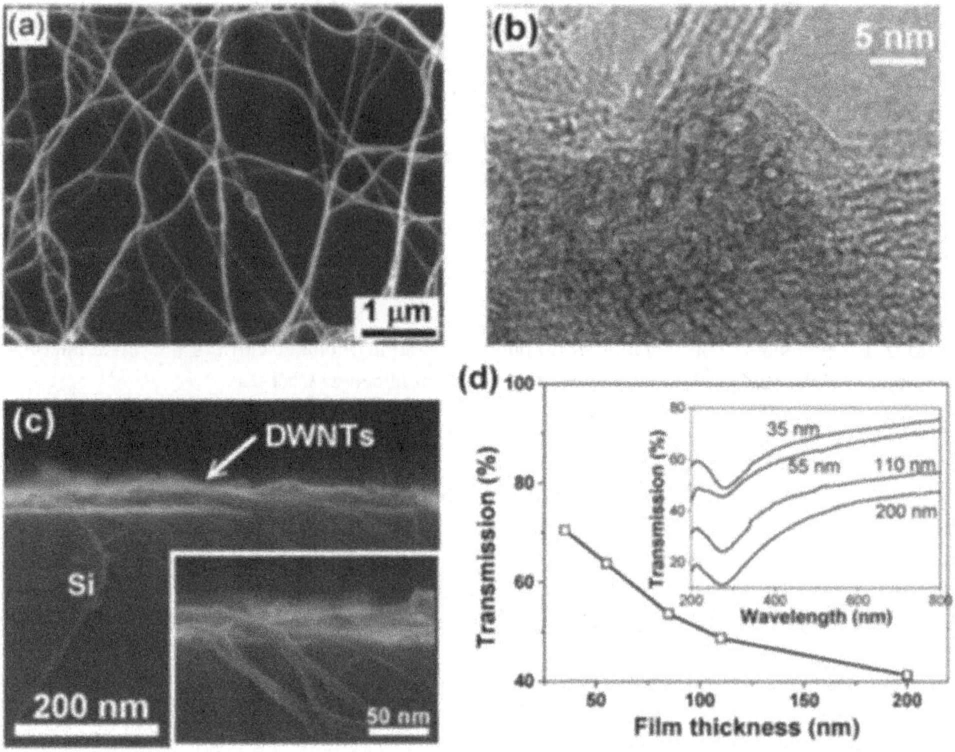

FIGURE 7.2 (a) SEM image of the DWCNTs film showing the porous network structure. (b) HRTEM image of the DWCNTs clearly showing the double walled structure. (c) Cross-sectional SEM image of the DWCNT film on the top of Si substrate. Inset shows the contact of the DWCNTs with Si surface. (d) Optical transmission (at 550 nm) of the DWCNTs films with different thicknesses. Inset shows the transmission of the DWCNT films in visible region. "Adapted with permission from [9]. Copyright (2007) American Chemical Society".

contacts were taken from front DWCNT and back Au/Ti films via Ag paste. In the spider-web-like structure the DWCNTs were highly interconnected and formed numerous percolating pathways on the Si surface. The DWCNT spider-webs were grown by CVD method and washed with H_2O_2 to remove amorphous carbon and then treated with hydrochloric acid to remove catalyst residue. A small piece of the bundle of DWCNTs was floated on distilled water and expanded into a thin uniform film which was directly transferred to the window of Si/SiO_2 wafer. The DWCNT film was quite uniform and transparent and allowed the incident light to reach underneath Si. Though DWCNTs film also absorbed some of the incident light but most of it was absorbed in the underlying Si substrate and the photogenerated charge carriers diffused to DWCNT-Si junction, which led to charge separation and their subsequent transportation and collection of electrons at bottom Au/Ti electrode and holes at DWCNTs. One of these devices exhibited V_{oc} of 0.54 V, J_{sc} of 26 mA/cm^2, FF of 53% and PCE of 7.4%. The authors also performed some studies with MWCNTs, which are metallic in nature, on Si substrates and found that with similar thickness of MWCNTs on Si exhibited comparable V_{oc} between 0.45–0.5 V but around 30 times lower J_{sc} of 0.8 mA/cm^2 and PCE of about 0.06%. It is worth mentioning here that the metallic MWCNTs form metal-semiconductor (M-S) Schottky contact instead of semiconductor-semiconductor (S-S) contact like in DWCNTs-Si. The MWCNTs were synthesized by vacuum filtration of a dispersed solution but the percolating network in the film was similar to that of DWCNT film. The MWCNTs formed a Schottky barrier with Si that resulted in reasonable V_{oc} but very poor charge extraction. To improve the efficiency of CNT-Si solar cells, researchers used SWCNTs instead of DWCNTs or MWCNTs.

Li et al. prepared SWCNT-Si solar cells via depositing horizontally aligned SWCNTs on n-type Si wafer following the superacid slide casting method and achieved very high FF of 73.8% and PCE of 11.5% [13]. Wang et al. prepared SWCNT-Si solar cells by depositing a high quality SWCNT network film grown by atmospheric pressure floating catalyst CVD (FC-CVD) technique on Si/SiO_2 substrate with a 1 mm window and 91% of transparency of SWCNT film and achieved PCE 12.5% without any post-production doping [14]. Au and In were respectively deposited on SWCNT and on the back side of Si substrate for anode and cathode contacts. The atmospheric pressure FC-CVD method enabled them to control the thickness of SWCNT and achieve different transparencies of SWCNT films varying from 70% to 97%. They also studied the effect of window size on the PCE of these solar cells and the effect of window size on the PCE for different transparencies of SWCNT films at 550 nm. Figure 7.3 shows schematically the structure of the SWCNT-Si solar cells and effect of window size on PCE for different transparencies of SWCNTs. The high PCE of these solar cells was attributed to low resistance and high transparency of SWCNT films. The post-production treatment of the optimized solar cells with HNO_3 resulted in doping of SWCNTs and a PCE of 14.5% [14]. Jia et al. also reported quite an enhancement in the PCE of SWCNT-Si solar cells by doping the SWCNT network by dilute HNO_3

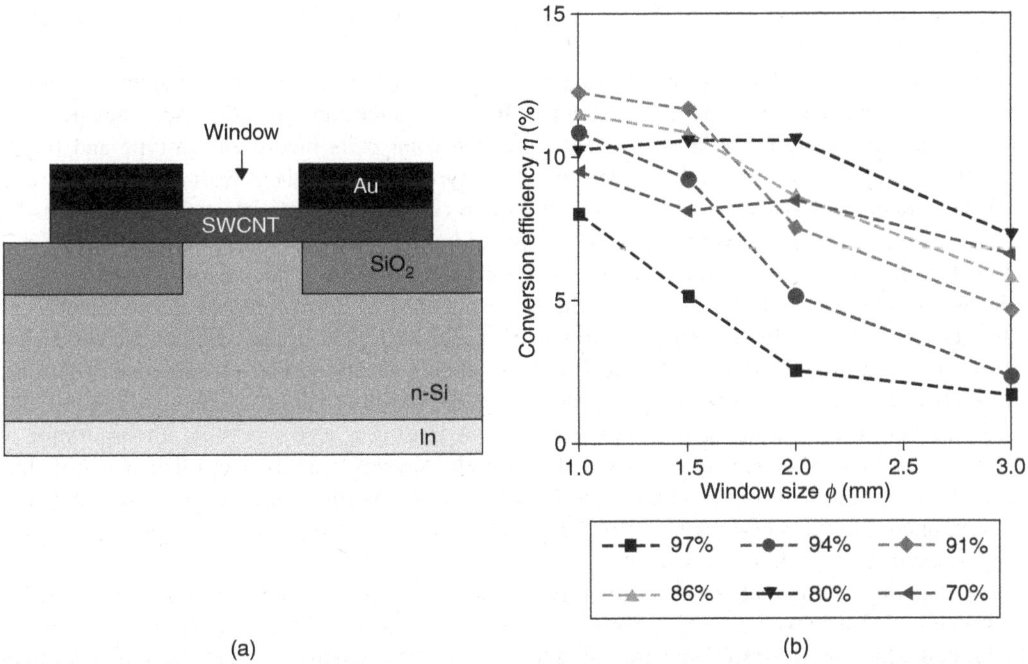

FIGURE 7.3 (a) Schematic representation of structure of SWCNT-Si solar cells. (b) Variation in PCE of SWCNT-Si solar cells as a function of window size in Si/SiO_2 substrate for different transparencies of the SWCNT films in the window. "Adapted with permission from [14]. Copyright (2014) American Chemical Society".

(0.5 M) [15]. The SWCNT films were grown on n-type Si wafer via CVD technique that exhibited a percolating network of semiconducting SWCNTs and formed p-n junction with n-type Si. The CNT network exhibited an optical transparency of over 85% and the sheet resistance less than 200 Ω/sq. The CNT network was quite porous with porosity of about 70% and formed numerous parallel junctions and photoelectrochemical units with Si wafer. On the average, where the initial efficiency of untreated SWCNT-Si solar cells was around 6–7%, the HNO_3 treatment enhanced their efficiencies to 11–13 %. One of the pristine SWCNT-Si solar cell, which exhibited a short circuit current density (J_{sc}) of 27.4 mA/cm^2, fill factor (*FF*) of 47% and PCE of 6.2%, the doping by infiltration of dilute HNO_3 in SWCNT film resulted in J_{sc} of 36.3 mA/cm^2 and *FF* of 72% that led to 13.8% PCE in that cell. Doping of SWCNT film with dilute HNO_3 reduced the internal series resistance significantly and enhanced charge carrier separation and transport. In addition to that HNO_3 infiltration formed Si-acid-CNT units in the areas not covered by CNTs, where Si and CNT worked as photo-electrodes and acid as electrolyte like in photoelectrochemical cells. The authors also tried infiltration of other liquids like NaCl solution as well but the effect was not similar to that with HNO_3.

Graphene, which is a single atomic 2D layer of carbon, where carbon atoms are connected with each other through covalent bonds in a honeycomb structure, is another important material used in solar cells. Graphene has been used in all types of solar cells to improve their performance. Applications of graphene could be similar to that of CNTs in solar cells, however the performance of graphene-based solar cells strongly depends upon the number of graphene layers and the doping of graphene-based materials. Graphene can be produced in single or multiple atomic layers via chemical vapor deposition (CVD) or Hummer's process [16]. Graphene has many advantages over CNTs, like high conductivity despite very low thickness, avoids interfacial resistance occurring between CNTs in the film, low porosity and provides extremely flat surface for device fabrication.

The sheet resistance and optical transparency of the graphene films depends upon the number of graphene layers in graphene films. Both the sheet resistance and optical transparency decrease as the number of graphene layers increases. As the solar cells incorporate n-type and p-type materials, graphene sheets can be made n-type or p-type by doping them with suitable materials. Since graphene is basically a sheet made of carbon (C) atoms and C is a tetravalent atom, for n-type doping it is doped with pentavalent atoms like phosphorous and for p-type doping it is doped with trivalent atoms like boron. The tetravalent atom takes an electron from graphene sheets and leaves a hole on the sheet whereas a pentavalent atom releases a free electron on the graphene sheet, which contributes to electrical conduction. Doping of graphene sheets has been observed to change their electrical, optical, chemical, and physical properties. There are a number of methods to dope the graphene sheets like plasma treatment, chemical doping, ball milling, thermal annealing, in-situ doping while growth of graphene sheets etc. Incorporation of graphene in Si solar cells is quite easy and it provides several benefits over other materials like complete substrate coverage and predetermined thickness. As the properties of graphene can be easily tuned, in graphene-Si solar cells they can be used to create p-type heterojunction, n-type heterojunction and Schottky contacts.

Li et al. deposited graphene sheets on n-type Si wafer to prepare graphene-Si Schottky junction solar cells and achieved around 1.5% efficiency at one sun illumination [17]. To fabricate the solar cells, they used Si/SiO_2 substrate with 0.1–0.5 cm^2 Si window and pre-deposited Au lines around the window on SiO_2 for contacts. Graphene sheets were prepared via CVD method on nickel foils. The freestanding films of graphene were achieved by detaching them from nickel in an acidic solution and then rinsing them with deionized water. The films were then dipped in a H_2O_2 solution for over 24 hrs to remove the amorphous carbon impurities and make them hydrophilic. The graphene films were then washed with deionized water and conformally transferred to the Si window on the Si/SiO_2 substrate. For the front contact Au was sputtered around the window whereas for the back contact Ti/Pd/Ag was sputtered on the back of n-Si substrate. In these solar cells graphene forms a Schottky junction with Si and serves as a top transparent electrode and antireflection layer. The light absorption takes place in n-Si and the photogenerated electrons in the vicinity of graphene-Si junction are transferred to graphene and holes diffuse to Ti/Pd/Ag electrode. The difference in the work functions of the graphene sheet and n-Si develops a built-in electric field in the junction. The solar cells with 0.1 cm^2 window area exhibited a PCE of 1.65% whereas the solar cell with 0.5 cm^2 window area showed a PCE of 1.34%. The average efficiency of unoptimized cells was around 1.5%. Though the efforts are being made to achieve high efficiency graphene-Si solar cells but unfortunately the efficiency of graphene-Si solar cells still remains lower than pure Si solar cells. Despite lower efficiencies the graphene-Si solar cells are expected to be more cost effective than pure Si solar cells.

Though over 90% of current PV market is dominated by Si solar cells, they are still out of reach of most people specially the poor. Si solar cells have some inherent problems like high cost, difficult to make in large area, bulky and fragile in nature. To make the most of PV technologies efforts are being made to make Si solar cells more cost effective and develop emerging PV technologies that include organic solar cells, perovskite solar cells, dye-sensitized solar cells, hybrid solar cells and quantum dot solar cells. To seek commercialization the emerging PV technologies still need to achieve the performance high enough for commercialization. Emerging PV technologies have shown several advantages over inorganic solar cells like highly cost effective, light weight, ease in production and can be produced roll to roll via conventional printing techniques. The power conversion efficiency of organic and dye-sensitized solar cells had been very low compared to Si solar cell because of low charge carrier mobilities in organic semiconductors. As the charge carrier mobility in organic semiconductors is relatively smaller than that in inorganic semiconductors, incorporation of CNTs and graphene in organic, dye sensitized and perovskite solar cells is expected to improve their performance manyfold, therefore incorporation of CNTs and graphene in such solar cells has been a highly researched area. We discuss here some of the applications of CNTs and graphene in emerging PV technologies and their latest developments.

7.3 APPLICATIONS OF CNTS AND GRAPHENE IN ORGANIC SOLAR CELLS

Organic solar cells (OSCs) are two electrode devices, where light absorbing materials are sandwiched between them. At least one of the electrodes in OSCs is optically transparent that allows the incident light to enter in the cell. OSCs incorporate organic semiconductors as the light absorbing medium. Organic semiconductors are the hydrocarbons that possess alternate single and double bonds between carbon atoms and the alternation of single and double bonds is knows conjugation. The conjugation is responsible for charge conduction in organic semiconductors. The first report on the organic solar cells was published in 1973 where organic semiconductor tetracene was sandwiched between Al and Au electrodes, and it exhibited power conversion efficiency of ~0.5% [18]. Since the electron and hole mobilities in organic semiconductors are very low compared to inorganic counterparts and the light absorption in them creates bound electron-hole pairs, called excitons, which have binding energy more than the thermal energy at the room temperature and they have the tendency to decay via electron-hole recombination. The excitons could be dissociated into free electron and hole to some extent at the electrodes but due to low diffusion lengths of excitons they decay before they dissociate, and the power conversion efficiency was very low. In 1986, C. W. Tang introduced the donor-acceptor concept, where excitons could easily dissociate into electrons and holes on the order of femto seconds [19]. He used perylene derivative 3,4,9,10-perylene tetracarboxylic bis-benzimidazole (PTCBI or Im-PTC) as acceptor and copper phthalocyanine (CuPc) as donor materials in bilayer configuration and the solar cell exhibited around 1% PCE. The energy difference between lowest unoccupied molecular orbitals (LUMOs) of the donor and acceptor materials imparts the required energy to break the excitons into free electrons and holes. The exciton dissociation takes place at donor-acceptor interface and after exciton dissociation the electrons transfer to LUMO of acceptor and holes remain in the highest occupied molecular orbital (HOMO) of the donor. For efficient exciton dissociation, the energy offset between LUMOs of donor and acceptor materials should be more than the binding energy of excitons. Since the exciton diffusion length in fullerene (C_{60}) is more than that in perylene derivatives, replacement of PTCDI by C_{60} with bathocuproine (BCP) as electron blocking layer resulted in a power conversion efficiency of 3.6% [20]. The concept of donor-acceptor combination in active layer proved to be a key finding for efficient OSCs.

As in bilayer structure the donor-acceptor interfacial area was just the area of the cell, Heeger et al. mixed the donor-acceptor materials together to form a single layer and this intermixing increased the donor-acceptor interfacial area to manyfold. The intermixing of donor-acceptor materials formed numerous donor-acceptor junctions in the bulk of the film therefore this concept was called bulk heterojunction (BHJ). In BHJ structures the donors and acceptors form different percolating pathways throughout the bulk film and electrons transport through acceptor channel whereas holes transport through donor channel to collect at the corresponding electrodes. Electrons collect at cathode whereas holes collect at anode. Heeger et al. used poly(2-methoxy-(5-ethylhexyloxy)-1,4-phenylene-vinylene) (MEH-PPV) as donor and phenyl [6,6′] C61 butyric acid methyl ester (PCBM) as acceptor to form the BHJ solar cells and these solar cells exhibited ~1.5% efficiency [21]. Later on, Shaheen et al. recorded a jump in efficiency to 2.5% when they used poly(2-methoxy-5-(3,7, dimethyloctyloxy)-1, 4-phenylenevinylene) (MDMO-PPV) as the donor along with PCBM as the acceptor [22]. BHJ structure also reduced the distance to be traveled by the excitons to dissociate into free holes and electrons and this helped a lot in improving the efficiency of OSCs. Padinger et al. fabricated OSCs using P3HT donor and PCBM acceptor and subjected them to thermal treatment under an externally applied voltage [23]. The devices were applied 2.7 V and then treated thermally at 75°C, and this treatment resulted in a PCE of 3.5% [23]. Since then, there have been lots of efforts made to improve the performance of OSCs, and the state-of-the-art certified efficiency in OSCs is over 18% [24]. More information about OSCs can be found in ref. [25]. Figure 7.4 shows schematic structure of a typical BHJ OSC and the molecular structures of P3HT donor and PCBM acceptor. Indium tin oxide (ITO) is an important transparent conducting oxide (TCO) and frequently used as optically transparent electrode in organic electronic devices like light emitting diodes and solar cells. Depending upon the structure of the OSC and the counter electrode, ITO could work as anode or cathode. Light enters in OSCs through ITO and get absorbed in active layer and generates excitons. The excitons diffuse to the donor-acceptor interface and dissociate into free electrons and holes. Electrons transport through acceptor network whereas holes transport through donor network. The electrons transport through electron transport layer (ETL) and collect at the cathode, whereas hole transport through hole transport layer (HTL) and collect at the anode. Under illumination of the solar cell, anode possesses positive charge and cathode possesses negative charge.

Though OSCs are supposed to be very cost effective, still some materials used in OSCs, like ITO electrode and barrier film, are very expensive and hold the control on the overall cost of these devices. Replacement of such materials with some other compatible and cheaper materials will make these devices more cost effective. Both CNTs and graphene have been used as alternate materials to ITO in organic solar cells [26, 27], however the trade-off among the transparency,

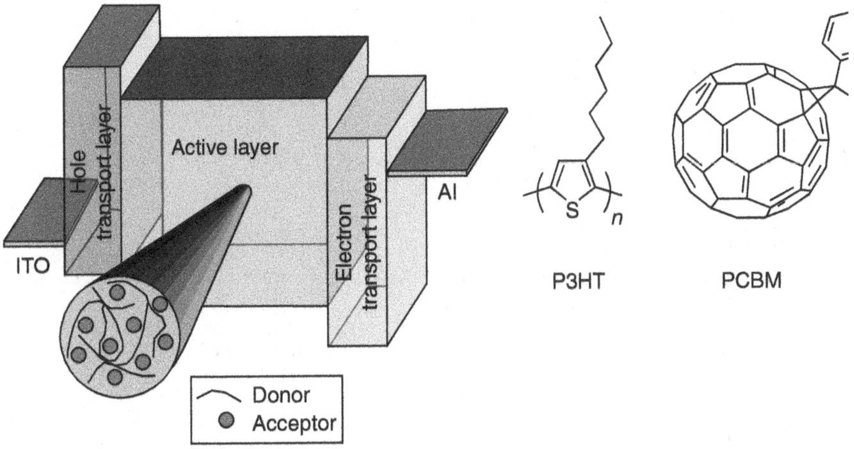

FIGURE 7.4 Schematic structure of a typical OSC. Here ITO works as anode and Al works as cathode. The molecular structures of P3HT (donor) and PCBM (acceptor) are also shown.

electrical conductivity and roughness with increased thickness limit their use in OSCs toward enhancing in their efficiency. CNT thin films have shown to have high electrical conductivity with optical transparency. As the prepared CNT films are quite rough, their roughness can be reduced by coating a layer of poly(ethylene-dioxythiophene):poly styrene sulfonate (PEDOT:PSS) on them. The conductivity of CNT films can be increased by chemical doping or by acid doping. Rowell et al. used SWCNT network to fabricate flexible transparent conducting electrode on polyethylene terephthalate (PET) sheet to replaced ITO and prepared organic solar cells on it that exhibited PCE similar to those prepared on ITO coated glass substrates [28]. They used a transfer printing method to print a smooth and uniform SWCNT network on the flexible plastic substrate. The SWCNT film exhibited transparency of over 85% in the visible range and sheet resistance of 200 Ω/sq. They used commercially available SWCNTs produced via arc discharge method and dissolved them in a solution with surfactants and sonicated. The well dispersed solution was then vacuum filtered over a porous alumina membrane and dried. The SWCNT film was then lifted off with a poly(dimethylsiloxane) (PDMS) stamp and transferred to the plastic substrate like stamp printing that resulted in a thickness of SWCNT network of around 30 nm. The organic solar cells were prepared in PET/SWCNT/PEDOT:PSS/P3HT:PCBM/Al structure and to compare the performance of SWCNT electrode with ITO, solar cells were also prepared in Glass/ITO/PEDOT:PSS/P3HT:PCBM/Al structure. The ITO coated glass had sheet resistance of 15 Ω/sq. The solar cells prepared on PET/SWCNT substrate exhibited PCE of 2.5% whereas those prepared on Glass/ITO substrate exhibited PCE of 3%. Figure 7.5(a) shows the *J-V* characteristics of the solar cells prepared on PET/SWCNT and Glass/ITO substrates. Inset shows the schematic of the device structure and photograph of the flexible solar cells on PET/SWCNT substrate. The photovoltaic parameters of both types of devices were similar except the *FF*, which was significantly lower in flexible solar cells based on SWCNT network and the reason behind was higher sheet resistance of SWCNT network compared to ITO. Zhang et al. used commercial CNTs prepared via HiPCO and arc discharge method and prepared their films and used them as transparent electrode in place of ITO [29]. The arc discharged-based CNTs showed better performance compared to those based on HiPCO CNTs in terms of conductivity, transparency, and surface roughness. The CNT films were doped with $SOCl_2$ to improve their conductivity and to smooth their roughness they were passivated with PEDOT. The sheet resistance and optical

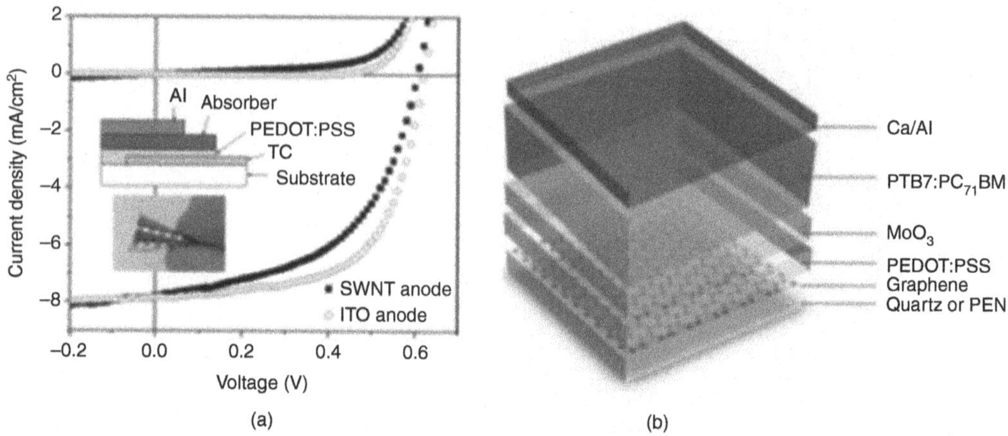

FIGURE 7.5 (a) Dark and illuminated *J-V* characteristics of the OSCs prepared on PET/SWCNT and Glass/ITO substrates by Rowell et al. Inset shows the device structure and photograph of the flexible OSCs on PET/SWCNT substrate. "Reproduced with permission from [28]. Copyright (2006) AIP Publishing". (b) Schematic structure of the OSCs prepared by Park et al. using graphene as the transparent conducting electrode. "Adapted with permission from [36]. Copyright (2014) American Chemical Society".

transparency of the optimized films were found to be ~160 Ω/sq and ~ 87% respectively. The conductivity of CNT films could be improved at a little cost of their transparency [30].

Similar to CNTs, graphene films have also shown excellent electrical, optical, mechanical and thermal properties, which make them suitable for transparent conducting electrode applications in OSCs [27, 31–33]. Graphene exhibits sheet resistance on the order of hundreds of ohms per square, high work function, optical transparency over 80% and mechanical flexibility, which make it a suitable replacement of ITO. Reina et al. prepared single and a few layer graphene sheets and studied their electrical and optical properties [34]. A 3 nm film of graphene having a few graphene layers showed optical transparency over 90% with sheet resistance of ~700 Ω/sq [34]. The doping of graphene sheets by acids can increase their conductivity many folds. Bae et al. reported a method for roll-to-roll fabrication and chemical doping of predominately monolayer of graphene films of 30-inch grown on flexible copper substrate [35]. The graphene films were treated with nitric acid, and they exhibited the sheet resistance as low as 125 Ω/sq and optical transparency of 97.4%. They further stacked four layers of doped graphene that led to get sheet resistance of 30 Ω/sq and transparency of ~90%, which were superior to the commercially available ITO.

Park et al. replaced ITO with graphene and prepared high efficiency flexible OSCs [36]. For flexible OSCs, they deposited graphene films on polyethylene naphthalate (PEN) substrates and used them as anode and cathode and fabricated both the normal and inverted OSCs. For comparison purpose they also deposited graphene on quartz substrates and prepared solar cells on them and compared the performance with those prepared on ITO coated glass substrates. As the active layer they used a blend of thieno[3,4-b]thiophene/benzodithiophene (PTB7) as donor and [6,6]-phenyl C71 butyric acid methyl ester ($PC_{71}BM$) as acceptor. Graphene sheets were prepared by low pressure CVD method on Cu foils and transferred to quartz or PEN sheets using poly methyl methacrylate (PMMA), which was then removed by acetone. For multiple layers of graphene, the same process was repeated on the substrate. The graphene electrode possessed three layers of single layer graphene and it exhibited a sheet resistance of ~300 Ω/sq and transparency of ~92% at 550 nm. The normal solar cells on the quartz substrate prepared in graphene/

PEDOT:PSS/MoO_3/PTB7:PC_{71}BM/Ca/Al and ITO/PEDOT:PSS/MoO_3/PTB7:PC_{71}BM/Ca/Al structures exhibited PCE of 6.1% and 6.7% respectively. Figure 7.5(b) shows schematically the structure of the solar cells prepared by Park et al. In case of inverted solar cells on quartz substrates the solar cells were prepared in the structures graphene/ZnO/PTB7:PC_{71}BM/MoO_3/Ag and ITO/ZnO/PTB7:PC_{71}BM/MoO_3/Ag and they exhibited PCE of 6.9% and 7.6%, respectively. In both the normal and inverted structures, the PCE of solar cells with graphene electrode was quite near to those based on ITO electrode. In normal design the graphene electrode works as anode whereas in inverted design graphene electrode works as cathode. The choice of an electrode to be used as anode or cathode depends upon the charge collecting layers used with it. If graphene is used in intimate contact with electron collecting layer it can work as cathode and if it is used in intimate contact with hole collecting layer it can work as anode. In case of flexible OSCs on PEN substrates the structures of normal and inverted solar cells were similar to those on the quartz substrates viz. PEN/graphene/PEDOT:PSS/MoO_3/PTB7:PC_{71}BM/Ca/Al and PEN/ graphene/ZnO/PTB7:PC_{71}BM/MoO_3/Ag. Both the normal and inverted flexible OSCs with graphene electrode exhibited excellent photovoltaic performance. The normal flexible solar cell exhibited PCE of 6.1% whereas inverted flexible solar cell exhibited PCE of 7.1%. Figure 7.6 shows the performance of flexible OSCs based on graphene electrode. These graphene electrodes were quite robust under mechanical bending and the solar cells did not show any significant change in their performance even after 100 bending cycles. This work by Park et al. was really appreciable as it successfully demonstrated high efficiency flexible OSCs with graphene electrodes used as both anode and cathode [36].

Jung et al. used graphene electrode as the replacement of ITO for OSCs and prepared inverted OSCs both on glass and plastic substrates and achieved high PCE [37]. The emphasis of their work was to fabricate low temperature processable flexible OSCs using graphene electrodes. They used ZnO-nanoparticles (ZnO-NP) as ETL that could be processed without annealing unlike sol-gel method and the solar cells were prepared in graphene/ZnO-NP/PTB7:PC_{71}BM/MoO_3/Ag structure. Graphene sheets were grown by CVD method and the electrode possessed three layers of graphene, which were transferred on the substrates using wet transfer method. The graphene electrode exhibited average sheet resistance of 305±17 Ω/sq and optical transparency of 92.9% at 550 nm. The PTB7:PC_{71}BM solar cells with ITO and graphene cathode showed PCE of 8.21% and 7.37% respectively. Along with PTB7:PC71BM active layer, they also used a blend of poly[4,8-bis(5-(2-ethylhexyl)thiophen-2-yl)benzo[1,2-*b*;4,5-*b'*]dithiophene-2,6-diyl-*alt*-(4-(2-ethylhexyl)-(3-fluorothieno[3,4-*b*]thiophene-)-2-carboxylate-2,6-diyl)] (PTB7-Th) donor and PC_{71}BM acceptor to prepare inverted OSCs on ITO and graphene electrodes. The solar cells prepared in ITO/ZnO-NP/PTB7-Th:PC_{71}BM/MoO_3/Ag structure on glass substrates exhibited PCE of 9.13% with J_{sc} of 16.96 mA/cm^2, FF of 68.5% and open circuit voltage (V_{oc}) of 0.79 V, whereas the solar cells prepared in graphene/ZnO-NP/PTB7-Th:PC_{71}BM/MoO_3/Ag structure on glass substrate exhibited PCE of 8.16% with J_{sc} of 16.26 mA/cm^2, FF of 66.4% and V_{oc} of 0.76 V. The flexible solar cells on PET substrate prepared in graphene/ZnO-NP/PTB7-Th:PC_{71}BM/MoO_3/Ag structure exhibited PCE of 7.41% with J_{sc} of 15.63 mA/cm^2, FF of 63% and V_{oc} of 0.76 V. The performance of graphene electrode-based solar cells was near to that of those based on ITO. Jung et al. also checked the robustness of flexible OSCs against the mechanical bending. Where the graphene on PET-based solar cells retained over 80% of their initial efficiency after 100 bending cycles, the ITO on PET-based solar cells degraded to 30% of their initial efficiency just after 20 bending cycles [37].

Du et al. reported 6.85% PCE in OSCs using graphene as the transparent conducting anode [38]. They used a blend of poly[(2,5-bis(2-hexyldecyloxy)phenylene)-alt-(5,6-difluoro-4,7-di(thiophen-2-yl)benzo[c]-[1,2,5]thiadiazole)] (PPDT2FBT): PC_{71}BM as the photoactive layer

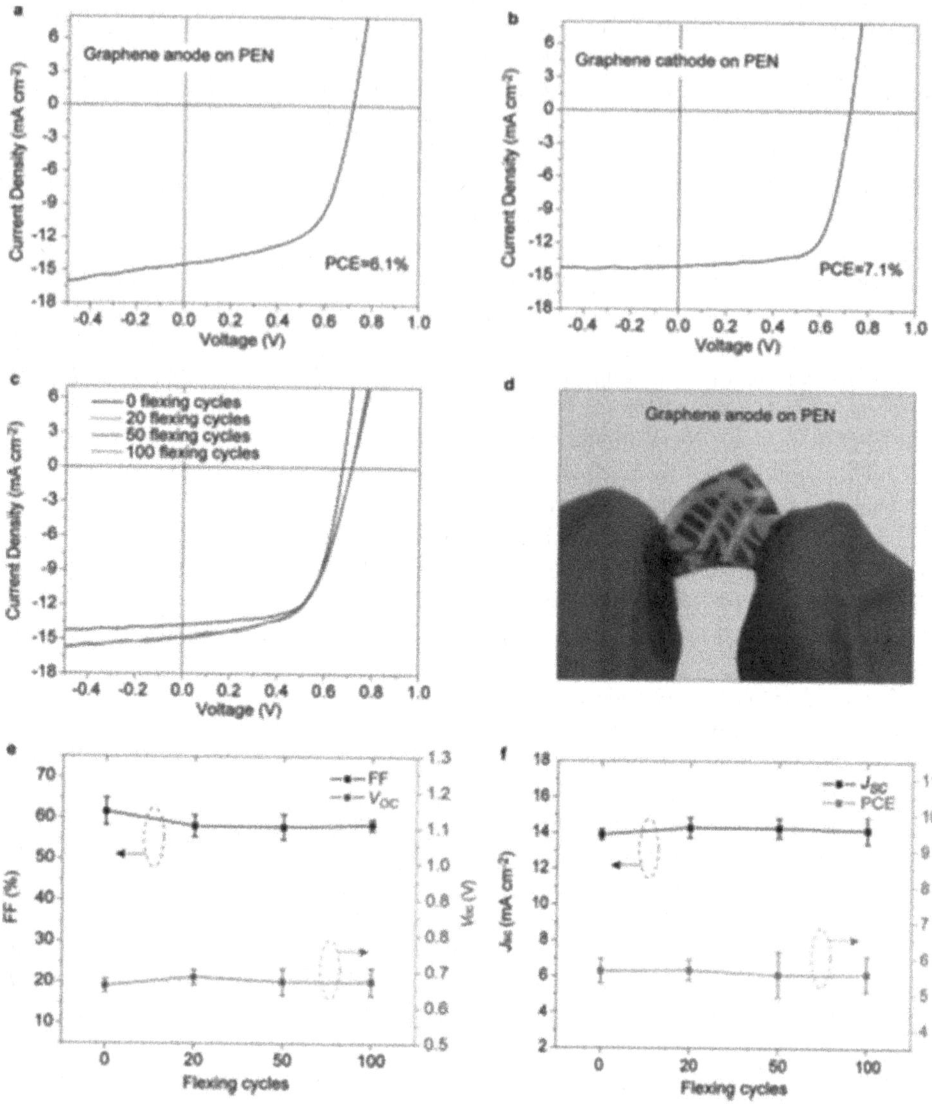

FIGURE 7.6 (a) *J-V* characteriatics of flexible OSC on PEN substrate with graphene as anode (normal design). (b) *J-V* characteristics of flexible OSC on PEN substrate with graphene as cathode (inverted design). (c) *J-V* characteristics of the graphene electrode-based champion solar cell after multiple bending cycles. (d) Photograph of the flexible OSC having graphene electrode as anode on PEN. (e) Variation in the *FF* and V_{oc} and (f) variation in the J_{sc} and PCE of the flexible solar cells after multiple bending cycles. "Adapted with permission from [36]. Copyright (2014) American Chemical Society".

and a benzimidazole (BI)-doped graphene as the transparent anode. The PET substrates with three layers of graphene were coated with Ag grids via thermal evaporation to reduce contact resistance and then 30 nm of MoO_3 was deposited on the graphene film via thermal evaporation. A 300 nm of PPDT2FBT:$PC_{71}BM$ film was spin coated onto MoO_3 and the 20 nm of Ca and 100 nm of Al were thermally deposited to complete the solar cells. These solar cells were mechanically robust as they exhibited only a little degradation on performance after 250 bending cycles. Interestingly the properties of graphene can be tuned in a number of ways to increase its acceptance in various electronic devices. For example, the conductivity of graphene can be increased by ozone treatment for a little while. The ozone treatment initially decreased the electrical resistance of graphene films by doping them p-type but after that the ozone exposure led to increment in the resistance and optical transparency due to surface oxidation and conversion into graphene oxide [39]. Ozone treatment also develops hydroxyl and carbonyl functional groups on graphene that helps in enhancement of V_{oc} in OSCs [40, 41]. The conductivity of ozone-treated graphene is also reduced because its sp^2 hybridized carbon-carbon network is disrupted into sp^3 hybridization at the sites of functionalization. For better suitability of graphene to work as anode or cathode in OSCs or in other optoelectronic devices, its work function can be tuned. Higher work function would make it more suitable for anode applications whereas lower work function would make it more suitable for cathode applications. Some of the important parameters and treatments that affect the work function of graphene are graphene film thickness, contact with metal or insulator, plasma treatment, UV treatment, and chemical doping. Naghdi et al. have discussed in detail several factors that can affect the work function of graphene [42].

Reduced graphene oxide (r-GO) is another important graphene derivative that has also been used as transparent conducting electrode in OSCs and other electronic devices [43]. Reduced graphene oxide can be derived from chemical or thermal reduction of graphene oxide (GO). Though graphene oxide is insulating in nature, reduced graphene oxide exhibits high conductivity and optically transparency. For thermal reduction of graphene oxide, the insulating graphene oxide film is deposited on the substrate by suitable means and then the film is annealed at high temperature in the presence of hydrazine vapor and graphene oxide converts into conducting reduced graphene oxide. The transparency and conductivity of reduced graphene oxide depends on its thickness. Yin et al. used reduced graphene oxide as the transparent conducting electrode to prepare flexible OSCs [43]. First of all, they prepared thin films of graphene oxide on SiO_2/Si substrate by spin coating method and then annealed those samples in Ar/H_2 environment at 1000°C to reduce graphene oxide. The reduced graphene oxide films were then transferred to PET substrates with the help of PMMA film. The reduced oxide films of PET substrates were coated with PEDOT:PSS HTL, followed by P3HT:PCBM active layer and TiO_2 nanoparticle ETL, and to complete the devices Al was deposited by thermal evaporation on the top. The optimized thickness of reduced graphene oxide electrode of 16 nm resulted in PCE of around 0.78%. The 16 nm of graphene oxide film exhibited sheet resistance of 3.2 kΩ/sq and transparency of 65%. The robustness of devices was also checked against the mechanical bending and the devices sustained over thousands of bending cycles. Zhao et al. prepared highly stable and transparent conductive films on PET substrates by spray coating of both SWCNTs and graphene oxide [44]. A mass ratio of GO to SWCNT of 0.2 in the SWCNT-GO transparent conducting films showed sheet resistance of 90.4 Ω/sq with an optical transmission of 87% at 550 nm. When a thin film of PEDOT:PSS was coated on SWCNT-GO films, it reduced the sheet resistance to 46.3 Ω/sq with optical transparency of 87.5% at 550 nm. In addition to increment in the conductivity of SWCNT-GO films, PEDOT:PSS increased the wettability and reduced roughness of the SWCNT-GO conducting film.

CNTs, graphene and their derivatives can also be used as interfacial layers between active layer and electrodes to improve charge collection. They can also be incorporated within the active layer and charge transport layers to improve photon harvesting, charge transport and collection [41, 45–49, 50]. Kim et al. used p-type SWCNTs as HTL between ITO and the active layer in OSCs. The active layer was made of the blend of poly(3-hexylthiophene) (P3HT) and PCBM and the solar cells were prepared in ITO/SWCNT/P3HT:PCBM/BCP/LiF/Al structure on the glass substrates [45]. Here BCP works as ETL and LiF works as electron injection layer (EIL).

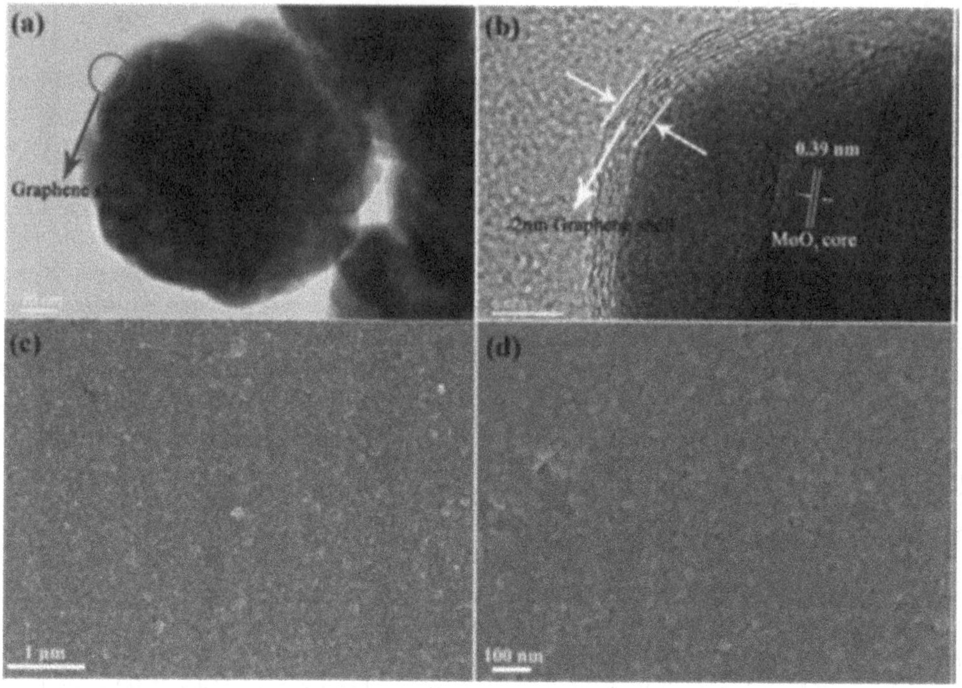

FIGURE 7.7 (a) TEM image of the graphene-MoO_3 nanoparticles. (b) HRTEM image of the MoO_3 particles covered by graphene. (c) Low magnification SEM image of the graphene-MoO_3 film on the glass substrate. (d) High magnification SEM image of the graphene-MoO_3 film on the glass substrate. Adapted with permission from [48].

The SWCNTs were dispersed in iso-propanol and deposited on the ITO coated glass substrate by spray process. The amount of SWCNTs on the substrate to manage the transparency and conductivity of SWCNT film was controlled by the frequency of spray coating. Initially the efficiency of OSCs increased and then decreased with increment in the substrate coverage or the frequency of spray coating. The optimized SWCNT coverage resulted in the peak PCE of 5.37%. Compared to the solar cells without SWCNTs, the SWCNTs provided better transportation of holes toward ITO via reducing the contact resistance and led to enhanced J_{sc} and FF. The reduction in the PCE for higher substrate coverage was attributed to reduction in the shunt resistance and increment in leakage current. Dang et al. prepared the hybrid graphene-MoO_3 particles and used them as HTL in OSCs [48]. The hybrid graphene-MoO_3 particles were prepared by chemical reaction of MoO_3 with graphene oxide. Figure 7.7 shows the SEM and TEM images of the graphene-MoO_3 particles. The solar cells were prepared on the ITO coated glass substrates, on which the aqueous hybrid solution of graphene-MoO_3 particles was spin coated and dried at 100°C for 10 min. Subsequently the active layer of poly[N-9″-hepta-decanyl-2,7-carbazolealt-5,5-(4′,7′-di-2-thienyl-2′,1′,3′-benzothiadiazole)] (PCDTBT) blended with $PC_{71}BM$ was spin coated and annealed at 70°C for 30 min. Finally, 0.8 nm of LiF and 100 nm of Al were deposited by thermal evaporation to complete the solar cells. Here graphene-MoO_3 layer served as HTL. Where the devices with MoO_3 HTL exhibited PCE of 5.71%, the solar cells with graphene-MoO_3 HTL showed the PCE of 7.07%. The graphene-MoO_3 HTL resulted in the enhancement in all the photovoltaic parameters like J_{sc}, V_{oc} and FF that all together resulted in enhancement in PCE. This enhancement in the performance has been attributed to the better hole transport properties of graphene-MoO_3 HTL. Liu et al. used sulfonated graphene oxide as HTL in OSCs [50]. For sulfonated graphene oxide, they introduced acidic moiety SO_3H into the graphene oxide and it exhibited high conductivity and led to interfacial doping of the donor. The OSCs based on sulfonated graphene oxide HTL exhibited PCE of 4.37% [50].

As graphene has high electron affinity and high charge carrier mobility that helps in dissociation excitons and extraction of charge carriers, it is also used as electron acceptor in OSCs [51–53]. The beauty of graphene is that it provides a large surface area to excitons to dissociate at the donor-graphene interface and provides continuous pathways for electron extraction. Liu et al. prepared a blend of poly(3-octylthiophene) (P3OT) as donor and organic solvent processable functionalized graphene as acceptor and prepared BHJ OSCs on the ITO coated glass substrates [51]. Graphene was functionalized to make it soluble in the common organic solvents and tune its highest occupied molecular orbital (HOMO) and lowest unoccupied molecular orbital (LUMO) levels to match with the HOMO and LUMO of donor and acceptor materials. For functionalized graphene, first of all the water soluble heavily oxidized graphene oxide was synthesized and then it was reacted with phenyl isocyanate to obtain functionalized graphene. The thin films of P3OT:graphene were prepared by spin coating of their solution in dichlorobenzene. For the optimum content of graphene for maximum efficiency the concentration of functionalized graphene was varied from 0 to 15 wt%. The solar cells were prepared in ITO/PEDOT:PSS/P3OT:graphene/LiF/Al structure and the optimized content on functionalized graphene of 5% in the active layer resulted in the maximum PCE of 1.4%. Though the BHJ devices having similar structure and same graphene content without thermal annealing exhibited very poor PCE of 0.32%, but the thermal annealing at 160°C for 20 min resulted in the devices having 1.4% PCE. Thermal annealing reorganized the nanoscale morphology of the P3OT:graphene active layer that enhanced J_{sc} from 2.5 mA/cm^2 to 4.2 mA/cm^2, V_{oc} from 0.56 V to 0.92 V and FF from 0.23 to 0.37. Bonaccorso et al. used functionalized graphene as the electron cascade acceptor in ternary OSCs [52]. They used poly[*N*-9′-heptadecanyl-2,7-carbazole-alt-5,5-(4′,7′-di-2-thienyl-2′,1′,3′-benzothiadiazole)] (PCDTBT) as the donor and $PC_{71}BM$ as the acceptor along with the functionalized graphene in the active layer. To prepare the blend solution of PCDTBT, $PC_{71}BM$ and functionalized graphene, PCDTBT and $PC_{71}BM$ were separately dissolved in 1:4 wt% ratio in a 1:3 mixture of dichlorobenzene and o-dichlorobenzene and mixed with graphene ink to prepare the ternary blend solutions with different graphene contents. The functionalized graphene ink was prepared separately by functionalizing the graphene nanoflakes by attaching the ethylenedinitrobenzoyl (EDNB) molecule to graphene nanoflakes. The functionalized graphene exhibited HOMO and LUMO levels favorable to HOMO and LUMO levels of PCDTBT and $PC_{71}BM$ and it worked as a bridge structure between PCDTBT and $PC_{71}BM$. Introduction of functionalized graphene in the PCDTBT:$PC_{71}BM$ active layer led to enhanced exciton dissociation and charge transport that resulted in enhanced PCE in the ternary blend. Functionalized graphene also provided an additional channel for transportation of electrons in active layer. BHJ solar cells were prepared in the structure ITO/PEDOT:PSS/PCDTBT:$PC_{71}BM$:functionalized graphene/TiO_2/Al structure on the glass substrates. Where the champion solar cells without graphene in the active layer exhibited the PCE of 5.59%, the champion solar cells with graphene in the active layer exhibited the PCE of 6.59% for 0.05% functionalized graphene in the active layer. Higher concentrations of graphene in the active layer led to reduction in the PCE mainly because of reduction in J_{sc} and FF. The reduction in J_{sc} and FF for higher concentration of graphene had been attributed to imbalance in the electron and hole collection because of high electron mobility in graphene. Yu et al. used a hybrid of graphene and fullerene (C_{60}) as the acceptor along with P3HT donor to significantly improve the PCE of BHJ OSCs [53]. C_{60} was covalently bonded to graphene sheets through lithiation reaction. Graphene sheets were prepared by chemical reduction of graphene oxide with hydrazine. For grafting of C_{60} onto graphene sheets, the graphene sheets were dispersed in toluene and then n-butyllithium in hexane was added dropwise under sonication and then C_{60} in toluene was added to the solution under sonication and to terminate the reaction methanol was added dropwise. The C_{60}-grafter graphene was washed with toluene and centrifuge to remove any residue of C_{60}. The final black powder was washed with methanol and dried in vacuum oven. For the active layer of BHJ solar cells, the C_{60}-grafter graphene (C_{60}-G) was blended with P3HT in 1:1 wt% ratio in chlorobenzene. The solar cells were prepared in ITO/PEDOT:PSS/P3HT:C_{60}-G/Al structure on glass substrate. Where the P3HT:C_{60} solar cells exhibited PCE of 0.47%, the P3HT:C_{60}-G solar cells resulted in the PCE of 1.22%. The enhanced PCE in P3HT:C60-G solar cells was attributed to improved electron transport properties in active layer. The photogenerated electrons at C_{60} are quickly transferred to graphene and graphene make them transport faster and collect at Al cathode.

Mixing of CNTs and graphene with most studied hole transport layer PEDOT:PSS has been observed to enhance the efficiency and stability of organic solar cells [54–57]. As graphene could be used as both anode and cathode in OSCs, it could be used as anode and cathode in the same device, but it would be tricky to deposit graphene on the top of active layer and make the graphene anode and cathode to work differently. There have been some reports where graphene and its derivatives have been combined with some other materials and used as both the anode and cathode materials in OSCs. Using graphene both as anode and cathode opens up the opportunities for transparent flexible OSCs and replacement of expensive and brittle ITO.

7.4 APPLICATIONS OF CNTS AND GRAPHENE IN DYE-SENSITIZED SOLAR CELLS

Dye-sensitized solar cells (DSSCs) possess two electrodes, one of which is optically transparent and usually it is fluorine-doped tin oxide (FTO) coated on glass substrate, and the other electrode is Pt or Pb coated on FTO on glass or some metal foils substrates. The transparent FTO electrode is coated with TiO_2 nanoparticles, which works as electron acceptor and transport the photogenerated electrons to FTO. DSSCs are a kind of electrochemical cells therefore the electron accepting TiO_2 coated FTO electrode called anode whereas electron donating Pt or Pb coated counter electrode called as cathode. The TiO_2 film is coated with light absorbing organic dyes and integrated with other Pb or Pt counter electrode. The two electrodes are integrated together and an electrolyte, usually a iodide/triiodide (I^-/I_3^-) redox couple is filled in between them. The electrolyte medium provides a path for ion migration from cathode to anode. Figure 7.8 shows the schematic of a typical DSSC. The incident light enters in the solar cells through TiO_2 coated FTO side and get absorbed in dye molecules. Light absorption causes electrons to excite to the LUMO of the dye from its HOMO and the electrons are transferred to TiO_2 and then to FTO electrode, whereas the holes get neutralized by the electrons from the oxidation of I^- into I_3^-. The I_3^- ions transport to the counter electrode through the electrolyte medium and get reduced there into I^- by taking electrons from counter electrode and then I^- ions transport back to the TiO_2 electrode. This way we get total negative charge on the TiO_2 coated FTO electrode and total positive charge on the counter electrode. The DSSCs were invented by Michael Gratzel in 1991, therefore these cells are also known as Gratzel cells. The first DSSCs prepared by Gratzel exhibited the PCE of 7.1–7.9% [58]. The state-of-the-art certified efficiency of DSSCs is 13% [3]. Initially these solar cells attracted the attention globally because of their low cost and high efficiencies, but for last two decades there has not been much progress in their performance. The reason behind performance stagnation in DSSCs is that there has not been much effort made in that direction. Actually, in 2009 a new kind of solar cell was invented which basically originated from DSSC, where

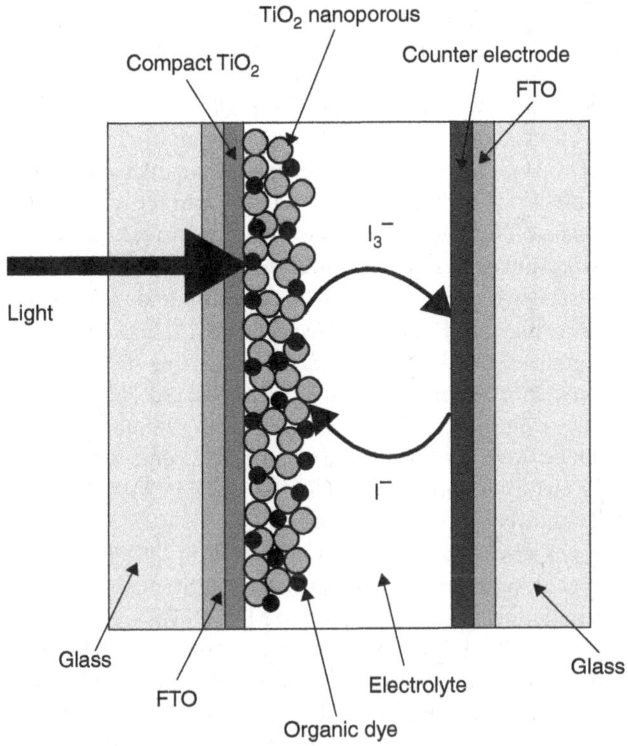

FIGURE 7.8 Schematic representation of the design of a typical DSSC.

instead of organic dye, some perovskite semiconductors were used as light absorbing medium and those solar cells are known as perovskite solar cells (PSCs). Since the discovery of PSCs, most of the groups working globally in DSSCs have switched to the development of PSCs rather than developing DSSCs. The PSCs and applications of CNTs and graphene in them are discussed in detail in the next section. In this section we discuss the applications of CNTs and graphene in DSSCs that played an important role in their progress.

CNTs and graphene have been explored for the replacement of counter electrode, use as photoelectrode interface layer and doping of the light absorbing medium to improve the performance of DSSCs and make them more cost effective [59–62]. Different interfaces in solar cells play crucial role in charge separation and transportation and controls the performance of the solar cells. Suitability of CNTs and graphene for the replacement of ITO in OSCs has already been discussed in the previous section. Similarly, CNTs and graphene have also been used to replace or modify the FTO/ITO and Pt counter electrode in DSSCs to provide better interfaces for charge separation and collection [63–66]. Kim et al. used graphene nanoplatelets (GnPs) to replace Pt counter electrode in DSSCs [64]. Since the adhesion of GnPs on the substrates was very poor, the GnPs were mixed with PEDOT:PSS to form a composite that retained the catalytic performance of GnPs and had good adhesion on the substrate through conductive PEDOT matrix. The PEDOT:PSS composite with GnPs (PPG) exhibited extremely low charge transfer resistance compared to Pt electrode and had strong electrochemical stability for $Co(bpy)^{2+/3+}$ redox couple. The solar cells were prepared in the designs Glass/FTO/compact TiO_2/TiO_2/Y123 dye/electrolyte/Pt/FTO/Glass and Glass/FTO/compact TiO_2/TiO_2/Y123 dye/electrolyte/PPG/FTO/Glass. Y123 dye was the light-absorbing photosensitive material and TiO_2 was the photoanode. The devices made of PPG counter electrode exhibited higher PCE of 8.33% compared to 7.99% of that made of Pt counter electrode. Xue et al. prepared 3D nitrogen doped graphene foams (N-GFs) by annealing of freeze-dried graphene oxide foams in ammonia and deposited them on FTO coated glass substrates to use them as the replacement of Pt counter electrode in DSSCs [67]. The solar cells with N-GFs counter electrode showed PCE of 7.07%, which was comparable to 7.44% from those based on Pt counter electrodes. The N-GFs provided high electrical conductivity and good electrocatalytic activity as the counter electrode and they facilitated high surface area, hydrophilicity, and well-defined porosity to enhance electrolyte-electrode interaction and electrolyte/reactant diffusion [67]. Ju et al. synthesized nitrogen doped graphene nanoplatelets (NGnP) and deposited them FTO coated glass substrates by electrospray coating method and used them as the counter electrode in DSSCs as a replacement of Pt counter electrode [68]. The solar cells were prepared using $Co(bpy)_3^{3+/2+}$ redox couple and O-alkylated-JK-225 organic dye. For comparison, the solar cells were also prepared using Pt counter electrode. The optimized NGnP electrode showed better photovoltaic performance compared to Pt electrode. NGnP counter electrode reduced the electrolyte/electrode resistance down to 1.73 Ω cm^2 from 3.15 Ω cm^2 with Pt electrode. The DSSCs based on NGnP counter electrode showed FF of 74.2% and a PCE of 9.05%, whereas those based on Pt counter electrode showed FF of 70.6% and PCE of 8.43%. These results showed that the graphene-based electrodes could outperform Pt electrode in DSSCs making them more stable and cost effective. Mei et al. reported high performance DSSCs using CNTs-based counter electrode [69]. The CNTs (including both SWCNTs and MWCNTs) were dispersed in poly(ethylene glycol) (PEG) and coated on the FTO coated glass substrates. The PEG was removed from the substrate by heating and that resulted in binder free CNTs film on FTO substrates. The DSSCs prepared with CNTs counter electrode exhibited comparable PCE with those based on Pt counter electrode. The DSSCs exhibiting CNT counter electrode showed a PCE of 7.81%, whereas those exhibiting the Pt counter electrode showed the PCE of 7.92%. The CNT electrode provided excellent electrochemical catalytic behavior with iodide/triiodide redox couple.

Song et al. deposited thin films of graphene like carbon (GLC) on FTO coated glass substrates by hot filament CVD method before TiO_2 photoanode and prepared DSSCs on them [94]. The GLC films exhibited good transparency and reduced the contact resistance. The thickness and transparency of GLC films was controlled by varying the deposition time. The solar cells with GLC films exhibited PCE of 6.9% which was higher than 5.9% without GLC film. Chen et al. introduced graphene between FTO and TiO_2 to enhance the electron transport from TiO_2 and then to FTO and reduced the charge recombination [61]. Graphene has work function (4.4 eV) similar to that of FTO therefore functionalization of FTO surface by graphene acts as the extension of the electrode and enhances the charge transport rate. In addition to that when the electrons are transferred to graphene from TiO_2, they can't transport back as TiO_2 conduction band is 4.2 eV, which provides an interface barrier of 0.2 eV for electron transfer from graphene to TiO_2. Along with functionalization of FTO surface by graphene, the TiO_2 was also functionalized to form TiO_2/graphene composite anode. For FTO functionalization graphene suspension in ethanol was spin coated on FTO, whereas for the functionalization of TiO_2, the TiO_2 nanoparticles and graphene sheets were mixed together for intimate contact between them and then a paste was formed. This paste was used to form TiO_2/graphane nanoporous films on bare FTO and graphene functionalized FTO substrates. To increase the substrate coverage by graphene on FTO, the graphene suspension was coated multiple times on the same substrate. Figure 7.9 shows schematically the devices with and without functionalization of FTO and TiO_2 with graphene and the electron transport pathways. N719 dye was used as the light-absorbing sensitizer, the counter electrode was made of Pt on FTO and the electrolyte was an iodide-based liquid (HL-HPE, 0.1 M, LiI, 50 mM I_2 and 0.5 M DMPII). The solar cells without any functionalization of FTO or TiO_2, exhibited PCE of 5.8%.

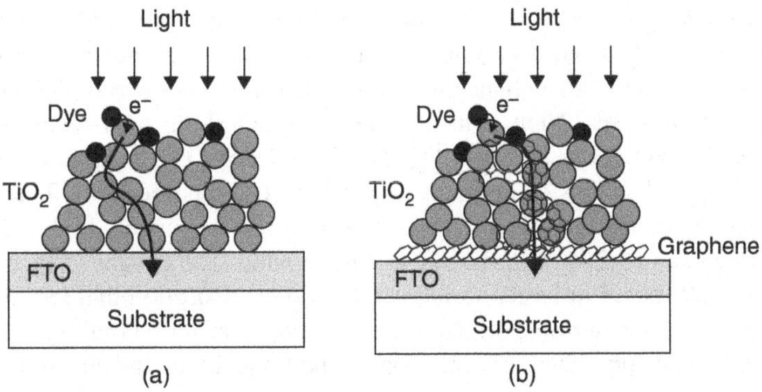

FIGURE 7.9 Schematic representation of the device designs and electron transport in the DSSCs (a) without and (b) with functionalization of FTO and TiO_2 with graphene [61].

A 0.02% content of graphene in TiO_2 only without any functionalization of FTO, resulted in a PCE of 7.2%. This enhancement in PCE was because of enhancement in all the photovoltaic parameter viz. J_{sc}, V_{oc} and FF. The functionalized FTO substrate with single coating of graphene with functionalized TiO_2 resulted in PCE of 7.74%. The enhanced PCE in case of functionalized FTO was mainly because of enhancement on J_{sc}. In case of the samples when graphene was coated twice on FTO the solar cell exhibited the PCE of 8.13%. The enhancement in PCE was mainly due to enhancement in J_{sc}. This enhancement in J_{sc} has been attributed to enhanced electron transport through graphene and reduced recombination. The bare nanoporous TiO_2 film allow the electrolyte to infiltrate and contact with FTO which results in direct charge carrier recombination but introduction of graphene layer in between FTO and TiO_2 prevented this direct contact and led to reduced recombination. These results clearly showed the contribution of graphene in the enhancement of PCE of DSSCs but as the thickness of graphene on FTO was increased further by three times spin coating of graphene the PCE decreased and this reduction in efficiency was attributed to reduction in transparency of the electrode due to thicker graphene film. Roh et al. used r-GO to modify the surface of FTO that led to enhanced charge transport, reduced series resistance and as a result to enhanced PCE [62]. Where the PCE with conventional FTO electrode was 7.12%, the surface modification of FTO by r-GO resulted in the enhanced PCE to 8.44%. This enhancement in PCE was mainly because of enhancement in J_{sc}, which was attributed to reduced series resistance and enhanced electron transport.

As TiO_2 and other semiconductors play a crucial role in controlling the PCE of DSSCs, efforts have also been made to improve the PCE of DSSCs via incorporating graphene into TiO_2 to improve the photon harvesting [70-73]. Manikandan et al. prepared a composite of TiO_2 and r-GO with different wt% composition of r-GO and used them to fabricate DSSCs [70]. For TiO_2/r-GO composites, TiO_2 nano-aggregates were grown and mixed with different amounts of GO powder and simultaneously GO was hydrothermally reduced to get TiO_2/r-GO composite. DSSCs were prepared using both TiO_2 nanoaggregates and TiO_2/r-GO composites. The solar cells having TiO_2/r-GO composite made of 0.5 mg of GO powder resulted in maximum PCE. Where the solar cells with TiO_2 photoanode showed PCE of 6.32%, the solar cells made of TiO_2/r-GO composite with 0.5 mg of GO showed PCE of 8.62%. The solar cells having TiO_2/r-GO composites made of 0.25 mg and 1.0 mg of GO powder exhibited PCE of 7.64 and 5.10% respectively. The enhancement in PCE was mainly because of enhancement in J_{sc} and that was caused by enhanced higher light scattering, enhanced dye loading, faster electron transport and reduced charge recombination. Similarly Kusumawati et al. [72] and Kazmi et al. [73] prepared the nanocomposites of

TiO_2 and graphene and used them as photoanode in DSSCs and achieved the PCE of 7.06% and 7.68% respectively. The TiO_2-graphene nanocomposites resulted in enhanced PCE compared to bare TiO_2 as photoanode and this enhancement was attributed to enhanced photon harvesting and reduced recombination losses. Nien et al. incorporated GO along with Ag into TiO_2 and used the composite as photoanode in DSSCs [74]. The modified TiO_2/GO/Ag composite resulted in enhanced PCE compared to TiO_2/GO composite and bare TiO_2 photoanodes. The DSSCs made of TiO_2/GO/Ag photoanode exhibted PCE of 5.33%, which was reasonably higher than 4.46% with TiO_2/GO photoanode and 3.79% with bare TiO_2 photoanode. The enhanced PCE with TiO_2/GO/Ag composite was attributed to larger surface area for dye adsorption that led to enhanced light absorption, enhanced electron transport and reduced charge carrier recombination.

Like in BHJ OSCs the photoactive layers possess both the donor and acceptor materials in the blend, a similar concept has also been applied in DSSCs to enhance their performance. The dye sensitizers have been modified to have to both the electron acceptor and electron donor entities to enhance photon harvesting. CNTs and graphene have been used as the sensitizer modifiers to enhance the PCE of DSSCs [75–78]. Volland et al. prepared a hybrid photosensitizer, which consisted of azulenocyanine as electron donor and few-layer graphene as electron acceptor and the hybrid photosensitizer exhibited wider absorption range with enhanced light absorption and better electron transport properties, which resulted in enhanced PCE [75]. The DSSCs having graphene/azulenocyanine/di-tetrabutylammonium cis-bis(isothiocyanato) bis(2,2'-bipyridyl-4,4'-dicarboxylato)ruthenium(II) (N719) sensitizer exhibited PCE of 8.32% which was better than 7.47% with graphene/N719 sensitizer-based solar cells. Yang et al. used graphene quantum dots (GQD) to incorporate with N719 dye to make a composite sensitizer to enhance photon harvesting and they achieved 8.9% efficiency in the GQD/N719 composite-based solar cells that was significantly higher than 7.6% efficiency of the N719-based solar cells [78].

7.5 APPLICATIONS OF CNTS AND GRAPHENE IN PEROVSKITE SOLAR CELLS

Perovskite solar cells (PSCs) are a new type of solar cells, which have basically originated from DSSCs. PSCs incorporate perovskite semiconductors as the light absorbers. Perovskite is a class of distinct materials which have ABX_3 formula, where A and B are cations and X is an anion. The two cations are usually different in size and the anion bonds both the cations. In 1839, Gustav Rose from Russia discovered calcium titanate ($CaTiO_3$) and named it as perovskite after L. A. Perovski. Since then, all the materials having ABX_3 formula and crystal structure like $CaTiO_3$ are known as perovskites materials. The perovskites include a range of materials from insulators to superconductors but nobody thought of using them in solar cells, until 2009 when Miyasaka et al. used synthetic perovskites methylammonium lead iodide ($CH_3NH_3PbI_3$) and methylammonium lead bromide $CH_3NH_3PbBr_3$ in DSSCs as the light absorber in place of organic dye [79]. These solar cells were not very promising as they exhibited PCE of around 3.8% (the DSSCs were already having around 11% PCE at that time) and degraded very rapidly as the electrolyte medium dissolved the perovskite crystals. They did not attract much attention but at the same time Michale Gratzel was working on the solid state DSSCs which include solid hole transport medium in place of electrolytes, and Gratzel et al. used $CH_3NH_3PbI_3$ perovskite semiconductors as the light absorber in solid state DSSCs and they got appreciable 9.7% PCE in 2012 [80]. This research got lots of attention worldwide and since then the research in perovskite-based solar cells got accelerated and these solar cells are being called perovskite solar cells (PSCs). Gratzel et al. modified the method of growing the crystals of perovskite semiconductors and achieved around 15% PCE in 2013 [81]. At the same time Snaith et al. used thermal evaporation method to grow the perovskite crystals and used them to fabricate PSCs and they also got over 15% PCE [82]. These works boosted global research in PSCs and in the last decade the number of research publications in PSCs have grown exponentially and the PCE in PSCs has reached certified 25.5%, which is nearly as good as that of traditional Si solar cells [24]. Just within a decade or so the PCE of PSCs has increased over 6 times and this is the only solar cell technology that has shown such an unprecedented progress in PCE in the history. Such a high performance of PSCs is the result of their excellent optical and electrical properties. These solar cells are prepared on FTO or ITO coated glass or plastic substrates. ITO or FTO here serves as transparent conducting electrode like in OSCs. The perovskite layer is sandwiched between FTO and the top metal electrode along with some HTL and ETL. PSCs can be prepared in normal as well as inverted geometries. The normal geometry could possess either planer heterojunction or mesoporous structure of the electron transport layer. The device structures, fabrication methods and working principal of PSCs are quite similar to those of OSCs. The only difference in OSCs and PSCs is that the former incorporates organic semiconductors as light absorbers, whereas latter incorporates perovskite semiconductors as the light absorbers. Figure 7.10 shows the schematic structures of typical PCSs with mesoporous and planer heterojunction structures with normal and inverted device designs. In brief, the normal device design possesses ITO or FTO as a cathode that is coated with an ETL, then perovskite, then HTL and finally top metal electrode, which works as an anode. In case of inverted device design the ITO or FTO is used as anode, which is coated with a HTL, then perovskite, then ETL and finally the top metal electrode, which works as cathode. In the mesoporous structure, the nanoparticles of $TiO_{2,}$ or some other suitable material, are deposited on hole blocking layer on FTO. TiO_2

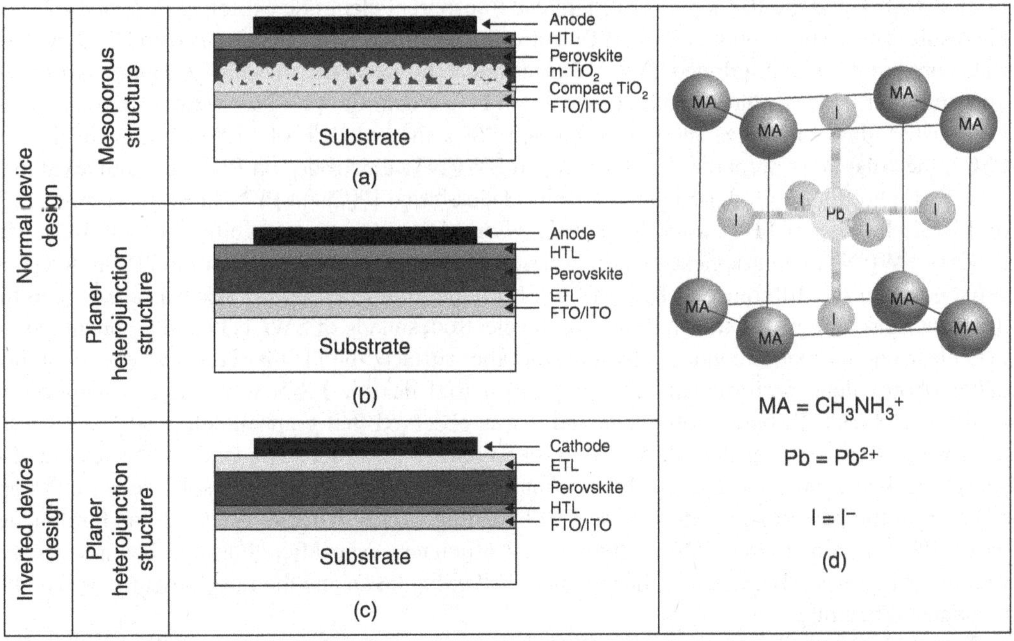

FIGURE 7.10 Schematic representation of normal (a) and (b) and inverted (c) device designs of typical PSCs. The normal PSCs could be prepared both in the mesoporous (a) and planer heterojunction (b) structures. (d) Crystal structure of a very common perovskite $CH_3NH_3PbI_3$ used in PSCs.

nanoparticles form a mesoporous nanostructured scaffold, which works as electron transport layer. The mesoporous scaffold is coated with perovskite semiconductor, and then a thin layer of HTL is coated on perovskite film to complete the solar cells the top metal electrode is deposited. Snaith et al. found that mesoporous scaffold structure in not compulsory to achieve high efficiency PSCs and they can also be prepared without them, and such structures were called planer structures [83]. In case of planer structure, FTO is coated with hole blocking compact film of ETL, which is followed by perovskite and then by HTL and then by top metal electrode. Though PSCs have shown very high PCE in all types of structures and high cost-effectiveness, but they are still far from commercialization because of their poor stability. PSCs have shown sensitivity to a number of parameters like moisture, oxygen, UV light and high temperature. They have also shown different materials in solar cell to react with each other and degrade cell performance very rapidly [84, 85].

CNTs and graphene have also been investigated to find different applications in PSCs toward improving their performance and make them more cost effective. One of the most studied applications of CNTs and graphene in PSCs is the replacement of brittle and expensive transparent ITO or FTO electrodes, like in OSCs [86]. Jeon et al. prepared inverted PSCs, using CNTs in particular the SWCNTs and graphene films as the transparent conducting electrodes in place of ITO on both the glass and plastic substrates [86]. Both the SWCNTs and graphene electrodes on glass and plastic substrates were doped with MoO_3 by depositing a thin film of MoO_3 on them. The MoO_3 doping of SWCNTs and graphene electrodes improved their conductivity and nanoscale morphology and a thin layer of PEDOT:PSS on MoO_3 doped SWCNTs or graphene electrodes reduced their roughness and contact resistances. The solar cells were prepared in the structure MoO_3 doped SWCNTs or MoO_3 doped graphene/PEDOT:PSS/$CH_3NH_3PbI_3$/C_{60}/BCP/LiF/Al on glass and PEN substrates. The PSCs made of SWCNTs transparent electrode exhibited PCE of 12.8% and the PSCs made of graphene transparent electrode exhibited PCE of 14.2% on glass substrates. The doping of SWCNT electrode was also done by treating it with HNO_3 and the PSCs prepared on HNO_3 doped SWCNT electrode exhibited PCE of 15.3% on glass substrates. To compare the performance with conventional ITO electrode the PSCs were also prepared on ITO coated glass substrates and the ITO-based PSCs showed PCE of 17.8%. In case of flexible PSCs, the solar cells prepared on MoO_3 doped SWCNTs electrodes on PEN substrate exhibited 11% PCE and the MoO_3 doped graphene electrodes-based PSCs on PEN substrate showed PCE of 13.3%. In case of ITO-based flexible PSCs on PEN substrates exhibited PCE of 16%. The PCEs of SWCNTs and graphene electrode-based PSCs were not much inferior to ITO-based solar cells, however the difference in PCE of SWCNTs and graphene-based PSCs was attribute to the different morphology and transmittance of the electrodes made of SWCNTs and graphene. Since ITO electrode has better conductivity and smoother surface, the ITO-based solar cells, exhibited better photovoltaic performance. The graphene-based flexible PSCs were more susceptible to strain than SWCNTs-based solar cells and it was observed that graphene electrode-based solar cells were slightly better than those based on SWCNTs electrode. The flexible PSCs were also tested for their robustness against the mechanical bending and ITO-based solar cells rapid degradation in the performance after around 200 bending cycles, whereas SWCNTs and graphene-based PSCs retained over 85% of their initial efficiency even after thousand bending cycles. Figure 7.11 shows schematically the structure of flexible PSCs and their performance against the mechanical bending.

You et al. prepared semitransparent PSCs, where one of the electrodes was FTO and the other electrode was made of graphene [87]. The solar cells were prepared in the structure FTO/TiO_2/$CH_3NH_3PbI_{3-x}Cl_x$/Spiro-OMeTAD/PEDOT:PSS/graphene/PMMA/PDMS on the glass substrates. Here the top transparent electrode was made of graphene, which was prepared by CVD method.

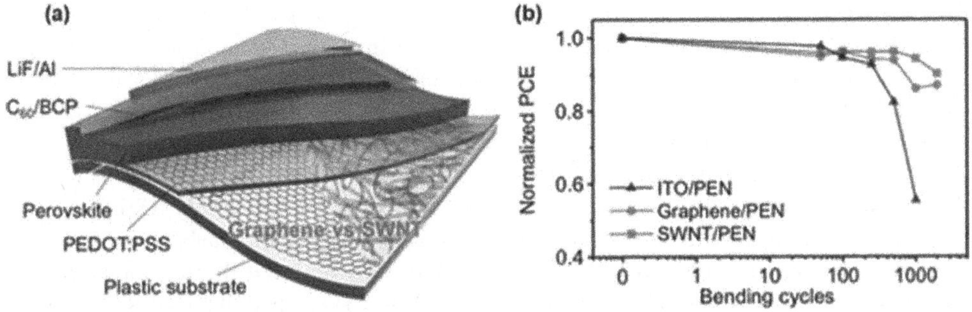

FIGURE 7.11 (a) Schematic of the flexible PSCs prepared by Jeon et al. and (b) their normalized PCE as a function of mechanical bending cycles. "Adapted with permission from [86]. Copyright (2017) American Chemical Society".

The CVD grown graphene sheets on Cu foil were coated with poly(methyl methacrylate) (PMMA) solution and annealed at 90°C for 30 min., which was followed by deposition of a freestanding PDMS film on PMMA and etching out of Cu foil in $FeCl_3$ solution. The graphene/PMMA/PDMS multilayer film was transferred to glass substrate with graphene side upwards. Now a thin layer of PEDOT:PSS was coated on graphene/PMMA/PDMS film on glass substrate and annealed at 120°C for 60 min. These PEDOT:PSS coated graphene electrodes were used to laminate FTO/TiO_2/ PEDOT:PSS/$CH_3NH_3PbI_{3-x}Cl_x$/Spiro-OMeTAD structure prepares separately on some other FTO coated glass substrates to obtain the final devices. PEDOT:PSS enhanced the conductivity of graphene electrode dramatically and it also supported the adhesion of graphene electrode on the Spiro-OMeTAD film. The optimized devices, when illuminated from FTO side exhibited average PCE of 12.02% and when illuminated from top graphene side they exhibited an average PCE of 11.65%. For comparison with conventional Au electrode the solar cells were also prepared in FTO/TiO_2/PEDOT:PSS/$CH_3NH_3PbI_{3-x}Cl_x$/Spiro-OMeTAD/Au structure on glass substrates and these solar cells showed PCE of 13.62%.

Apart from replacing the conducting electrode in PSCs, CNTs and graphene have also been used to modify both the HTL and ETL to improve the charge collection toward improving their PCE and lifetime. As both the CNTs and graphene have been observed to have hydrophobic behavior, therefore have been incorporated in charge transport layers of PSCs to prevent the diffusion of moisture and oxygen into the devices and improve their lifetime. Habisreutinger et al. doped HTL with CNTs in PSCs and the doping of HTL by CNTs resulted in highly stable PSCs [88]. The solar cells were prepared in the structure FTO/TiO_2/m-Al_2O_3/$CH_3NH_3PbI_{3-x}Cl_x$/P3HT/SWCNTs/PMMA/Ag on glass substrates, where P3HT was used as HTL, which was doped with SWCNTs and embedded in insulating polymer matrix. Such PSCs exhibited PCEs up to 15.3% with an average PCE of 10±2%. The doping of P3HT with SWCNTs led to strong retardation in thermal degradation compared to bare P3HT and also the resistance to water ingress was remarkably enhanced. For solar cells fabrication, TiO_2 was deposited on FTO coated glass substrates and followed by coating of Al_2O_3 nanoparticles from their colloidal dispersion in isopropanol. After that, the perovskite layer was deposited on Al_2O_3 mesoporous (m-Al_2O_3) scaffold and then HTL was deposited. For comparison, a number of materials like P3HT, Spiro-MeOTAD, PTAA and SWCNTs doped P3HT in PMMA matrix were used as HTL. For SWCNTs dope P3HT layer, the SWCNTs were functionalized by P3HT and dispersed in chloroform. The solution of functionalized SWCNTs was spin coated on the perovskite layer and then immediately PMMA or poly(bisphenol A carbonate) (PC) solution in chlorobenzene was spin coated onto

P3HT/SWCNTs film. Finally, to complete the solar cells Ag or Au electrodes were thermally evaporated onto HTL. As PSCs show rapid degradation in the moist environment, one of the PSCs employing P3HT/SWCNTs in PC matrix was exposed to continuous water flow for 60 s and this PSC showed remarkable stability in PCE even after direct exposure to water flow. The PCE was measured before and after the water exposure, where the PCE before water exposure was 12.9% and the PCE after water exposure for 60 s was measured to be 12.7%. This work was quite encouraging as it showed that the use of functionalized CNTs in highly resilient polymer matrix in PSCs can improve their thermal and moisture stability tremendously.

Li et al. also prepared semitransparent PSCs, where they replaced the top metal electrode with CNTs electrode [89]. The solar cells were prepared on FTO coated glass substrates. A compact layer of TiO_2 was deposited on FTO and then followed by a thin layer of mesoporous TiO_2. $CH_3NH_3PbI_3$ perovskite film was deposited on mesoporous TiO_2 and then a CNTs network was laminated onto perovskite film as the top electrode. There was no HTL in this device and it exhibited PCE of up to 6.87%. For comparison with conventional Au electrode, the solar cells were also prepared with deposition of Au as the top electrode on the perovskite film, and the solar cells with Au electrode showed PCE of 5.14%. One of the PSCs with CNTs electrode was illuminated from both the CNT and FTO electrode sides and tested for its PCE. The solar cell showed 6.29 % efficiency when illuminated from FTO side and 3.88% efficiency when illuminated from CNT side. When Spiro-MeOTAD (used as HTL in PSCs) was added to the top CNT electrode, it formed a composite electrode with CNTs and the PCE of the solar cells improved to 9.9%. This enhancement in PCE was because of reduced recombination and enhanced hole extraction in solar cells due to Spiro-MeOTAD. Semitransparent PSCs with the use of CNTs and graphene electrodes opens up the opportunity of making transparent PSCs for windows applications in buildings and vehicles. There are a number of reports where r-GO has been used as HTL and ETL in PSCs [90, 91]. r-GO has also been mixed with TiO_2 to form electron transport layer in PSCs and the electron transport layer made of the composite of TiO_2 and r-GO has shown promising efficiency of over 19% in perovskite solar cells [92]. Ammonia modified GO was mixed with PEDOT:PSS and used as HTL in inverted PSCs and the incorporation of ammonia modified GO into PEDOT:PSS not only improved the PCE of PSCs but also enhanced their stability [93].

7.6 CONCLUSION

CNTs and graphene are excellent materials and have found various applications in all sorts of photovoltaic devices ranging from first generation to third generation. In this chapter I have discussed the applications of CNTs and graphene in Si solar cells from first generation and dye sensitized, organic and perovskite solar cells from third generation. CNTs and graphene are purely made of carbon and are completely renewable, their incorporation in solar cells makes them more cost effective and gives the opportunity to make flexible and semitransparent solar cells, which would be not only be efficient but also stable compared to conventional materials used. CNTs and graphene have found applications in all the components of a solar cell including electrodes, active light-absorbing layers and charge transport layers. Use of CNTs and graphene in solar cells is very important from both the academic and technological aspects.

REFERENCES

1. Fritts, C., 1885. On the Fritts selenium cell and batteries, *Van Nostrands Engineering Magazine* 32: 388–395.
2. Chapin, D., Fuller C., and Pearson, G. 1954. A new silicon p-n junction photocell for converting solar radiation into electrical power. *J. Appl. Phys.* 25: 676–677.

3. https://upload.wikimedia.org/wikipedia/commons/a/aa/CellPVeff%28rev210104%29.png.
4. Jorio, A., Dresselhaus, G., and Dresselhaus, M. 2008. *Carbon Nanotubes: Advanced Topics in the Synthesis, Structure, Properties and Applications*. Springer.
5. Wilder, J.W.G., Venema, L.C., Rinzler, A.G., Smalley, R.E., and Dekker, C. 1998. Electronic structure of automatically resolved carbon nanotubes, *Nature* 391: 59–62.
6. Lee, J.U., 2005. Photovoltaic effect in ideal carbon nanotube diodes. *Appl. Phys. Lett.* 87: 073101.
7. Wei, W., Bao, X.Y., Soci, C., Ding, Y., Wang, Z.-L., and Wang, D. 2009. Direct heteroepitaxy of vertical InAs nanowires on SI substrates for broad band photovoltaics and photodetection. *Nano Lett.* 9: 2926–2934.
8. Li, Z., Kunets, V.P., Saini, V., Xu, Y., Dervishi, E., Salamo, G.J., Biris, A.R., and Biris, A.S. 2009. Light-harvesting using high density p-type single wall carbon nanotube/n-type silicon heterojunctions. *ACS Nano* 3: 1407–1414.
9. Wei, J., Jia, Y., Shu, Q., Gu, Z., Wang, K., Zhuang, D., Zhang, G., Wang, Z., Luo, J., Cao, A., and Wu, D. 2007. Double-walled carbon nanotube solar cells. *Nano Lett.* 7: 2317–2321.
10. Mintmire, J.W., and White, C.T., 1995. Electronic and structural properties of carbon nanotubes. Carbon 33: 893.
11. Liu, P., Sun, Q., Zhu, F., Liu, K., Jiang, K., Liu, L., Li, Q., and Fan, S. 2008. Measuring the Work-Function of Carbon Nanotubes with Thermionic Method. *Nano Lett.* 8: 647.
12. Jia, Y., Wei, J., Wang, K., Cao, A., Shu, Q., Gui, X., Zhu, Y., Zhuang, D., Zhang, G., Ma, B., Wang, L., Liu, W., Wang, Z., Luo, J., and Wu, D. 2008. Nanotube-Silicon Heterojunction Solar Cells. *Adv. Mater.* 20: 4594–4598.
13. Li, X., Jung, Y., Sakimoto, K., Goh, T.H., Reed, M.A., and Taylor, A.D. 2013. Improved efficiency of smooth and aligned single walled carbon nanotube/silicon hybrid solar cells. *Energy Environ. Sci.* 6: 879–887.
14. Wang, F., Kozawa, D., Miyauchi, Y., Hiraoka, K., Mouri, S., Ohno, Y., and Matsuda, K. 2014. Fabrication of Single-Walled Carbon Nanotube/Si Heterojunction Solar Cells with High Photovoltaic Performance. *ACS Photonics* 1: 360–364.
15. Jia, Y., Cao, A., Bai, X., Li, Z., Zhang, L., Guo, N., Wei, J., Wang, K., Zhu, H., Wu, D., and Ajayan, P.M. 2011. Achieving High Efficiency Silicon-Carbon Nanotube Heterojunction Solar Cells by Acid Doping. *Nano Lett.* 11: 1901–1905.
16. Gao, H., and Duan, H., 2015. 2D and 3D graphene materials: Preparation and bioelectrochemical applications. *Biosen. Bioelectron.* 65: 404.
17. Li, X., Zhu, H., Wang, K., Cao, A., Wei, J., Li, C., Jia, Y., Li, Z., Li, X., and Wu, D. 2010. Graphene on Silicon Schottky junction solar cells. *Adv. Mater.* 22: 2743.
18. Gosh, A.K., and Feng, T., 1973. Rectification, space-chargelimited current, photovoltaic and photoconductive properties of Al/tetracene/Au sandwich cell. *J. Appl. Phys.* 44: 2781.
19. Tang, C.W. 1986. Two layer organic photovoltaic cell. *Appl. Phys. Lett.* 48: 183.
20. Peumans, P., and Forrest, S.R., 2001. Very-high-efficiency double-heterostructure copper phthalocyanine/C60 photovoltaic cells. *Appl. Phys. Lett.* 79: 126–128.
21. Yu, G., Gao, J., Hummelen, J., Wudl, F., and Heeger, A.J. 1995. Polymer photovoltaic cells: enhanced efficiencies via a network of internal donor-acceptor heterojunctions. *Science* 270: 1789–1791.
22. Shaheen, S.E., Brabec, C.J., Sariciftci, N.S., Padinger, F., Fromhert, T., and Hummelen, J.C. 2001. 2.5% efficient organic plastic solar cells. *Appl. Phys. Lett.* 78: 841–843.
23. Padinger, F., Rittberger, R.S., and Sariciftci, N.S. 2003. Effects of postproduction treatment on plastic solar cells. *Adv. Funct. Mater.* 13: 85–88.
24. https://en.wikipedia.org/wiki/Solar_cell_efficiency#/media/File:CellPVeff(rev210104).png.
25. Kumar, P. 2016. *Organic Solar Cells: Device physics, processing, degradation and prevention,* USA, CRC Press.
26. Pasquier, A.D., Unalan, H.E., Kanwal, A., Miller, S., and Chhowalla, M. 2005. Conducting and transparent single-wall carbon nanotube electrodes for polymer-fullerene solar cells. *Appl. Phys. Lett.* 87: 203511.
27. Park, H., Chang, S., Zhou, X., Kong, J., Palacios, T., and Gradecak, S. 2014. Flexible Graphene Electrode-Based Organic Photovoltaics with Record-High Efficiency. *Nano Lett.* 14: 5148.

28. Rowell, M.W., Topinka, M.A., McGehee, M.D., Prall, H.J., Dennler, G., Sariciftci, N.S., Hu, L., and Gruner, G. 2006. Organic solar cells with carbon nanotube network electrodes. *Appl. Phys. Lett.* 88: 233506.
29. Zhang, D., Ryu, K., Liu, X., Polikarpov, E., Ly, J., Tompson, M.E., and Zhou, C. 2006. Transparent, Conductive, and Flexible Carbon Nanotube Films and Their Application in Organic Light-Emitting Diodes. *Nano Lett.* 6: 1880.
30. Geng, H.Z., Kim, K.K., So, K.P., Lee, Y.S., Chang, Y., and Lee, Y.H. 2007. Effect of Acid Treatment on Carbon Nanotube-Based Flexible Transparent Conducting Films. *J. Am. Chem. Soc.* 129: 7758.
31. Das, S., Pandey, D., Thomas, J., and Roy, T. 2019. The Role of Graphene and Other 2D Materials in Solar Photovoltaics. *Adv. Mater.* 31: 1802722.
32. Jung, S., Lee, J., Seo, J., Kim, U., Choi, Y., and Park, H. 2018. Development of Annealing-Free, Solution-Processable Inverted Organic Solar Cells with N-Doped Graphene Electrodes using Zinc Oxide Nanoparticles. *Nano Lett.* 18: 1337.
33. Yang, M.K., and Lee, J.K. 2020. CNT/AgNW multilayer electrodes on flexible organic solar cells. *Electron. Mater. Lett.* 16: 573–578.
34. Reina, A., Jia, X., Ho, J., Nezich, D., Son, H., Bulovic, V., Dresselhaus, M.S., and Kong, J. 2008. Large Area, Few-Layer Graphene Films on Arbitrary Substrates by Chemical Vapor Deposition. *Nano Lett.* 9: 30.
35. Bae, S., Kim, H., Lee, Y., Xu, X., Park, J.S., Zheng, Y., Balakrishnan, J., Lei, T., Kim, H.R., Song, Y.I., Kim, Y.J., Kim, K.S., Ozyilmaz, B., Ahn, J.H., Hong, B.H., and Iijima, S. 2010. Roll-to-roll production of 30-inch graphene films for transparent electrodes. *Nat. Nanotechnol.* 5: 574.
36. Park, H., Chang, S., Zhou, X., Kong, J., Palacios, T., and Gradecak, S. 2014. Flexible graphene electrode-based organic photovoltaics with record-high efficiency. *Nano Lett.* 14, 9: 5184–5154.
37. Jung, S., Lee, J., Seo, J., Kim, U., Choi, Y., and Park, H. 2018. Development of Annealing-Free, Solution-Processable Inverted Organic Solar Cells with N-doped Graphene Electrodes using Zinc Oxide Nanoparticles. *Nano Lett.* 18: 1337.
38. Du J., Zhang, D., Wang, X., Jin, H., Zhang, W., Tong, B., Liu, Y., Burn, P.L., Cheng, H.M., Ren, W. 2021. Extremely efficient flexible organic solar cells with a graphene transparent anode: Dependence on number of layers and doping of graphene. *Carbon* 171: 350–358.
39. Yuan, J., Ma, L.P., Pei, S., Du, J., Su, Y., Ren, W., and Cheng, H.M. 2013. Tuning the Electrical and Optical Properties of Graphene by Ozone Treatment for Patterning Monolithic Transparent Electrodes. *ACS Nano* 7: 4233–4241.
40. Du, T., Adeleye, A.S., Zhang, T., Yang, N., Hao, R., Li, Y., Song, W., and Chen, W. 2019. Effects of ozone and produced hydroxyl radicals on the transformation of graphene oxide in aqueous media. *Environ. Sci.: Nano* 6: 2484–2494.
41. Yang, D., Zhou, L., Chen, L., Zhao, B., Zhang, J., and Li, C. 2012. Chemically modified graphene oxides as a hole transport layer in organic solar cells. *Chem. Commun.* 48: 8078–8080.
42. Naghdi, S., Arriaga, G.S., and Rhee, K.Y. 2019. Tuning the work function of graphene toward application as anode and cathode. *J. Alloy Comp.* 805: 1117.
43. Yin, Z., Sun, S., Salim, T., Wu, S., Huang, X., He, Q., Lam, Y.M., and Zhang, H. 2010. Organic Photovoltaic Devices Using Highly Flexible Reduced Graphene Oxide Films as Transparent Electrodes. *ACS Nano* 4: 5263–5268.
44. Zhao, H., Geng, W., Cao, W.W., Wen, J.G., Wang, T., Tian, Y., Jing, L.C., Yuan, X.T., Zhu, Z.R., and Geng, H.Z., 2020. Highly stable and conductive PEDOT:PSS/GO-SWCNT bilayer transparent conductive films. *New J. Chem.* 44: 780–790.
45. Kim, D.H., and Park, J.G. 2012. Polymer photovoltaic cell embedded with p-type single walled carbon nanotubes fabricated by spray process. *Nanotechnology* 23: 325401.
46. Reyes, O.A., Quintana, I.C., Maldonado, J.L., Collazo, J.N., and Borja, D.R. 2020. Single graphene derivative layer as a hole transport in organic solar cells based on PBDB-T:ITIC. *Appl. Opt.* 59: 8285.
47. Abdulrazzaq, O.A., Bourdo, S.E., Saini, V., and Biris, A.S. 2020. Acid-free polyaniline:graphene-oxide hole transport layer in organic solar cells. *J. Mater. Sci.: Mater. Electron.* 31: 21640–21650.

48. Dang, Y., Wang, Y., Shen, S., Huang, S., Qu, X., Pang, Y., Ravi, S., Silva, P., Kang, B., and Lu, G., 2019. Solution processed hybrid Graphene-MoO_3 hole transport layers for improved performance of organic solar cells. *Org. Electron.* 67: 95–100.
49. Li, S.S., Tu, K.H., Lin, C.C., Chen, C.W., and Chhowalla, M. 2010. Solution-Processable Graphene Oxide as an Efficient Hole Transport Layer in Polymer Solar Cells. *ACS Nano* 4: 3169–3174.
50. Liu, J., Xue, Y., and Dai, L., 2012. Sulfated Graphene Oxide as a Hole Extraction Layer in High Performance Polymer Solar Cells. *J. Phys. Chem. Lett.* 3: 1928–1933.
51. Liu, Z., Liu, Q., Huang, Y., Ma, Y., Yin, S., Zhang, X., Sun W., and Chen, Y. 2008. Organic Photovoltaic Devices Based on a Novel Acceptor Material: Graphene. *Adv. Mater.* 20: 3924–3930.
52. Bonaccorso, F., Balis, N., Stylianakis, M.M., Savarese, M., Adamo, C., Gemmi, M., Pellegrini, V., Stratakis, E., and Kymakis, E. 2015. Functionalized Graphene as an Electron-Cascade Acceptor for Air-Processed Organic Ternary Solar Cells. *Adv. Funct. Mater.* 25: 3870–3880.
53. Yu, D., Park, K., Durstock M., and Dai, L. 2011. Fullerene-Grafted Graphene for Efficient Bulk Heterojunction Polymer Photovoltaic Devices. *J. Phys. Chem. Lett.* 2: 1113–1118.
54. Pathak, C.S., Singh, J.P., Singh, R. 2018. Preparation of novel graphene-PEDOT:PSS nanocomposite films and fabrication of heterojunction diodes with *n*-Si, *Chem. Phys. Lett.* 694: 75–81.
55. Wang, Y., Wang, H., Xu, J., He, B., Li, W., Wang, Q., Yang, S., and Zou, B., 2018. PEDOT:PSS Modification by blending graphene oxide to improve the efficiency of organic solar cells. *Poly. Compos.* 39: 3066–3072.
56. Wageh, S., Raïssi, M., Berthelot, T., Laurent, M., Rousseau, D., Abusorrah, A.M., Al-Hartomy, O.A., and Al-Ghamd, A.A. 2021. Digital printing of a novel electrode for stable fexible organic solar cells with a power conversion efficiency of 8.5%. *Sci. Reports* 11: 14212.
57. Kymakis, E., Klapsis, G., Koudoumas, E., Stratakis, E., Kornilios, N., Vidakis, N., and Franghiadakis, Y., 2006. Carbon nanotube/PEDOT:PSS electrodes for organic photovoltaics. *Eur. Phys. J. Appl. Phys.* 36: 257–259.
58. O'Regan, B., and Gratzel, M. 1991. A low-cost, high-efficiency solar cell based on dye-sensitized colloidal TiO_2 films. *Nature* 353: 737–740.
59. Kavan, L., 2014. Exploiting Nanocarbons in Dye-Sensitized Solar Cells. *Top. Curr. Chem.* 348: 53–94.
60. Janani, M., Srikrishnarka, P., Nair, S.V., and Nair, A.S. 2015. An in-depth review on the role of carbon nanostructures in dye-sensitized solar cells. *J. Mater. Chem. A* 3: 17914–17938.
61. Chen, T., Hu, W., Song, J., Guai, G.H., and Li, C.M. 2012. Interface Functionalization of Photoelectrodes with Graphene for High Performance Dye-Sensitized Solar Cells. *Adv. Funct. Mater.* 22: 5245–5250.
62. Roh, K.M., Jo, E.H., Chang, H., Han, T.H., and Jang, H.D. 2015. High performance dye-sensitized solar cells using graphene modifiedfluorine-doped tin oxide glass by Langmuir–Blodgett technique. *J. Solid State Chem.* 224: 71–75.
63. Dong, P., Zhu, Y., Zhang, J., Peng, C., Yan, Z., Li, L., Peng, Z., Ruan, G., Xiao, W., Lin, H., Tour, J.M., and Lou, J. 2014. Graphene on Metal Grids as the Transparent Conductive Material for Dye Sensitized Solar Cell. *J. Phys. Chem.* C118: 25863–25868.
64. Kim, J.C., Rahman, Md. M., Ju, M.J., and Lee, J.J., 2018. Highly conductive and stable graphene/PEDOT:PSS composite as a metal free cathode for organic dye sensitized solar cells. *RSC Adv.* 8: 19058–19066.
65. Pang, B., Dong, L., Ma, S., Dong, H., and Yu, L. 2016. Performance of FTO-free conductive graphene-based counter electrodes for dye-sensitized solar cells. *RSC Adv.* 6: 41287–41293.
66. Rahman, M.Y.A. 2021. Review of graphene and its modification as cathode for dye-sensitized solar cells. *J Mater Sci: Mater Electron.* https://doi.org/10.1007/s10854-021-06898-z.
67. Xue, Y., Liu, J., Chen, H., Wang, R., Li, D., Qu, J., and Dai, L. 2012. Nitrogen-Doped Graphene Foams as Metal-Free Counter Electrodes in High-Performance Dye-Sensitized Solar Cells. *Angew. Chem., Int. Ed.* 51: 12124–12127.
68. Ju, M.J., Kim, J.C., Choi, H.J., Choi, I.T., Kim, S.G., Lim, K., Ko, J., Lee, J.J., Jeon, I.Y., and Baek, J.B. 2013. N-Doped Graphene Nanoplatelets as Superior Metal-Free Counter Electrodes for Organic Dye-Sensitized Solar Cells. *ACS Nano* 7: 5243–5250.

69. Mei, X., Cho, S.J., Fan, B., and Ouyang, J. 2010. High-performance dye-sensitized solar cells with gel-coated binder-free carbon nanotube films as counter electrode. *Nanotechnology* 21: 395202.
70. Manikandan, V.S., Palai, A.K., Mohanty, S., and Nayak, S.K. 2019. Hydrothermally synthesized self-assembled multi-dimensional TiO_2/Graphene oxide composites with efficient charge transfer kinetics fabricated as novel photoanode for dye sensitized solar cell. *J. Alloys Compd.* 793: 400–409.
71. Chong, S.W., Lai, C.W., Juan, J.C., and Leo, B.F. 2019. An investigation on surface modified TiO_2 incorporated with graphene oxide for dye-sensitized solar cell. *Sol. Energy* 191: 663–671.
72. Kusumawati, K., Daoud, S.K., and Pauporte, T. 2016. TiO_2/graphene nanocomposite layers for improving the performances of dye-sensitized solar cells using a cobalt redox shuttle. *J. Photochem. Photobiol. A* 329: 54–60.
73. Kazmi, S.A., Hameed, S., Ahmed, A.S., Arshad, M., and Azam, A. 2017. Electrical and optical properties of graphene-TiO_2 nanocomposite and its applications in dye sensitized solar cells (DSSC). *J. Alloys Compd.* 691: 659–665.
74. Nien, Y.H., Chen, H.H., Hsu, H.H., Kuo, P.Y., Chou, J.C., Lai, C.H., Hu, G.M., Kuo, C.H., and Ko, C.C. 2019. Enhanced photovoltaic conversion efficiency in dye-sensitized solar cells based on photoanode consisting of TiO_2/GO/Ag nanofibers. *Vacuum* 167: 47–53.
75. Volland, M., Lennert, A., Roth, A., Ince, M., Torres, T., and Guldi, D.M. 2019. Azulenocyanines immobilized on graphene; on the way to panchromatic absorption and efficient DSSC blocking layers. *Nanoscale* 11: 10709–10715.
76. Zamiri, G., and Bagheri, S. 2018. Fabrication of green dye-sensitized solar cell based on ZnO nanoparticles as a photoanode and graphene quantum dots as a photo-sensitizer. *J. Colloid Interface Sci.* 511: 318–324.
77. Jahantigh, F., Ghorash, S.M.B., and Mozaffari, S. 2020. Orange photoluminescent N-doped graphene quantum dots as an effective co-sensitizer for dye-sensitized solar cells. *J. Solid State Electrochem.* 24: 883–889.
78. Yang, W., Park, I.W., Lee, J.M., Choi, H. 2020. Influence of Oxidized Graphene Quantum Dots as Photosensitizers. *J. Nanosci. Nanotechnol.* 20: 3432–3436.
79. Kojima, A., Teshima, K., Shirai, Y., and Miyasaka, T., 2009. Organometal Halide Perovskites as Visible-Light Sensitizers for Photovoltaic Cells. *J. Am. Chem. Soc.* 131(17): 6050–6051.
80. Kim, H.S., Lee, C.R., Im, J.H., Lee, K.B., Moehl, T., Marchioro, A., Moon, S.J., Baker, R.H., Yum, J.H., Moser, J.E., Grätzel, M., and Park, N.G. 2012. Lead Iodide Perovskite Sensitized All-Solid-State Submicron Thin Film Mesoscopic Solar Cell with Efficiency Exceeding 9%. *Sci. Reports* 2: 591.
81. Burschka, J., Pellet, N., Moon, S.J., Baker, R.H., Gao, P., Nazeeruddin, Md. K., and Grätzel, M. 2013. Sequential deposition as a route to high-performance perovskite-sensitized solar cells. *Nature* 499: 316.
82. Liu, M., Johnston, M.B., Snaith, H.J. 2013. Efficient planar heterojunction perovskite solar cells by vapour deposition. *Nature* 501: 395–398.
83. Lee, M.M., Teuscher, J., Miyasaka, T., Murakami, T.N., and Snaith, H.J. 2012. Efficient hybrid solar cells based on meso-superstructured organometal halide perovskites. *Science* 338: 643.
84. Chauhan, A.K., Kumar, P., 2017. Degradation in perovskite solar cells stored under different environmental conditions. *J. Phys. D: Appl. Phys.* 50: 325105.
85. Chauhan, A.K., and Kumar, P. 2019. Photo-stability of perovskite solar cells with Cu electrode. *J. Mater. Sci.: Mater. Electron.* 30: 9582–9592.
86. Jeon, I., Yoon, J., Ahn, N., Atwa, M., Delacou, C., Anisimov, A., Kauppinen, E.I., Choi, M., Maruyama, S., and Matsuo, Y. 2017. Carbon Nanotubes versus Graphene as Flexible Transparent Electrodes in Inverted Perovskite Solar Cells. J. Phys. Chem. Lett. 8: 5395–5401.
87. You, P., Liu, Z., Tai, Q., Liu, S., and Yan, F. 2015. Efficient Semitransparent Perovskite Solar Cells with Graphene Electrodes. *Adv. Mater.* 27: 3632–3638.
88. Habisreutinger, S.N., Leijtens, T., Eperon, G.E., Stranks, S.D., Nicholas, R.J., and Snaith, H.J. 2014. Carbon Nanotube/Polymer Composites as a Highly Stable Hole Collection Layer in Perovskite Solar Cells. *Nano Lett.* 14: 5561–5568.

89. Li, Z., Kulkarni, S.A., Boix, P.P., Shi, E., Cao, A., Fu, K., Batabyal, S.K., Zhang, J., Xiong, Q., Wong, L.H., Mathews, N., and Mhaisalkar, S.G. 2014. Laminated Carbon Nanotube Networks for Metal Electrode-Free Efficient Perovskite Solar Cells. *ACS Nano* 8: 6797–6804.
90. Palma, A.L., Cina, L., Pescetelli, S., Agresti, A., Raggio, M., Paolesse, R., Bonaccorso, F., and Carlo, A.D. 2016. Reduced graphene oxide as efficient and stable hole transporting material in mesoscopic perovskite solar cells. *Nano Energy* 22: 349.
91. Tavakoli, M.M., Tavakoli, R., Hasanzadeh, S., and Mirfasih, M.H., 2016. Interface Engineering of Perovskite Solar Cell Using a Reduced-Graphene Scaffold. J. *Phys. Chem. C* 120: 19531.
92. Cho, K.T., Grancini, G., Lee, Y., Konios, D., Paek, S., Kymakis, E., and Nazeeruddin, M.K. 2016. Beneficial Role of Reduced Graphene Oxide for Electron Extraction in Highly Efficient Perovskite Solar Cells, *Chem. Sus. Chem* 9: 3040.
93. Feng, S., Yang, Y., Li, M., Wang, J., Cheng, Z., Li, J., Ji, G., Yin, G., Song, F., and Wang, Z., 2016. High-Performance Perovskite Solar Cells Engineered by an Ammonia Modified Graphene Oxide Interfacial Layer. *ACS Appl. Mater. Interfaces* 8: 14503.
94. Song, M., Seo, H.K., Ameen, S., Akhtar, M.S., and Shin, H.S. 2014. Low Resistance Transparent Graphene-Like Carbon Thin Film Substrates for High Performance Dye Sensitized Solar Cells. *Electrochimica Acta* 115: 559–565.

8 Carbon Nanotubes and Graphene for Supercapacitors Applications

Harsharaj S. Jadhav, Ranjit S. Kate, Suyog A. Raut, Ramchandra S. Kalubarme, and Bharat B. Kale

CONTENTS

8.1 INTRODUCTION

In recent times, energy storage and conversion systems (EES) have attracted much attention due to the worldwide increased alarming situation with the use of fossil fuels. In EES, especially electrical energy storage technology is a top choice because of its widespread application in consumer electronic products such as wearable electronics and electric vehicles (EVs). Especially rapid development in the EVs creates soaring demand for high energy density, power density, and long-life EES. Among different EES, electrochemical capacitors (ECs) and secondary batteries such as Li-ion batteries (LIBs) with characteristics of high efficiency, flexibility, and durability have gained much attention and are widely explored in practical applications [1–5]. The Ragon plot, shown in Figure 8.1, shows the comparison of different energy storage technologies [6]. In comparison with other EES, the supercapacitor has gained much interest in academic research and industries due to its unique advantages such as high power density and long cycle life.

ECs, also commonly known as capacitors, supercapacitors, or ultracapacitors, are electrochemical devices that exhibit greater energy density than conventional dielectric capacitors and

DOI: 10.1201/9781003231943-8

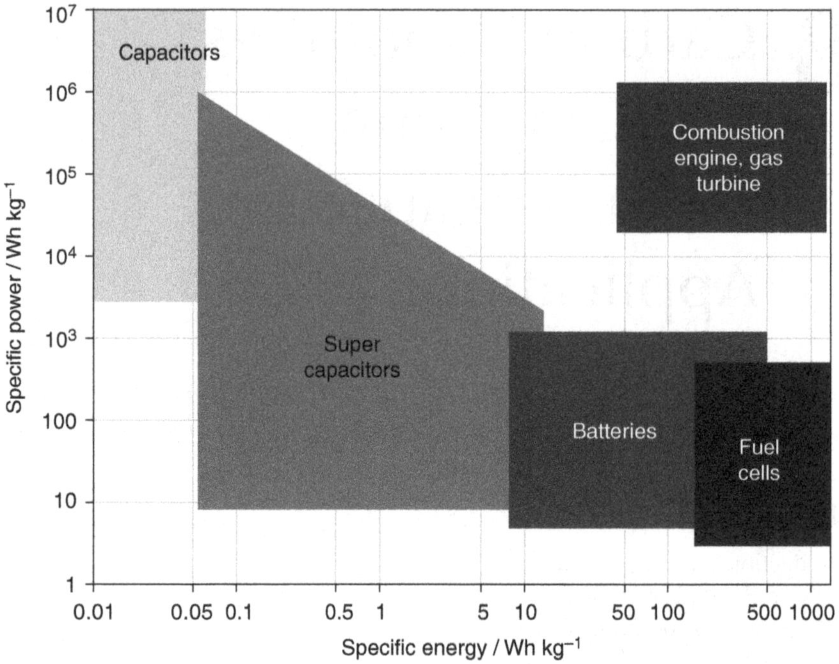

FIGURE 8.1 Ragone plot of various energy storage and conversion systems. Reprinted with permission from [6]. Copyright {2004} American Chemical Society.

higher power density than secondary LIBs. Based on the charge storage mechanism and the nature of electrode materials, ECs can be classified into three main types namely electrochemical double-layer capacitors (EDLCs), pseudocapacitors, and hybrid capacitors [7–8]. In EDLCs, electrostatic charge accumulates at the interface between the electrode surface and the electrolyte. In contrast, pseudocapacitors work on the principle of faradic redox reactions occurring on the surface of electrodes. As the name indicates, hybrid capacitors work on the combined principle of EDLC and pseudocapacitor. Figure 8.2 summarizes the different electrode configurations of ECs.

8.2 WORKING PRINCIPLE OF EDLCS

The EDLC is mainly composed of an electrolyte, separator, and two carbon-based materials as electrodes. In EDLCs charge storage takes place either electrostatically or via a non-faradic process. The charge storage in EDLCs is mainly based on the principle of the electric double layer, formed at the interface between active carbon electrode and electrolyte. With an external applied electric field, there is an accrual of electric charge at the electrode and electrolyte interface, where excess or deficit charges on the electrode surface trigger the movement of ions in the electrolyte to the respective electrodes to build the charge across the electrode/electrolyte interface to maintain electroneutrality [10]. This charge formation includes the ions adsorption as well as surface dissociation from both electrolyte and surface defects, which exclusively involve the electrostatic accrual of charge. In EDLCs, cations move toward the negative electrode whereas anions move toward the positive electrode surface. There is no charge transfer of ions exchange that happens across the electrode/electrolyte interface. Figure 8.3 reflects the schematics of the charge-discharge mechanism of EDLCs.

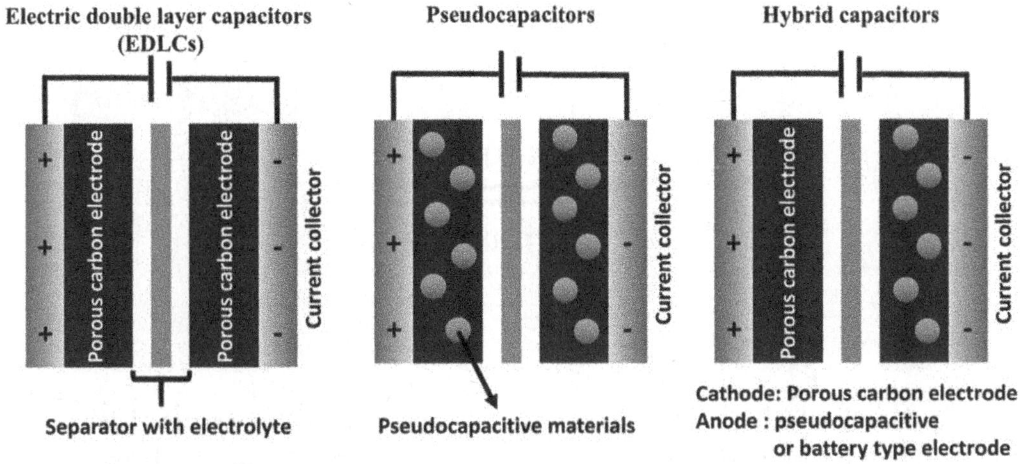

FIGURE 8.2 Schematic illustration of ECs with different electrode configurations. Reprinted with permission from [9]. Copyright {2016} IOP Science.

The surface available for electrolyte ions to diffuse can be responsible for charge storage, therefore it is important to optimize the desired pore size, pore volume, surface properties, and conductivity of electrode material [12]. In EDLCs, electrode material with growth in surface area and reduction in the distance between electrodes can significantly enhance its energy density than conventional capacitors [13,14]. The chemical or composition stability of electrodes during electrostatic charge storage is mainly attributed to no charge transfer between electrode and electrolyte. Therefore, the charge storage process in EDLCs is highly reversible and fast which permits achieving ultralong cyclability cycling stability with better power performance. The electrochemical performance of EDLCs can be adjusted by changing the nature of electrolytes such as H_2SO_4 or KOH, which generally possess lower equivalent series resistance (ESR) and required the least pore size as associated with organic electrolytes [15]. Additionally, an aqueous electrolyte exhibits a lower range of breakdown voltages. The tradeoffs between capacitance, ESR, and voltages need to be taken into account while choosing electrolytes for EDLCs. Overall, the nature of electrolyte and electrode material is of great importance in high-performance EDLC design. As an electrode, carbon-based materials are cheaper, exhibit high surface area, and have well-established fabrication techniques as compared to other materials, such as metal oxides/phosphide/sulfides and polymers.

8.3 KEY PARAMETERS FOR SUPERCAPACITORS

1. **Capacitance (C)**: The stored charge i.e., capacitance of SCs depends on the nature of electrode active material and the electrolyte used. In the case of aqueous electrolytes, carbon-based electrode materials can deliver 250–350 F/g, whereas, in the organic electrolyte, the capacitance can reach 150–200 F/g. These different capacitance values are mainly ascribed to the porous nature of carbon structure and the dielectric constant of solvents.
2. **Cell voltage (V)**: The maximum potential difference (V max) at the SC terminals depends on the nature of the current collector and electrolyte used. The theoretical thermodynamic stability domain for aqueous electrolytes is 1.2 V. However, non-aqueous/organic electrolytes exhibit a much greater stable domain of ~3 V, and possess much lower ionic conductivity than aqueous electrolytes.

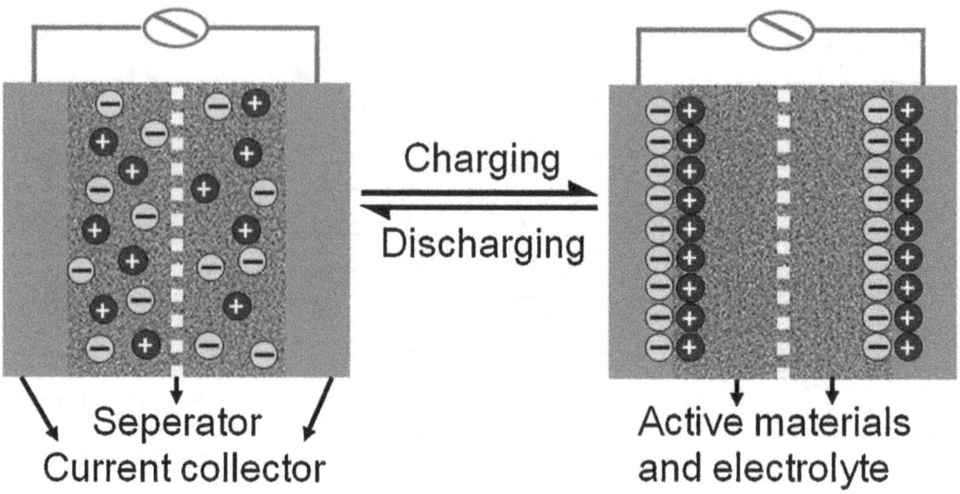

FIGURE 8.3 Charge storage mechanism in EDLC. Reprinted from [11] with permission from Elsevier.

3. **Equivalent Series Resistance (ESR)**: The equivalent series resistance (ESR) mainly consists of electrolyte resistance and resistance that arises from electrodes such as intrinsic resistance of the material and contact resistance. The lower series resistance of aqueous electrolytes than non-aqueous/organic electrolytes is mainly ascribed to the greater ionic conductivity. This resistance can significantly affect the power performance of SC.
4. **Energy density (E)**: The energy stored in SCs can be determined using the following equation:

$$E = \frac{1}{2} C V^2$$

where E is the energy (J), C is the capacitance (F) and V is the voltage (V). The energy of the SCs depends on the operating window used; therefore, organic electrolytes-based SCs with higher potential stability windows (~2.7–3 V) can store more energy than the aqueous system.

5. **Power (P_{max})**: This is the most important parameter used to characterize the SCs. The power is nothing but the product of voltage, current, or the energy delivered per unit of time and determined using the following equation:

$$P_{max} = \frac{V_{max}^2}{4ESR}$$

Where P_{max} is the maximum power for SC in W and V_{max} is the maximum cell voltage, and ESR is the equivalent series resistance. This power is a theoretical value that corresponds to the power that would be delivered during a discharge at a nominal time. The energy density or power improvement of SC can be achieved by maximizing the specific capacitance or cell voltage.

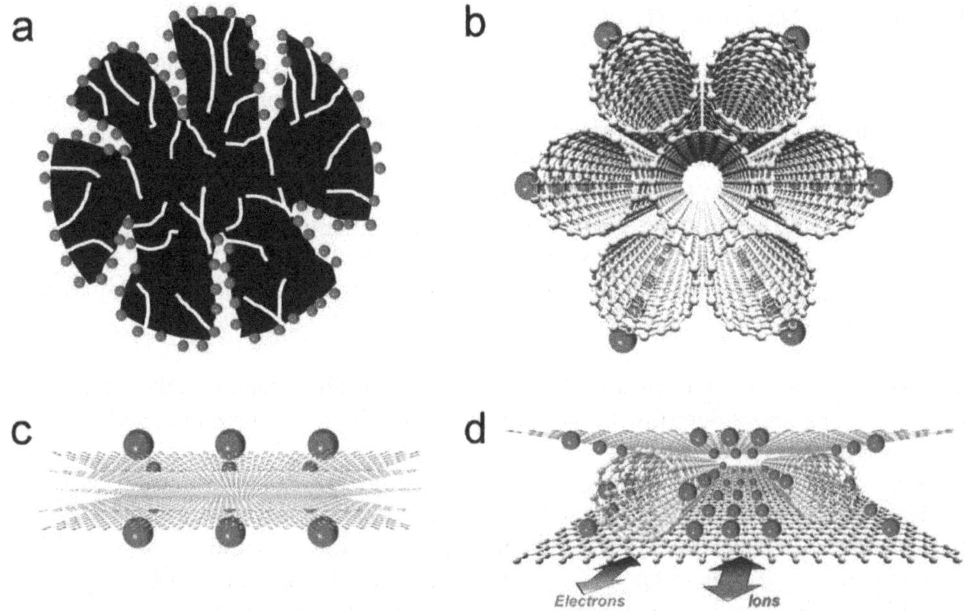

FIGURE 8.4 Comparison of different carbon materials as an electrode for supercapacitors. (a) activated carbon (b) single-walled carbon nanotubes (SWCNTs) bundles (c) pristine graphene (d) graphene/CNT composite. Reprinted with permission from [18]. Copyright 2011, The Royal Society of Chemistry.

8.4 THE DIFFERENT FORMS OF CARBON

In the case of EDLCs, carbon materials are the first choice owing to their larger surface area, better electrical conductivity, and extended electrochemical stability. As shown in Figure 8.4, to date, activated carbon (AC), single-walled carbon nanotubes (SWCNTs), graphene sheets, and graphene/CNT composites are widely used as carbon-based materials for SCs applications. The AC was widely used as electrode material in industries for many years because of its low cost, high surface area, and high packing density [16]. The pore structure and surface area of AC can be controlled by using various chemical activation methods [17]. However, in AC lot of carbon atoms cannot be accessed by the electrolyte ions, which limits their specific capacitance. Additionally, the lower electrical conductivity of AC results in low specific capacitance per area, which restricts its application in high power density applications [18]. On the other hand, CNTs and graphene with exohedral pores allowed the easy transport of ions, and the quick transfer of electrons is mainly attributed to the intrinsic covalent bonds of sp^2 hybridized carbon atoms. The covalent bonds of sp^2 hybridized carbon atoms in CNTs and graphene make it highly electrical conductive with extremely high mechanical strength. Additionally, sp^2 hybridized carbon exhibited higher crystallinity than the sp^3 hybridized carbon (AC), allowing its stable use in the high voltage system. This chapter exclusively covers the detailed discussion on the use of CNTs and graphene for SCs application based on a literature survey.

A graph showing the possibilities of charge storage in different types of carbon materials used in supercapacitors (a) activated carbon (b) single-walled carbon nanotubes (SWCNTs) (c) pristine graphene and (d) Graphene/CNT composite.

8.4.1 Carbon Nanotubes (CNTs)

Since the discovery of CNTs by Iijima in 1991, it has attracted considerable attention in various research areas including EES [19]. CNTs can be viewed as a graphene sheet rolled up into a nanoscale tube to form so-called single-wall CNTs (SWCNTs) (Figure 8.5(a)). The additional one or more graphene tubes around a core of SWCNTs form the double-wall CNT (DWCNTs) or multi-walled CNTs (MWCNTs) (Figure 8.5(b–c)). The diameter of CNTs ranges from a few angstroms to tens of nanometers, with a length of several micrometers to centimeters [20]. The SWCNTs show a theoretical specific surface area value of 1315 m^2/g. However, in the case of MWCNTs, it would be lower, which can be determined by the diameter of the tubes and the number of graphene walls [21].

Synthesis of CNTs: Until now plenty of methods have been developed to synthesize CNTs such as chemical vapor deposition (CVD), arc discharge, pulsed laser vaporization, and pyrolysis of hydrocarbons or carbon monoxide.

Chemical vapor deposition (CVD): The CVD process ensures the synthesis of SWCNTs, however, poses some challenges such as selection of catalyst and growth conditions. Some researchers have proposed the use of high melting point alloy nanoparticles as a catalyst for SWCNT synthesis. Due to its high melting point, the nanoparticles remain in a solid state and ensure crystalline growth during the CVD process [23]. The next challenge is the synthesis of selective chiral chemistry and diameter control for CNTs to have better applicability. The electronic properties of the SWCNTs are sensitive to the structure; even a slight change in the chirality will result in a change in the band structure. Enormous efforts have been put to synthesize the SWCNTs with the selective chiral structure to tune the electronic properties. One of the many methods is to use

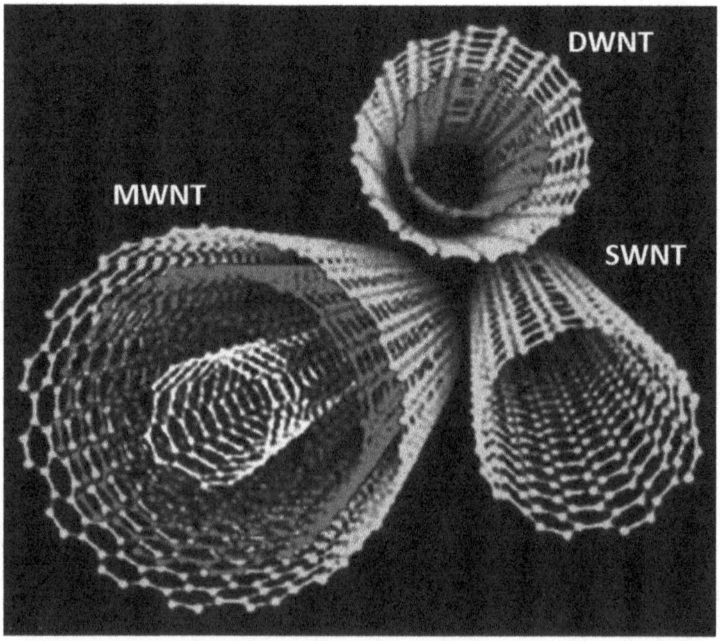

FIGURE 8.5 (a) SWCNTs with different helicities, (b) DWCNTs and (c) MWCNTs. Reprinted from [22] with permission of Taylor & Francis Online.

well-controlled and high-temperature stabilized catalyst nanoparticles. Having all these methods like selective etching, laser ablation, flame synthesis, pyrolysis, electrolysis, electron or ion beam irradiation, template methods, arc discharge[19, 24–26], and CVD in hand the large-scale synthesis of controlled structure of CNTs remains a challenge for many applications. The method named floating catalyst chemical vapor deposition (FCCVD) has been in use for continuous, high-yielding SWCNTs production while maintaining intrinsic electronic properties and reducing the processing time [27].

Arc discharge: The arc discharge method is used for the synthesis of CNTs having fewer structural defects in comparison with other routes. The arc discharge can produce temperature well above 2000°C which help to eliminate the structural defects. The synthesis of CNTs by using arc discharge is very simple, efficient, and timesaving. The DC arc discharge method is widely used and utilizes the two graphitic electrodes usually water-cooled to produce a CNTs chamber filled with He gas below the atmospheric pressure. The arc produced in between anode and cathode consumes the electrodes as a carbon source. Apart from He gas, some reports use hydrogen or methane to produce the CNTs. The purity and yield of the CNTs are sensitive to the gas pressure in the reaction chamber. Also, the CNTs produced by using high-pressure methane gas and high arc current produce thick nanotubes along with many nanoparticles.

Hata and Futaba et al. have demonstrated the water-assisted CVD for the synthesis of highly pure CNTs (99.98%) with 2.5 mm of height over the substrate in 10 min. of time [28]. The well-aligned, highly pure CNTs have been synthesized by selective etching of carbon atoms with water assistance. The etching process occurs at CNT caps as well as at the interface between CNTs and a metal catalyst. The layer-by-layer CNTs have been grown over the silicon substrate using iron as a catalyst by a CVD method. The selective etching by water helps to remove the catalyst from the cap and opens the ends of the CNTs [29].

8.4.2 Graphene

Graphene is nothing but graphite-derived monolayer of carbon atoms with atomic thickness, which displays outstanding electrical conductivity, thermal conductivity, strong mechanical strength, and chemical stability. Another advantage of graphene is the easy modification of its structure by adding different functional groups to induce a wide array of electrical and mechanical properties. Further, graphene allows to the creation of flexible structures (0D, 1D, 2D, and 3D), providing a fine-tuned surface area for enhancing the desired structure as shown in Figure 8.6. Some other different examples include graphene foam, graphene films, yarn-like graphene, and graphene quantum dots. With these several superior properties, it would be better to construct a supercapacitor to achieve high power density and long-life cycle performance [30, 31]. Some chemical and physical techniques for the synthesis of graphene have been developed as discussed in the following section.

Synthesis of graphene: Similar to CNTs there are several methods adopted for the synthesis of graphene. The physicochemical properties of graphene highly depend on the preparation technique. In general, graphene can be synthesized in two ways, the first way involves the synthesis of graphene sheets by dismantling from graphite, and the second involves the synthesis of carbon from other sources such as CNTs [32]. However, in the case of dismantling from graphite synthesis of high-quality graphene is a key challenge. Because if the sheets of graphene are not separated adequately from each other, then these sheets will endure irreversible accumulation to restock through Van der Waals interactions which will result in the creation of a 3D graphite

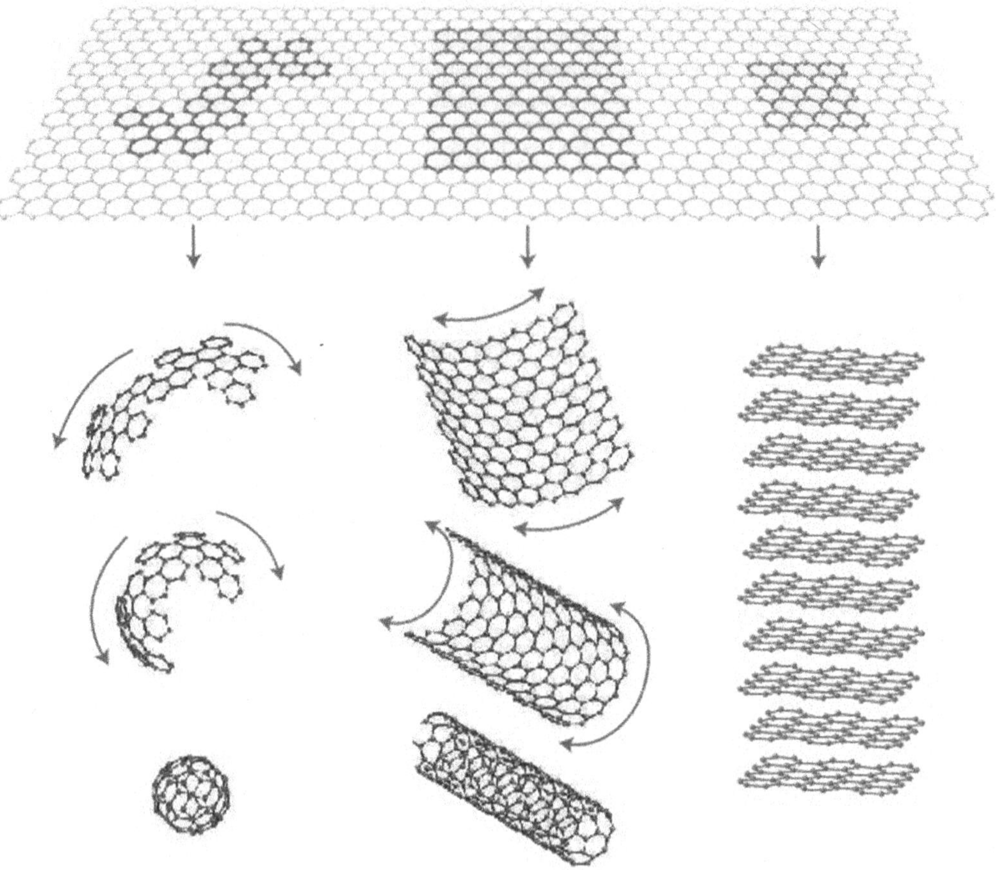

FIGURE 8.6 Schematic diagrams of graphene as the backbone for various graphitic materials. Reprinted from [30]. Copyright 2007, Nature Publishing Group.

structure [33]. Therefore, formation of graphene with individual single sheets is a high priority. Following are some well-known effective synthesis techniques for graphene.

Mechanical exfoliation: In which an adhesive scotch tape was used for the stick and pills process several times on the commercially available highly oriented pyrolytic graphite (HOPG). This method produces a monolayer thin graphene sample exfoliated from 1 μm-thick graphite flakes. However, during this process contamination of graphene from glue residues of scotch tape limits the mobility of its electric charge. The post-heating treatment is useful to remove the residues of scotch tape from the obtained graphene sheets [34].

Chemical vapor deposition: This is a well-established synthesis approach used for the growth of monolayer or a few layers of graphene on a large scale [35]. The CVD offers graphene with better properties, such as monolayer structure, large crystal domains, and fewer defective sheets, which enhances its carrier mobility [36]. CVD is a chemical process in which gaseous reactants are allowed to deposit on a substrate. For example, gaseous ethylene can be deposited on a hot Ir surface to produce graphene sheets with superior morphological properties [37]. Similarly, epitaxial growth of crystalline monolayer graphene on Ruthenium substrate was demonstrated by

Sutter et al. [38]. 3D graphene network with a large surface area could be synthesized by using Ni foam as a template and ethanol as a carbon source, which allows rapid access to electrolytes [39].

Unrolling of CNTs: The CNTs can be unrolled and exfoliated to produce graphene. In the electrostatic synthesis approach, Li ions and NH_3 solvent were intercalated within the layers of MWCNTs through electrostatic attraction between the positive Li-ions and negative MWCNTs, resulting in the formation of ruptured MWCNTs which can be further exfoliated by acid and heat treatment [40].

Liquid phase exfoliation: The solvent-based exfoliation can produce graphene at a large scale efficiently. With the use of appropriate surfactants graphite can be exfoliated in water or organic solvents. For example, graphene oxide formed from graphite oxide in water solvent possesses poor electrical conductivity, because of disruption in the π orbital structure during oxidation [41]. This limitation can be resolved by exfoliation of graphite in an organic solvent instead of graphite oxide. For example, the similar surface energies of nanotubes and graphene suggest that the exfoliation of graphite to graphene could be achieved by using the organic solvent *N*-methyl-2-pyrrolidone (NMP) [42].

Thermal reduction: The thermal reduction involves acid and heat treatment, where sufficient oxidation is needed during the acid treatment while an appropriate (greater than van der Waals forces exist between graphite or graphene oxide layers) pressure is required during heat treatment. As a result of oxidation, the formed functional group (epoxide and hydroxyl) decomposes and builds up pressure between the graphite layers to separate them into graphene sheets [43].

Chemical reduction: Chemical reduction also involves redox reactions to produce graphene sheets from graphite, where firstly graphite can be oxidized to graphite oxide and then exfoliated into graphene oxide sheets followed by its reduction to graphene [44]. The selection of an appropriate amount of reducing agent, solvent, and surfactant is important to get graphene of good quality. The hydrazine hydrate is a well-known reducing agent, however, its toxic nature and re-amalgamation of separated graphene sheets forced researchers to find a non-hazardous way to reduce graphene oxide to graphene. In this regard, S. Perera et al. reported a process to reduce graphene oxide using an alkaline hydrothermal technique which exhibits similar properties to graphene produced via the hydrazine reduction route [44].

8.5 CARBON NANOTUBES-BASED SUPERCAPACITOR

The 1D CNTs with a high surface area have been widely used for supercapacitors because of their unique physical and chemical properties [45]. The CNTs synthesized via different synthesis methods in different forms such as compressible foam [46], stretchable non-woven cloth [47], vertically aligned CNT (VACNT film) [48], and aligned network of CNTs [49], with high conductivity and the large specific area, can be directly used as electrodes for EDLCs. The capacitance of bare CNTs can be enhanced by surface functionalization, doping heteroatoms, or making composites with other carbon sources. Additionally, the large aspect ratio and ease of surface functionalization make CNTs a highly apposite material to chemically integrate various nanomaterials to achieve high capacitance and long-term life. This section includes a comprehensive view of the recent progress in CNTs and modified CNTs-based supercapacitors.

It is well known that the poor mesoporosity of activated carbons is the main bottleneck to achieving high capacitance. This issue can be solved by using SWCNTs or MWCNTs. The unique mesoporous nature of CNTs allows easy penetration of electrolytes, i.e., more electrolyte penetration may expose more surfaces to achieve capacitance. In general, CNTs exhibit lower ESR than bare CNTs, the reported capacitances ranged from 5 to 200 F/g having a surface area in the range of 120 to 500 m^2/g. In 1997, for the first time, Niu et al. tested nitric-acid-functionalized CNTs for supercapacitor application which displays a specific capacitance value of 102 and 49 F/g at 1 and 100 Hz in 38 wt% H_2SO_4 electrolyte [50]. Similarly, An et al. investigated the capacitive performance of DC arc discharge-derived SWCNTs, which deliver a maximum specific capacitance of 180 F/g and measured power density of 20 kW/kg at energy densities in the range from 7 to 6.5 Wh/kg at 0.9 V in 7.5 N KOH electrolyte solution [51]. This group claimed that the enhanced capacitance was mainly ascribed to the increased specific surface area and lower pore size distribution in the range of 30–50 Å. The applicability of MWCNTs synthesized by different methods such as catalytic decomposition of acetylene, CVD for supercapacitor, and its correlation with microstructure and elemental composition of the material was investigated by Frackowiak et al. Depending on the type of CNTs and/or their post-treatments, the specific capacitance was varied from 4 to 135 F/g [52]. The higher surface area does not always guarantee to delivery of higher capacitance because capacitive performance also depends on the pore size distribution and conductivity of electrode material. Therefore, higher capacitance can be achieved by optimizing all these factors. For example, vertically aligned CNTs with a specific area of 69.5 m^2/g exhibit better specific capacitance of 14.1 F/g with excellent rate capability than those of entangled CNTs, mainly because of its larger pore size with regular pore structure and more conductive paths [53].

The chemical activation [51], functionalization, and heat/surface treatment [54] of CNTs can boost their specific capacitance. For example, after the chemical activation of CNTs using nitric acid, the value of capacitance increases drastically due to increased functional groups on the surface of CNTs. Also, chemical KOH activation is the best approach to increase supercapacitor performance by developing beneficial microporous nature in CNTs [55]. The hydrophilicity of MWCNTs was enhanced by introducing surface carboxyl groups to get a 3.2 times larger capacitance in an aqueous electrolyte [56]. The capacitance of functionalized CNTs(c-MWNT) showed capacitance of 51.3 F/g than that of bare CNTs (16.1 F/g). On the other hand, the longer functionalization of CNTs with alkyl groups increases its hydrophobicity and blocks the proton access to CNTs, resulting in the complete disappearance of capacitance (Figure 8.7(a–b)).

Furthermore, heteroatom (N, S, P, B) doping in CNTs is the best strategy to enhance capacitance, surface wettability, and electronic conductivity. These heteroatoms can be introduced in CNTs either with post-modification or *in-situ* functionalization [57, 58]. In this regard, Sevilla et al. show the successful synthesis of N-doped carbon-coated CNTs via hydrothermal carbonization in the presence of N-containing carbohydrates. The N-doped carbon-coated CNTs exhibited a 2- to 4-fold increase in capacitance (~50–60 F/g) as compared to bare CNTs (~20 F/g) and only 4–6% loss of capacitance after 10000 cycles at 10 A/g [59]. Similarly, the effectiveness of S-doping in CNTs was investigated in detail by Kim et al. [60]. The S-doped CNTs (S-CNTs) were successfully synthesized via chemical vapor deposition using dimethyl disulfide as the carbon source. The purified S-CNTs(P-S-CNTs) showed the highest specific capacitance of 120.2 F/g at a high discharge current density of 100 mA/cm^2, which is superior to those of MWCNTs (87 F/g), super-P (83.9 F/g), and De-P-S-CNTs (90 F/g). The effect of co-doping on capacitive activity was investigated by Paul et al. [61]. The spray pyrolysis techniques were used to synthesize nanoporous brush-like CNTs grown on individual carbon fiber of a carbon cloth (CC) substrate with simultaneous co-doping with B and N heteroatom (BNCNT-CC) as shown in Figure 8.7(c). The flexible solid-state symmetrical supercapacitor constructed with BNCNT-CC electrodes demonstrated high aerial capacitance of 106.8 mF/cm2 (~21.4 F/cm^3), with attractive

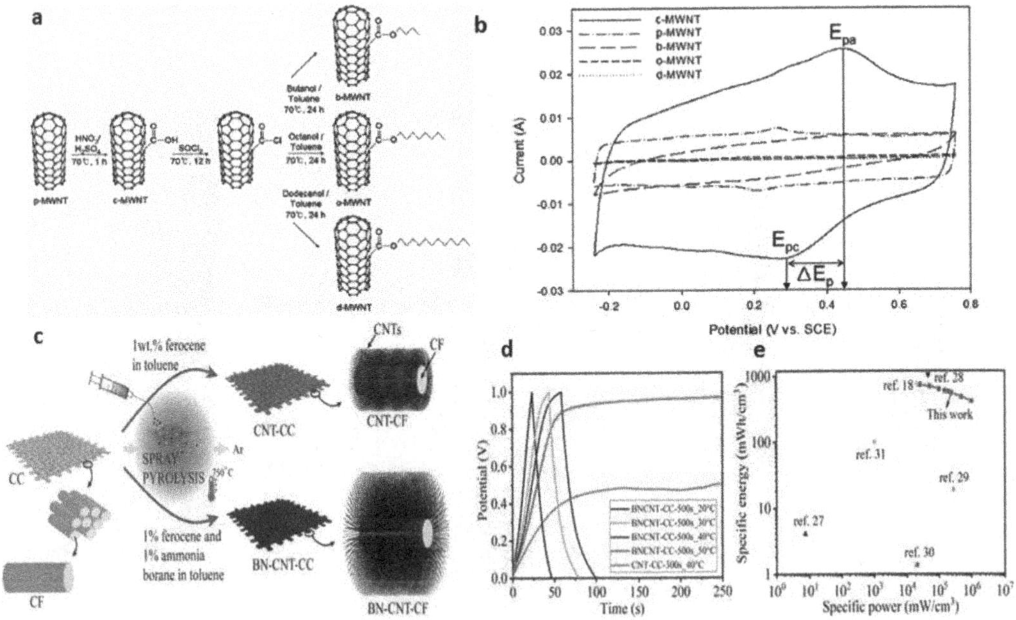

FIGURE 8.7 (a) Schematic diagram of synthetic route for surface functionalization. (b) CV curves of various functionalized MWNTs recorded in H_2SO_4 aqueous electrolyte with a scan rate of 100 mV/s [56]. (c) Synthesis process of radially grown CNTs and BN-CNTs on carbon fiber (CF) of carbon cloth (CC) with a brush-like morphology using one-step spray pyrolysis inside a horizontal quartz tube. (d) Temperature-dependent charge-discharge curves for BNCNT-CC-500s and CNT-CC-300s based symmetric supercapacitor. (e) Ragone plot of the specific energy vs. specific power, comparing some other reported supercapacitors. Reprinted from [61] with permission of Elsevier.

capacitance retention (86.4%) after 5000 charge-discharge cycles, specific energy of 741.8 mWh/cm^3 at 25 W/cm^3 specific power at room temperature (Figure 8.7(d–e)).

Yarn or fiber-based supercapacitors proved their applicability over conventional 3D or 2D energy devices for powering wearable electronics. The cylindrical geometry and electrical conductivity of yarns warrant the easy fabrication of devices through the direct process and their incorporation into wearable devices/parts. The fabrication of a CNT yarn (CNT-yarn) based flexible all-solid-state coaxial-type supercapacitor was reported by Jha et al. [62]. The meticulous thermal engineering of the electrode-electrolyte interface introduces selective chemical functionalities across the interface, which results in accessing a new domain of ultrafast scan rate capability with solid-state electrolyte (Figure 8.8(a)). The supercapacitor device was operable at a scan rate (25000 mV/S), exhibiting high energy (6.2 mWh/cm^3) and power density (4465 mW/cm^3) (Figure 8.8(b)). The capacitive performance retention of the device was tested by its seamless integration through sewing on nitrile gloves (Figure 8.8(c)).

8.6 GRAPHENE-BASED SUPERCAPACITOR

Graphene-based supercapacitor electrodes exhibit moderate specific capacitance of (~200 F/g) due to stacking, aggregation, and poor wetting capability between graphene and electrolyte [63]. Therefore, in the recent research approaches, the use of modified graphene for supercapacitors

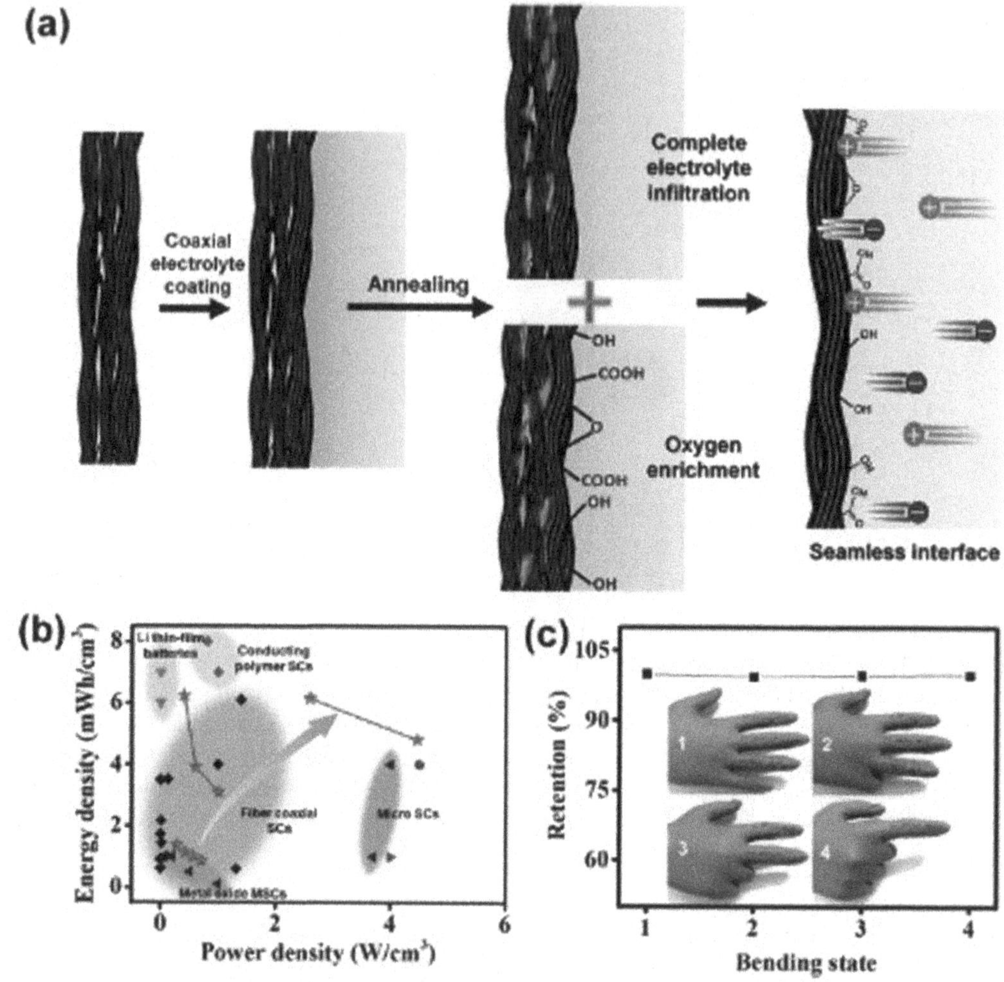

FIGURE 8.8 (a) Schematic representation of the mechanism for obtaining ultra- high scan rate operability of CNT coaxial supercapacitor by thermal engineering the electrode- electrolyte interface. (b) Ragone plot, comparing the energy and power density of this report with other literature reports. Points marked in red asterisk are values obtained in this work. (c) Retention of the capacitance of CCS with gradual bending. Inset shows pictures of the integration of CCS onto a nitrile glove and its subsequent mechanical flexing at different bending states.Reprinted from [62].Copyright {2020} American Chemical Society.

has been explored widely. This modifications strategy includes the idea of the use of exfoliated graphene, corrugated graphene sheets, preparation of thin layer graphene sheets, and doping of heteroatoms, which can help to improve capacitance by increasing specific surface area and reducing graphene aggregation [64]. For example, Li et al. synthesized few-layer graphene (FLG) via intercalation of graphite and pre-graphite oxide followed by reduction of graphite oxide. The as-prepared FLG electrode exhibits a specific surface area of 1400 m^2/g and capacitance of 180 F/g in 1M Na_2SO_4 electrolyte [65]. In another study, the use of highly corrugated graphene sheets (HCGS) to prevent aggregation of graphene sheets was demonstrated [66]. These corrugated graphene sheets were synthesized via simple, cheap, efficient, and scalable thermal expansion

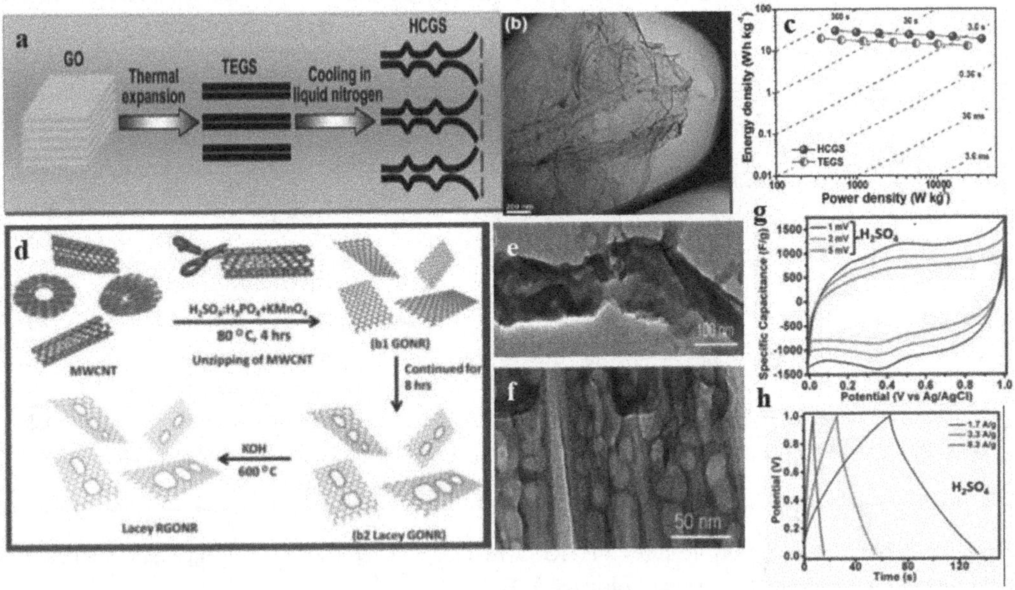

FIGURE 8.9 (a) Schematic illustration for the possible formation of HCGS. (b) TEM image of asprepared HCGS. (c) Ragone plot related to energy and power densities of the HCGs and TEGS electrode. Reprinted from [66] with permission of Elsevier (d) Schematic for the formation of LRGONR. (e) TEM image of MWCNT showing the initial stage of nanotube opening in axial direction. (f) LRGONR showing the holes (10 – 50 nm) in individual graphene sheet. (g) CV of LRGONR in 2M H_2SO_4 (3 electrode cell). (h) LRGONR galvanic charge- discharge (2 electrode cell). Reprinted from [67]. Copyright {2015} American Chemical Society.

and subsequent nitrogen cooling method (Figure 8.9(a–b)). The corrugated graphene sheets with a large surface area showed capacitance of 349 F/g in 6M KOH electrolytes. The HCGS exhibits a higher energy density of 30.4 Wh/kg at a power density of 547 W/kg, which is superior to thermally expanded graphene sheets (TEGS) (19.9 Wh/kg) (Figure 8.9(c)). Additionally, graphene nanoribbons (GNRs) have been widely explored owing to their beneficial properties such as edge-dependent electronic properties, and variation in band-gap due to electron confinement. The synthesis of lacey reduced graphene oxide nanoribbons (LRGONR) through chemically unzipping the MWCNTs using a strong oxidizing agent was reported by Sahu et al., as shown in Figure 8.9(d) [67]. In the initial stage of MWCNT opening, chemical moiety (MnO_4-) attacks in axial direction along the bond strain and opens it layer by layer (Figure 8.9(e)). The chemical moiety (MnO_4-) enters between two GONR layers separating them as a destacked LGONR with random hole formation (Figure 8.9(f)). While LRGONR was tested for performance, CV showed mixed EDLC and pseudocapacitive behavior with high capacitance of 1042, 817, and 621 F/g at scan rates of 1, 2, and 5 mV/s in three-electrode cells, respectively (Figure 8.9(g)). The two-electrode charge-discharge characteristics curves of LRGONR showed identical nature with a steady performance of 97% capacity holding over 3000 cycles at current density of 8.3 A/g (Figure 8.9(h)) [67].

Holey graphene (HG) based nanomaterials offer several structural virtues as compared to graphene nanosheets (GNs) for supercapacitor application. In HG, holes not only distort GNs in geometry but also weaken the van der Waals force between GNs. The several advantages of HG over GNs are summarized in Figure 8.10 [68].

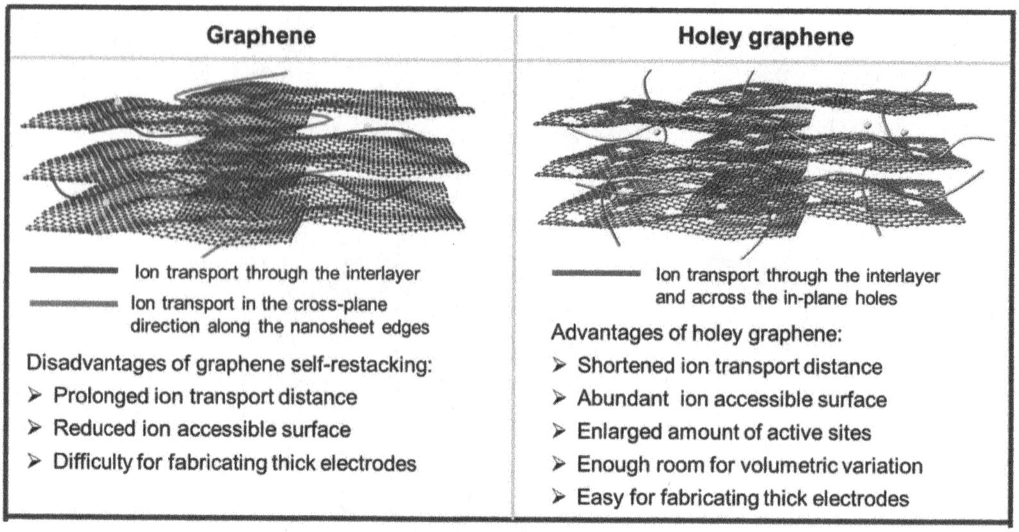

FIGURE 8.10 Advantageous features of HG- based nanomaterials for supercapacitor applications. Reprinted from [68] with permission of Elsevier.

The doping of heteroatoms (e.g., N, S, B, P) into graphitic structure have been attracted as an efficient strategy for improving electrochemical performance for supercapacitor because of the improved hydrophilicity between electrode/electrolyte and enhanced conductivity of materials. For example, N-doped HG (NHG) was synthesized via the wet-chemical etching method using GO as a graphene source and urea as a nitrogen source (Figure 8.11(a–b)). The NHG with a surface area of 1216 m^2/g exhibited a higher specific capacitance of 343 F/g at 0.3 A/g than N-graphene (NG, 296 F/g at 0.3 A/g) with a surface area of 630 m^2/g (Figure 8.11(c)) [69]. The enhanced capacitance was mainly associated with N-doping in graphitic structure and the formation of nanoholes in GNs, which increases the number of active sites and facilitated the diffusion of electrolyte ions. In recent times, co-doping of heteroatoms in graphitic structures has also attracted much attention due to its synergistic effect on electrochemical performance. The N-doping can enhance performance by accelerating the charge mobility of negative charges on carbon surface and fluorination can increase electrical conductivity, refine pore structure as well as enhance polarization from the highly electronegative F functional groups. Chen et al. reported the synthesis of N, F co-doped HG hydrogel via hydrothermal route using GO and H_2O_2 to form HG and pyridine hydro fluoride as a source of N and F (Figure 8.11(d)). The as-synthesized N, F-doped HG hydrogel showed a specific capacitance of 345.4 and 197.9 F/g at 1 A/g and 100 A/g, respectively [70]. While, a symmetric solid-state device using PVA-KOH gel electrolyte and N, F-doped HG displays a maximum energy density of 7.99 Wh/kg and power density of 10.08 kW/kg with longer cycle life (Figure 8.11(e–f)). Further, N, S, and P tri-doped graphene synthesized by hydrothermal reaction delivered high gravimetric capacitance of 295 F/g at 1 A/g, excellent rate capability, and outstanding cycling stability [71].

A graph shows synthesis route of doped holey graphene and their electrochemical performance (a) schematic for the synthesis of N-doped holey graphene (NHG), (b) TEM image of N-HG, (c) comparison of specific capacitance of N-graphene (NG) and NHG under various current densities, (d) schematic illustration of the synthesis of N, F-doped holey graphene hydrogel (NF-HGH), (e) Ragone plots with a comparison with other symmetric flexible solid-state supercapacitors, (f) the cycle stability after 10–000 cycles at a current density of 2 A/g (inset shows the LED illuminated by the devices connected in series).

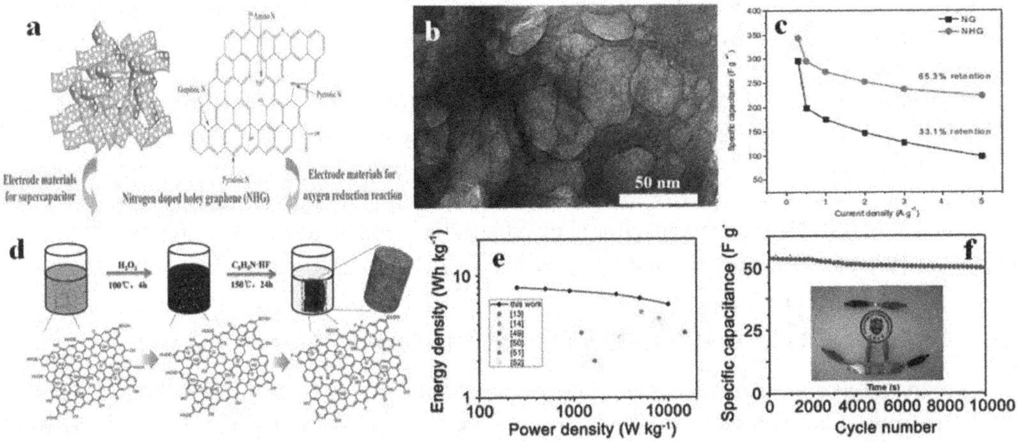

FIGURE 8.11 (a) Schematic for the synthesis of N-doped holey graphene (NHG). (b) TEM image of N-HG. (c) Comparison of specific capacitance of NG and NHG under various current densities. Reprinted from [69] with permission of Elsevier. (d) Schematic illustration of the synthesis of N, F-doped holey graphene hydrogel (NF-HGH). (e) Ragone plots with a comparison with other symmetric flexible solid-state supercapacitors. (f) The cycle stability after 10000 cycles at a current density of 2 A/g (inset shows the LED illuminated by the devices connected in series). Reprinted from [69]. Copyright 2019, The Royal Society of Chemistry.

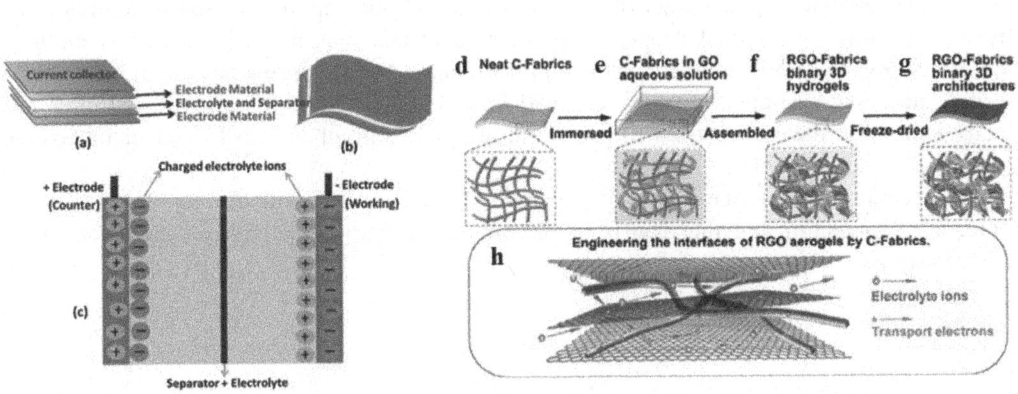

FIGURE 8.12 (a) Schematic illustration of the conventional supercapacitor. (b) Schematic illustration of the typical flexible supercapacitor. (c) Schematic representation of the two-electrode supercapacitor system. Reprinted with permission from [72]. Copyright {2020}, American Chemical Society. (d-g) Fabrication of the binary 3D architecture. (h) Engineering the interface of rGO aerogel with Cfabrics, showing effective charge diffusion and transfer in the binary 3D architectures.Reprinted with permission from [73]. Copyright {2015}, American Chemical Society.

8.7 GRAPHENE-BASED FLEXIBLE SUPERCAPACITORS

In recent times, the development of flexible supercapacitor devices is in high demand. There are several challenges such as providing enough flexibility with low device weight, low cost, and high performance to meet the current demand for electronic devices for consumer and industrial applications. In this regard, as an electrode material graphene is one of the best choices for the production of flexible supercapacitors due to its high surface area, superior thermal characteristics, and mechanical properties (Figure 8.12(a–c)). The use of rGO in graphene-based devices is of foremost choice to overcome the current challenge of low electrochemical stability.

The synthesis of the 3D porous architecture of rGO aerogel and carbon fabrics (RGO fabrics) as an electrode for flexible energy storage devices. The synthesized 3D architecture exhibited a high surface area (1111 m^2/g) as compared to bare rGO (without carbon fabrics), which improves its micropore size and channels for ionic diffusion and electronic transportation (Figure 8.12(d–h)). When it is tested for supercapacitor showed specific capacitance of 391 and 195 F/g at 0.1 and 5 A/g, respectively with outstanding cycle life [73].

8.8 CARBONOUS COMPOSITES FOR SUPERCAPACITOR

For carbonaceous supercapacitor materials, carbon nanotubes (CNTs) and graphene are promising electrode materials due to their excellent electrical conductivity and efficient electron/ion diffusion that will enhance the effective ion accessible surface area [74, 75]. Even though graphene and CNTs show better performance as supercapacitors, there are some drawbacks of pristine graphene and CNTs such as a limited surface area for the formation of double-layer and poor electrical conductivity. The hybrid graphene–CNT composites were demonstrated to be favorable for various applications by combining both strategic properties of graphene and CNTs [76, 77]. Figure 8.13 shows quick diffusion pathways for the electrolyte ions where CNTs can serve as a spacer between the graphene nanosheets. Also, they can improve electrical conduction for the electrons. CNTs also used as a binder to prevent the collapsed structure of the graphene into the electrolyte and also show more highly opened structures [18]. Given outstanding individual properties of these carbonaceous materials provides mutualistic effects to improve the supercapacitor performance. Many approaches, including hydrothermal processes, chemical vapor deposition, cathodic electrophoretic deposition, and electrodeposition have been tried to prepare high-quality graphene/CNTs and metal oxides hybrid electrodes.

In this section, we describe the use of graphene/CNT composites with different types of metal oxides and conducting polymer matrix as the efficient electrode material for electrochemical supercapacitors.

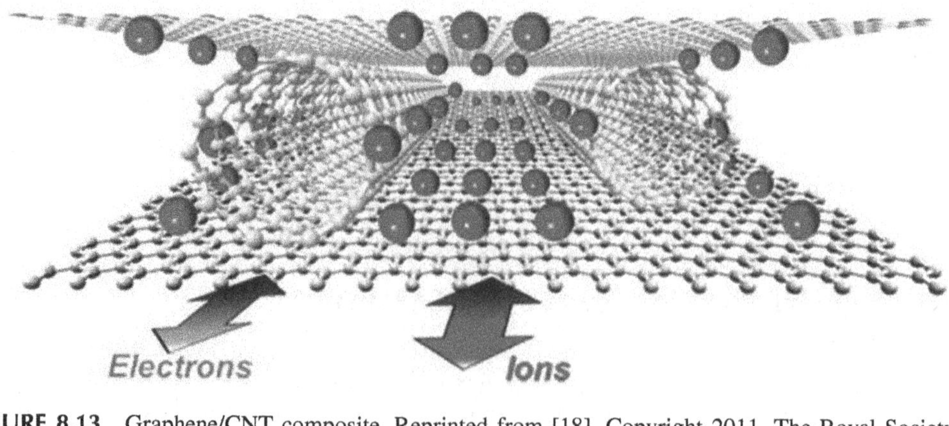

FIGURE 8.13 Graphene/CNT composite. Reprinted from [18]. Copyright 2011, The Royal Society of Chemistry.

8.8.1 Graphene/CNT-Based Supercapacitors

Due to the mutualistic effect, the conductive graphene-CNTs network such as it provides channels for electron transport and reduces the influence of defects in the graphene can enhance the performance of the electrode.

Mao et al. reported a (CNT)/CGB hybrid by direct growth of CNTs on the crumpled graphene ball surface which bridges the active material and the current collector to improve the affinity for

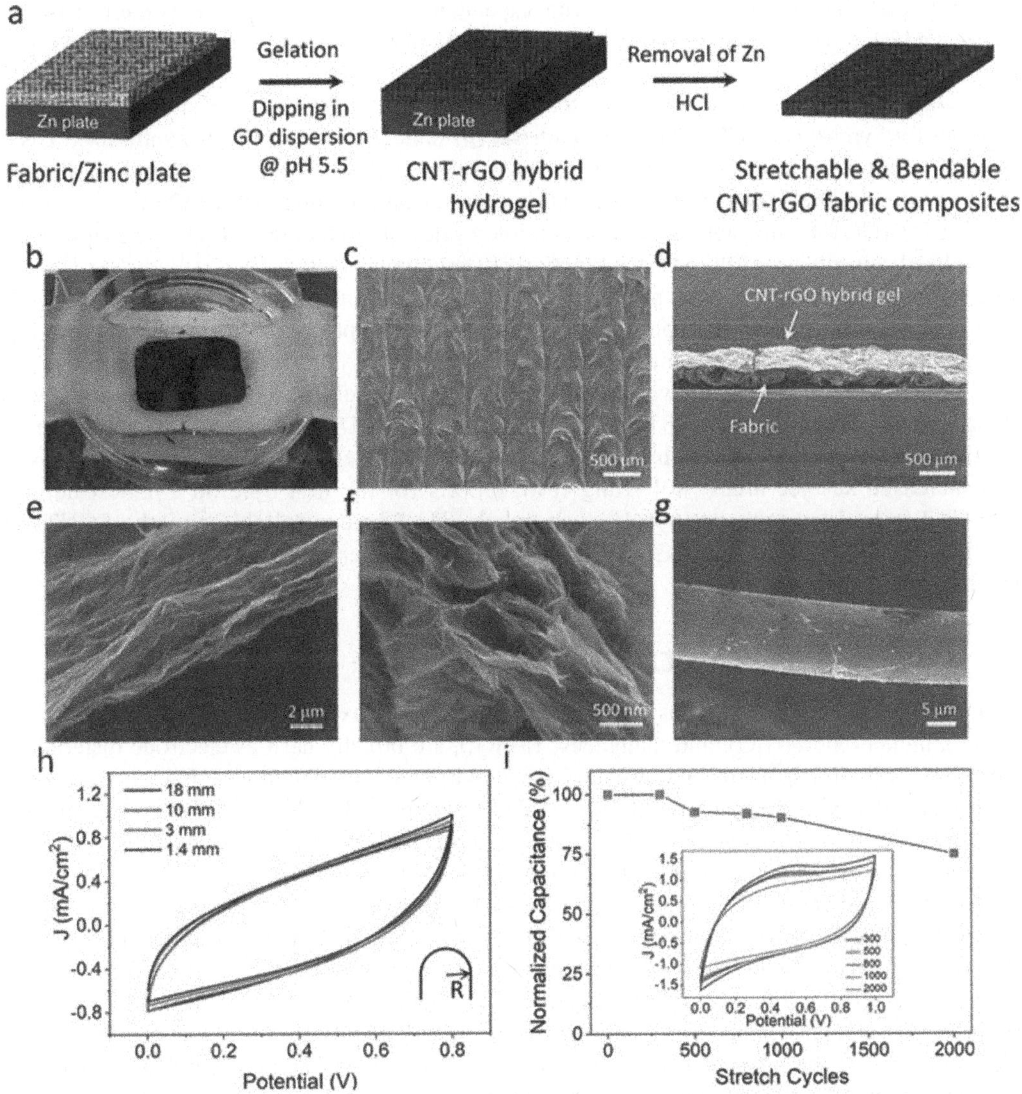

FIGURE 8.14 (a) Schematic diagram for the synthesis of CNT-rGO@F synthesis. (b) Optical image of the as-synthesized CNT-rGO@F composite. (c-g) SEM images of the freeze-dried CNT-rGO@F composite. (h) CV measurements of the CNT-rGO@F-based SC bent to various bending radii. (i) CV measurements of the CNT-rGO@F supercapacitor after multiple cycles of stretching and releasing (inset) and normalized capacitance of the SC as a function of stretching cycles. Reprinted from [79]. Copyright {2021}, American Chemical Society.

supercapacitor applications [78]. The CNT porous networks and 3D CGB both synergistically improve specific capacitance and stability in supercapacitors. Kang et al. demonstrate a flexible and stretchable supercapacitor with a combination of reduced graphene oxide (rGO) and CNT [79]. Figure 8.14(a) shows the schematic representation of the synthesis of the CNT-rGO@F composite. Figure 8.14(b–g) shows the optical and SEM images of CNT-rGO@F. The synergistic effect of porous structure and one-dimensional structure of the CNT-rGO framework and CNT with a large surface area suitable for stretchable energy storage applications. CNT-rGO@F SC exhibits superior capacitance retention with different bending radii. Capacitive performance remains nearly unchanged on folding with capacitance retention of 95.2% (Figure 8.14(h)). CNT–rGO@F stretchable supercapacitor showed capacitance of 10.13 mF/cm^2 with good retention (Figure 8.14(i)).

A graph showing the schematics for the synthesis of CNT/RGO composite. (a) Schematic diagram for synthesis of CNT-rGO@F synthesis (b) optical image of the as-synthesized CNT-rGO@F composite (c-g) SEM images of the freeze-dried CNT-rGO@F composite (h) CV measurements of the CNT-rGO@F-based SC bent to various bending radii (i) CV measurements of the CNT-rGO@F supercapacitor after multiple cycles of stretching and releasing (inset) and normalized capacitance of the SC as a function of stretching cycles.

Cheng et al. describes a graphene and SWCNT composite electrode for use in supercapacitors by a blending process [18]. The obtained specific capacitance is 290.4 Fg^{-1} for graphene/CNT supercapacitors with a two-electrode system. Zang et al. synthesized MWCNT forests by CVD followed by electroplating of nickel onto CNTs. As compared with pristine VACNT capacitance of the CNT/Ni/CNT and CNT/Ni/Graphene increases significantly with 90% retention which is mainly attributed to the enhanced electrical conductivity and increased surface areas [80]. Yang et al. reports for the first time on a nanostructured GN-CNT hybrid by a one-pot pyrolysis approach [81]. The controllable growth of CNTs on GN in GN-CNT hybrids exhibits remarkable physical properties resulting in excellent electrochemical performances.

8.8.2 Metal Oxides@CNT/GN-Based Supercapacitors:

Transition metal oxides, such as iron oxide (Fe_2O_3), zinc oxide (ZnO), cobalt oxide (CoO, Co_3O_4), nickel oxide (NiO), and manganese (MnO_2), are broadly used as electrode materials in supercapacitors due to their natural abundance, environmental friendliness, low cost, and high capacitance [82, 83]. However, rapid capacity decline upon cycling is one of the most commonly known limitations of metal oxides. Fabricating hybrid metal oxides with conductive materials, such as CNT and GN is the most studied solution for this limitation.

Zhang et al. demonstrated a facile two-step synthesis approach to synthesize nitrogen-doped rGO/MWCNTs/manganese dioxide (NGC/MnO_2) for flexible solid-state supercapacitors [84]. Firstly, they have synthesized NGC fiber by incorporating MWCNTs and urea into graphene oxide (GO) aqueous by a convenient hydrothermal method. Secondly, to further enhance the properties, MnO_2 was coated on a surface of NGC fiber. The prepared MnO_2-coated NGC fiber electrode showed the highest specific capacitance of 367.7 F/cm^2 at a current density of 0.5 A/cm^3. The prepared supercapacitor showed outstanding mechanical flexibility and electrochemical stability, which suggests its potential application as a flexible energy storage device. Kong et al. designed a 3D binder-free RuO_2-GNS-CNT (RuO_2-GC) electrode by cathodic electrophoretic deposition and electrodeposition [85]. CFC (carbon fiber cloth) has been used as a supporter of GNS and CNT composites and the surface of GNS and CNT is anchored with RuO_2 nanoparticles through the electrodeposition process.

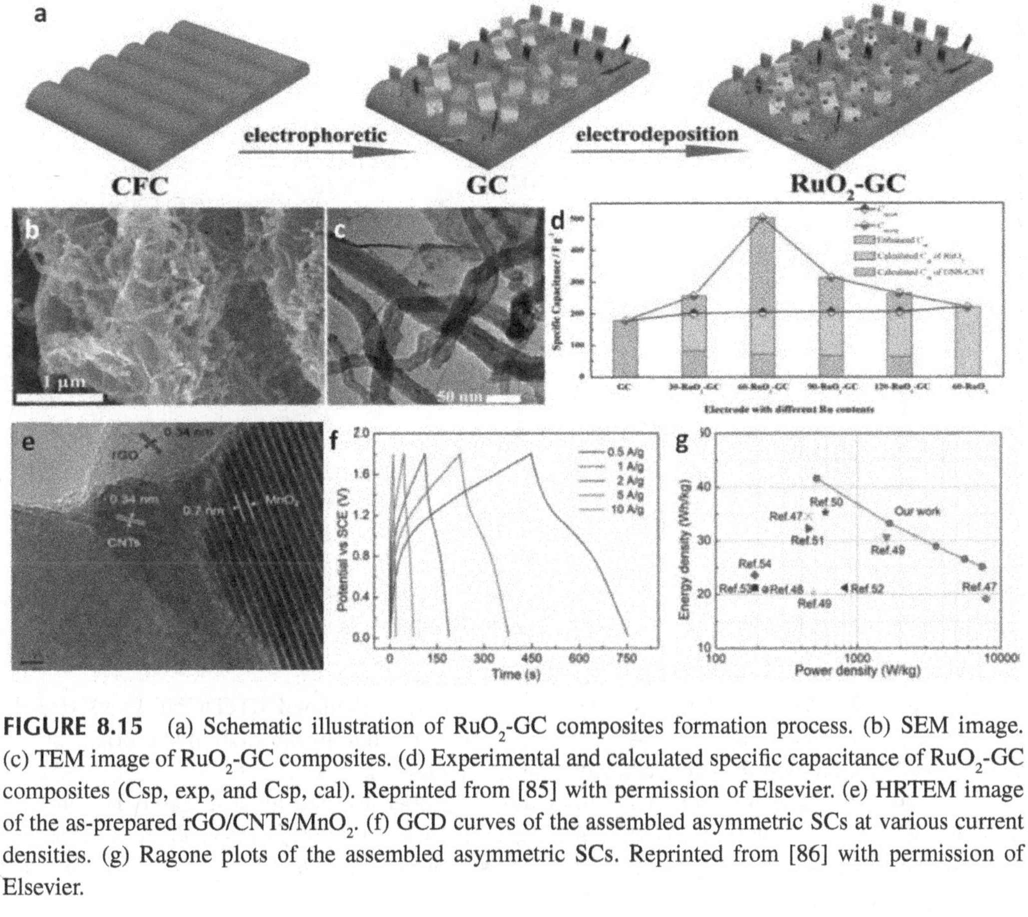

FIGURE 8.15 (a) Schematic illustration of RuO_2-GC composites formation process. (b) SEM image. (c) TEM image of RuO_2-GC composites. (d) Experimental and calculated specific capacitance of RuO_2-GC composites (Csp, exp, and Csp, cal). Reprinted from [85] with permission of Elsevier. (e) HRTEM image of the as-prepared rGO/CNTs/MnO_2. (f) GCD curves of the assembled asymmetric SCs at various current densities. (g) Ragone plots of the assembled asymmetric SCs. Reprinted from [86] with permission of Elsevier.

Figure 8.15(a) shows the synthesis process of the RuO_2-GC electrode on the surface of CFC. FESEM and TEM image (Figure 8.15(b–c)) shows that RuO_2 nanoparticles are homogeneously distributed on the surface of GNS and CNTs. Their results suggest that for a superior electrochemical performance, hybrid electrodes are well-suited. The plot of specific capacitance vs Ru mass loading clearly shows that the capacitance of RuO_2-GC electrodes is higher than the sum of the intrinsic capacitance of the pure GNS-CNT as EDLC and RuO_2 pseudocapacitors (Figure 8.15(d)). Similar work was carried out by tang et al. whereby combining EPD, CVD, and ECD methods hierarchical CF templated rGO/CNTs/MnO_2 electrode was successfully fabricated [86]. The MnO_2 nanoparticles were observed in the CNTs forest with discrete distribution (Figure 8.15(e)). This unique structure inherits the good pseudocapacitance of MnO_2 nanoparticles as well as excellent EDLC behavior of graphene, and 1D CNTs. The specific capacitance was found to be 54.4 F/g at 0.5 A/g and retain the value of 37.8 F/g@20 A/g (Figure 8.15(f)). A maximum energy density of 41.6 Wh/kg at a corresponding power density of 513.7 W/kg was delivered for CF-rGO/CNTs/MnO_2//AC (Figure 8.15(g)). Mohammadi et al. synthesized rGO-CNT-Co_3S_4 nanocomposite via a hydrothermal approach (Figure 8.16) [87].

Here, the synergetic effect among rGO, CNT, and Co_3S_4 components makes rGO, CNT, and Co_3S_4 nanocomposite a favorable electrode material for supercapacitor applications. Ramli et al. demonstrated CNT/GNF/Fe_2O_3 ternary composites via a simple hydrothermal route. The effect of

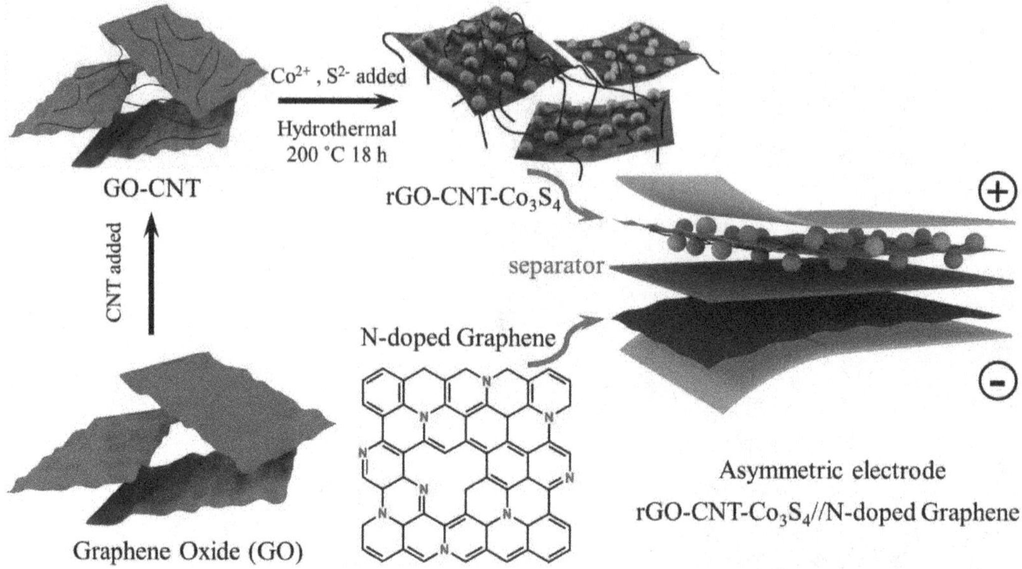

FIGURE 8.16 Fabrication process of asymmetric rGO-CNT-Co_3S_4//N-doped graphene supercapacitor electrode and the preparation process of rGO-CNT-Co_3S_4. Reprinted from [87] with permission of Elsevier.

Fe_2O_3 addition ratios was studied, and their capacitance was calculated. CNT/GNF/Fe_2O_3 reveals an enhanced Cs value of up to 307 F/g which is attributed due to the mutualistic effect of the EDLC of CNT/GNF and the pseudocapacitance of Fe_2O_3 [88].

In nutshell, all these materials have their unique properties, and most researchers combine these properties by synthesizing new hybrid materials. By combining the graphene, CNTs, and metal oxides; in this manner, the synergistic effects between individual components can be fully realized.

8.8.3 Conducting Polymer/Graphene/CNT-Based Supercapacitors

The composite materials of graphene, carbon nanotubes, and conducting polymers have been considered for SC electrodes. The conductive polymers have advantages like high low production cost and specific capacitance through these SCs usually have poor stability because of the unstable polymeric backbone structure. Carbon nanotubes and graphene exhibit a high and stable double-layer capacitance due to superior electric properties and nanoscale structure, such as low cost, large specific area, and cycle stability. However, the commercial application for SCs is limited due to the restricted specific capacitance of CNTs. On the other hand, conducting polymers have been widely used as the electrode material in SCs. There are various conducting polymer materials, namely, polypyrrole (PPy), polyaniline (PANI), and ethylene-vinyl acetate copolymer (EVA) have been used in SCs.

Yuan et al. synthesized 3D free-standing graphene aerogel film as a scalable and cost-effective method [89]. In this study, to construct a 3D porous structure GO/CNT gel was coated on electrospun polyacrylonitrile (PAN) nanofiber membrane flexible substrate. In the GCP aerogel film, the bottom PAN nanofiber membrane and the upper interconnected GO/CNT aerogel provide support for the upper aerogel and macroporous structure that provides a rapid transmission channel for the electrolyte ions and electrons. Further, introducing pseudocapacitance surface modification on the GCP aerogel film was done by polypyrrole (PPy), and further, by using pyrolysis nitrogen-doped RGO/CNT/carbon nanofiber (NRCC) composite aerogel film was obtained. This approach provides a method to prepare high energy density flexible electrode materials for supercapacitors on large scale. Liu et al. reported polyaniline coated rGO/CNT composite fibers ((RGO/CNTs)@PANI, RCP) for high-performance solid-state symmetric supercapacitor [90]. The designed high-performance supercapacitor can be used as a promising wearable energy storage device due to its excellent mechanical and electrochemical performance.

8.9 CONCLUSION

This chapter summarizes the recent advancement in the synthesis and application of carbon nanotube, graphene and their composites with metal oxide, and conducting polymers as electrode material in supercapacitors. In brief, carbonous materials like carbon nanotubes and graphenes with high conductivity and high surface area can be prepared with controlled microstructures to enhance their electrochemical performance. As explained all through this chapter, carbonous materials have proven to be beneficial in increasing the performance of supercapacitors because of their novel geometrical characteristics and unique electronic properties. Further, CNT and graphene assist supercapacitors by (i) boosting the electrochemical reaction or molecular adsorption occurring at the electrode-electrolyte interface by providing a large surface area, and (ii) giving rise to high conductivity and/or porous structure to facilitate both electron and ion transport and electrolyte diffusion, to ensure the electrochemical process occurs highly efficiently.

REFERENCES

1. Yang, Z., Ren, J., Zhang, Z., Chen, X., Guan, G., Qiu, L., Zhang, Y., and Peng H. 2015. Recent advancements in nanostructured carbon for energy applications. Chem Rev 115: 5159–223.
2. Cao, X., Yin, Z., and Zhang, H. 2014. Three-dimensional graphene materials: preparation, structures, and application in supercapacitors. Energy Environ Sci 7: 1850–1865.
3. Kyeremateng, N.A., Brousse, T., Pech, D. 2017. Microsupercapacitors as miniaturized energy-storage components for on-chip electronics. Nature Nanotechnology 12: 7–15.
4. He, H., Fu, Y., Zhao, T., Gao, X., Xing, L., Zhang, Y., Xue, X. 2017. All-solid-state flexible self-charging power cell basing on piezo-electrolyte for harvesting/storing body-motion energy and powering wearable electronics. Nano Energy 39: 590–60.
5. Qi, D., Liu, Y., Liu, Z., Zhang, L., Chen, X. 2017. Design of Architectures and Materials in In-Plane Micro-supercapacitors: Current Status and Future Challenges. Adv. Mater. 29: 1602802–1602821.
6. Winter, M., Brodd, R.J. 2004. What Are Batteries, Fuel Cells, and Supercapacitors? Chem. Rev. 104: 4245–4270.
7. Zhang, J., Zhao, X. 2012. On the Configuration of Supercapacitors for Maximizing Electrochemical Performance. ChemSusChem 5: 818–841.
8. Chen, T., Dai, L. 2013. Carbon nanomaterials for high-performance supercapacitors. Mater. Today 16: 272–280.
9. Yoon, Y., Lee, K., Lee, H. 2016. Low-dimensional carbon and MXene-based electrochemical capacitor electrodes. Nanotechnology 27: 172001–172022.
10. Simon, P., Gogotsi, Y. 2009. Materials for electrochemical capacitors. Nat. Mater. 330–336.
11. Chen. T., Dai, L. M. 2013. Carbon nanomaterials for high-performance supercapacitors. Mater. Today 16: 272–280.
12. Zhang, X., Zhang, H., Li, C., Wang, K., Sun, X., Ma, Y. 2014. Recent advances in porous graphene materials for supercapacitor applications. RSC Adv. 4: 45862–45884.
13. Burke, A. J. 2000. Ultracapacitors: why, how, and where is the technology. Power Sources 91: 37–50.

14. Kotz, R., Carlen, M. 2000. Principles and applications of electrochemical capacitors. Electrochim. Acta 45: 2483–2498.
15. Pandolfo, A. G., Hollenkamp, A. F. 2006. Carbon properties and their role in supercapacitors. J. Power Sources 157: 11–27.
16. Zhang, L. L., Zhao, X. S. 2009. Carbon-based materials as supercapacitor electrodes. Chem. Soc. Rev. 38: 2520–2531.
17. Xing, W., Huang, C. C., Zhuo, S. P., Yuan, X., Wang, G. Q., Hulicova-Jurcakova, D., Yan, Z. F., Lu, G. Q. 2009. Hierarchical porous carbons with high performance for supercapacitor electrodes. Carbon 47: 1715–1722.
18. Cheng, Q., Tang, J., Ma, J., Zhang, H., Shinyaa, N., Qin, L.C. 2011. Graphene and carbon nanotube composite electrodes for supercapacitors with ultra-high energy density. Phys. Chem. Chem. Phys. 13: 17615–17624.
19. Iijima, S. 1991. Helical microtubules of graphitic carbon. Nature 354: 56–58.
20. Harris, P.F. 1999. *Carbon Nanotubes and Related Structures: New Materials for the Twenty-first Century*, Cambridge University Press: Cambridge.
21. Peigney, A., Laurent, Ch., Flahaut, E., Bacsa, R.R., Rousset, A. 2001. Specific surface area of carbon nanotubes and bundles of carbon nanotubes. Carbon 39: 507–514.
22. I. Rafique, A. Kausar, Z. Anwar, B. Muhammad, 2016. Exploration of epoxy resins, hardening systems, and epoxy/carbon nanotube composite designed for high performance materials: A Review, Polymer-plastics Technology and Engineering 55: 312–333.
23. Jia, X. & Wei, F. 2017. Advances in Production and Applications of Carbon Nanotubes. Top.Curr. Chem 375: 1–35.
24. Iijima, S. & Ichihashi, T. 1993. Single-shell carbon nanotubes of 1-nm diameter. Nature 363: 603–605.
25. Lee, R., Nikolaev, P., Dai, H., Petit, P., Robert, J., Xu, C., Lee, Y., Kim, S., Smalley, R. 1996. Crystalline Ropes of Metallic Carbon Nanotubes. Science 273: 483–487.
26. Ren, Z. F. 1998. Synthesis of Large Arrays of Well-Aligned Carbon Nanotubes on Glass. Science 282: 1105–1107.
27. Hussain, A., Ding, Er., Mclean, B., Mustonen, K., Ahmad, S., Tavakkoli, M., Page, A., Zhang, Q., Kotakoski, J., Kauppinen, E. 2020. Scalable growth of single-walled carbon nanotubes with a highly uniform structure. Nanoscale 12: 12263–12267.
28. Hata, K., Futaba, D. N., Mizuno, K., Namai, T., Yumura, M., Iijima, S. 2010. Water-Assisted Highly Efficient Synthesis of Impurity-Free Single-Walled Carbon Nanotubes. Science 306: 1362–1364.
29. Zhu, L., Xiu, Y., Hess, D. W. & Wong, C.-P. 2005. Aligned Carbon Nanotube Stacks by Water-Assisted Selective Etching. Nano Lett. 5: 2641–2645.
30. Geim, A. K., Novoselov, K. S. 2007. The rise of graphene. Nat. Mater. 6: 183–191.
31. Najib, S., Erdem, E.2019. Current progress achieved in novel materials for supercapacitor electrodes: mini review. Nanoscale Adv. 1: 2817–2827.
32. Edwards, R. S., Coleman, K. S. 2013. Graphene synthesis: relationship to applications. Nanoscale 5: 38–51.
33. Zhang, L., Zhou, R., Zhao, X.S. 2010. Graphene-based materials as supercapacitor electrodes. J Mater Chem. 20: 5983–5992.
34. Yoon, S.M., Choi, W.M., Baik, H. Shin, H., Song, I., Kwon, M., Bae, J., Kim, H., Lee, Y., Choi, J. 2012. Synthesis of multilayer graphene balls by carbon segregation from nickel nanoparticles. ACS Nano 6: 6803–6811.
35. Wu, Z., Winter, A., Chen, L., Sun, Y., Turchanin, Andrey., Feng, Xinliang., Müllen, K. 2012. Three-dimensional nitrogen and boron co-doped graphene for high-performance all-solid-state supercapacitors. Adv Mater. 24: 1–6.
36. Zhang, L., Zhou, R., Zhao, X.S. 2010. Graphene-based materials as supercapacitor electrode. J Mater Chem 20: 5983–5992.
37. Coraux, J., N'Diaye, A., Busse, C., Michely, T. 2008. Structural coherency of graphene on Ir (111). Nano Lett. 8: 565–570.
38. Sutter, P. W., Flege, J. I., Sutter, E. A. 2008. Epitaxial graphene on ruthenium. Nat. Mater. 7: 406–411.

39. Cao, X., Shi, Y., Shi, W., Lu, G., Huang, X., Yan, Q., Zhang, Q., Zhang, H. 2011. Preparation of Novel 3D Graphene Networks for Supercapacitor Applications. Small, 7: 3163–3168.
40. Cano-Márquez, A., Rodríguez-Macías, F., Campos-Delgado, J., Espinosa-González, C., Tristán-López, F., Ramírez-González, D., Cullen, D., Smith, D., Terrones, M., Vega-Cantú, Y. 2009. Ex-MWNTs: Graphene Sheets and Ribbons Produced by Lithium Intercalation and Exfoliation of Carbon Nanotubes. Nano Lett. 9: 1527–1533.
41. Stankovich, S. D., Dikin, A., Piner, R. D., Kohlhaas, K. A., Kleinhammes, A., Jia, Y., Wu, Y., Nguyen, S. T., Ruoff, R. S. 2007. Synthesis of graphene-based nanosheets via chemical reduction of exfoliated graphite oxide. Carbon 45: 1558–1565.
42. Coleman, J. N. 2009. Liquid-Phase Exfoliation of Nanotubes and Graphene. Adv. Funct. Mater. 19: 3680–3695.
43. Hou, J., Shao, Y., Ellis, M. W., Moored, R. B., Yie, B. 2011.Graphene-based electrochemical energy conversion and storage: fuel cells, supercapacitors and lithium-ion batteries. Phys. Chem. Chem. Phys. 13: 15384–15402.
44. Perera, S. D., Mariano, R. G., Nijem, N., Chabal, Y., Ferraris, J. P., Balkus, K. J. 2012. Alkaline deoxygenated graphene oxide for supercapacitor applications: An effective green alternative for chemically reduced graphene. J. Power Sources 215: 1–10.
45. Zhou, Y. S., Zhu, Y. C., Xu, B. S., Zhang, X. J. 2019. High electroactive material loading on a carbon nanotube/carbon nanofiber as an advanced free-standing electrode for asymmetric supercapacitors. Chem. Commun. 55: 4083–4086.
46. X, Gui., J, Wei., K, Wang., A, Cao., H, Zhu., Y, Jia., Shu, Q., Wu, D. 2009. Carbon Nanotube Sponges. Adv. Mater. 21: 1–5.
47. Song, L., Ci, L., Lu, L., Zhou, Z., Yan, X., Liu, D., Yuan, H., Gao, Y., Wang, J., Liu, L., Zhao, X., Zhang, Z., Dou, X., Zhou, W., Wang, G., Wang, C., Xie, S. 2004. Direct synthesis of macroscale single-walled carbon nanotube non-woven material. Adv. Mater. 16: 1529–34.
48. Di, J.T., Yong, Z., Yang, X.J., Li, Q.W. 2011. Structural and morphological dependence of carbon nanotube arrays on catalyst aggregation. Appl. Surf. Sci. 258: 13–8.
49. Liu, W., Zhang, X., Xu, G., Bradford, P.D., Wang, X., Zhao, H., Zhang, Y., Jia, Q., Yuan, F., Li, Q., Qiu, Y., Zhu, Y. 2011. Producing superior composites by winding carbon nanotubes onto a mandrel under a poly (vinyl alcohol) spray. Carbon 49: 4786–91.
50. Niu, C., Sichel, E.K., Hoch, R., Moy, D., Tennent, H. 1997. High power electrochemical capacitors based on carbon nanotubes electrodes. Appl. Phys. Lett. 70: 1480.
51. An, K. H., Kim, W. S., Park, Y. S., Moon, J.M., Bae, D. J., Lim, S. C., Lee, Y. S., Lee, Y.H. 2001. Electrochemical Properties of High-Power Supercapacitors Using Single-Walled Carbon Nanotube Electrodes. Adv. Funct. Mater. 11: 387–392.
52. Frackowiak, E., Metenier, K., Bertagna, V., Beguin, F. 2000. Supercapacitor electrode from multiwalled carbon nanotubes. Appl. Phys. Lett. 77: 2421–2425.
53. Zhang, H., CaO, G.P., Yang, Y.S. 2007. Using a cut–paste method to prepare a carbon nanotube fur electrode. Nanotechnol. 18: 195607–195611.
54. Zhou, C., Kumar, S., Doyle, C.D., Tour, J. 1997. Functionalized Single Wall Carbon Nanotubes Treated with Pyrrole for Electrochemical Supercapacitor Membranes. J.M. Chem. Mater. 17: 1997–2002.
55. Frackowiak, E., Delpeux, S., Jurewicz, K., Szostak, K., Cazorla Amoros, D., Beguin, F. 2002. Enhanced capacitance of carbon nanotubes through chemical activation, Chem. Phys. Lett. 261: 35–41.
56. Kim, Y.T., Ito, Y., Tadai, K., Mitani, T., Kim, U.S., Kim, H.S., Cho, B.W. 2005. Drastic change of electric double layer capacitance by surface functionalization of carbon nanotubes. Appl. Phys. Lett. 87: 234106–234109.
57. Kim, N. D., Kim, W., Joo, J. B., Oh, S., Kim, P., Kim, Y., Yi, J. 2008. Electrochemical capacitor performance of N-doped mesoporous carbons prepared by ammoxidation. J. Power Sources 180: 671–675.

58. Sevilla, M., Yu, L., Fellinger, T. P., Fuertes, A. B., Titirici, M.-M. 2013. Polypyrrole-derived mesoporous nitrogen-doped carbons with intrinsic catalytic activity in the oxygen reduction reaction. RSC Adv. 3: 9904–9910.
59. Sevilla, M., Yu, L., Zhao, L., Ania, C.O., Titiricic, M.M. 2014. Surface Modification of CNTs with N-Doped Carbon: An Effective Way of Enhancing Their Performance in Supercapacitors. ACS Sustainable Chem. Eng. 2: 1049–1055.
60. Kim, J.H., Ko, Y., Kim, Y., Kim, K., Yang, C-M. 2021. Sulfur-doped carbon nanotubes as a conducting agent in supercapacitor electrodes. J. Alloys and compounds 855: 157282–157320.
61. Paul, R., Roy, A. 2021. BN-co doped CNT based nanoporous brushes for all-solid-state flexible supercapacitors at elevated temperatures. Electro. Acta. 365: 137345–137369.
62. Jha, M., Ball, R., Seelaboyina, R. 2020. Subramaniam, All Solid-State Coaxial Supercapacitor with Ultrahigh Scan Rate Operability of 250–000 mV/s by Thermal Engineering of the Electrode–Electrolyte Interface. C. ACS Appl. Energy Mater. 3: 3454–3464.
63. Zhang, X., Sui, Z., Xu, B., Yue, S., Luo, Y., Zhan, W., Liu, B. J. 2011. Mechanically strong and highly conductive graphene aerogel and its use as electrodes for electrochemical power sources. Mater. Chem. 21: 6494–6497.
64. Li, Z. J., Yang, B. C., Zhang S. R., Zhao, C. M. 2012. Graphene oxide with improved electrical conductivity for supercapacitor electrodes. Appl. Surf. Sci. 258: 3726–3731.
65. Kim, Y.S., Kumar, K., Fisher F.T., Yang, E.-H. 2012. Out-of-plane growth of CNTs on graphene for supercapacitor applications. Nanotechnology 23: 015301–015308.
66. Yan, J., Liu, J., Fan, Z., Wei T., Zhang, L. 2012. High-performance supercapacitor electrodes based on highly corrugated graphene sheets. Carbon 50: 2179–2188.
67. Sahu, V., Shekhar, S., Sharma, R., Singh, G. 2015. Ultrahigh Performance Supercapacitor from Lacey Reduced Graphene Oxide Nanoribbons. ACS Appl. Mater. Interfaces 7 : 3110–3116.
68. Liu, T., Zhang, L., Cheng, X., Hu, X. 2020. Holey graphene for electrochemical energy storage. Cell reports physical Science 1: 100215.
69. Jiang, Z-J., Jiang, Z., Chen, W. 2014. The role of holes in improving the performance of nitrogen-doped holey graphene as an active electrode material for supercapacitor and oxygen reduction reaction. J. Power Sources 251: 55–65.
70. Chen, Y., Li, Y., Yao, F., Peng, C., Cao, C., Feng, Y., and Feng, W. 2019. Nitrogen and fluorine co-doped holey graphene hydrogel as a binder-free electrode material for flexible solid-state supercapacitors. Sustain. Energy Fuels 3: 2237–2245.
71. Liu, J., Zhu, Y., Chen, X., and Yi, W. 2020. Nitrogen, sulfur and phosphorus tri-doped holey graphene oxide as a novel electrode material for application in supercapacitor. J. Alloys Compounds 815: 152328–152339.
72. Singh, L., Mahapatra, D. 2020. Adapting 2D nanomaterials for advanced applications. ACS symposium series: ACS publication.
73. Song, W. L.; Song, K.; Fan, L. Z. 2015. A Versatile Strategy toward Binary Three-Dimensional Architectures Based on Engineering Graphene Aerogels with Porous Carbon Fabrics for Supercapacitors. ACS Appl. Mater. Interfaces 7: 4257–4264.
74. Dong, L., Liang, G., Xu, C., Ren, D., Wang, J., Pan, Z.Z., Li, B., Kang, F., Yang, Q.H. 2017. Stacking up layers of polyaniline/carbon nanotube network inside papers as highly flexible electrodes with large areal capacitance and superior rate capability. J. Mater. Chem. A 5: 19934–19942.
75. Huang, L., Li, C., Shi, G. 2014. High-performance and flexible electrochemical capacitors based on graphene/polymer composite films. J. Mater. Chem. A 2: 968–974.
76. Tamailarasan, P., Ramaprabhu, S. 2012. Carbon Nanotubes-Graphene-Solid like Ionic Liquid Layer-Based Hybrid Electrode Material for High Performance Supercapacitor. J. Phys. Chem. C 116: 14179–14187.
77. Cheng, Y., Lu, S., Zhang, H., Varanasi, C.V., Liu, J. 2012. Synergistic Effects from Graphene and Carbon Nanotubes Enable Flexible and Robust Electrodes for High-Performance Supercapacitors. Nano Lett. 12: 4206–4211.

78. Mao, S., Wen, Z., Bo, Z., Chang, J., Huang, X., Chen, J. 2014. Hierarchical Nanohybrids with Porous CNT-Networks Decorated Crumpled Graphene Balls for Supercapacitors. ACS Appl. Mater. Interfaces 6: 9881–9889.
79. Kang, S. H., Lee, G.Y., Lim, J., Kim, S.O. 2021. CNT–rGO Hydrogel-Integrated Fabric Composite Synthesized via an Interfacial Gelation Process for Wearable Supercapacitor Electrodes. ACS Omega 6: 19578–19585.
80. Zang, X., Jiang, Y., Sanghadasa, M., Lin, L. 2020. Chemical vapor deposition of 3D graphene/carbon nanotubes networks for hybrid supercapacitors. Sensors and Actuators A: Physical 304: 111886–111907.
81. Yang, Z.Y., Zhao, Y.F., Xiao, Q.Q., Zhang, Y.X., Jing, L. Yan, Y.M., Sun, K.N. 2014. Controllable Growth of CNTs on Graphene as High-Performance Electrode Material for Supercapacitors. ACS Appl. Mater. Interfaces 6: 8497–8504.
82. Zhi, M., Xiang, C., Li, J., Li, M., Wu, N. 2013. Nanostructured carbon-metal oxide composite electrodes for supercapacitors: a review. Nanoscale 5: 72–88.
83. Hu, Y., Guo, C. 1991. Carbon Nanotubes and Carbon Nanotubes/Metal Oxide Heterostructures: Synthesis. Characterization and Electrochemical Property. in M. Naraghi (ed.), *Carbon Nanotubes—Growth and Applications*, IntechOpen, London.
84. Zhang, L., Tian, Y., Song, C., Qiu, H., Xue, H. 2021. Study on preparation and performance of flexible all-solid-state supercapacitor based on nitrogen-doped RGO/CNT/MnO_2 composite filers. Journal of Alloys and Compounds 859: 157816–157847.
85. Kong, S., Cheng, K., Ouyang, T., Gao, Y., Ye, K., Wang, G., Cao, D. 2017. Facile electrodepositing processed of RuO_2-graphene nanosheets-CNT composites as a binder-free electrode for electrochemical supercapacitors. Electrochimica Acta 246: 433–442.
86. Tang, C., Zhao, K., Tang, Y., Li, F., Meng, Q. 2021. Forest-like carbon foam templated rGO/CNTs/MnO_2 electrode for high-performance supercapacitor. Electrochemica Acta 375: 137960–137968.
87. Mohammadi, A., Arsalani, N., Tabrizi, A., Moosavifard, S., Naqshbandi, Z., Ghadimi, L. 2018. Engineering rGO-CNT wrapped Co_3S_4 nanocomposites for high-performance asymmetric supercapacitors. Chemical Engineering Journal 334: 66–80.
88. Ramli, N., Rashid, S., Mamat, Md., Sulaiman, Y., Krishnan, S. 2018. Incorporation of iron oxide into CNT/GNF as a high-performance supercapacitor electrode. Incorporation of iron oxide into CNT/GNF as a high-performance supercapacitor electrode, Materials Chemistry and Physics 212: 318–324.
89. Yuan, S., Fan, W., Jin, Y., Wang, D., Liu, T. 2021. Free-standing flexible graphene-based aerogel film with high energy density as an electrode for supercapacitors. Nano Materials Science 3: 68–74.
90. Liu, D., Du, C.P., Wei, W., Wang, H., Wang, Q., Liu, P. 2018. Skeleton/skin structured (RGO/CNTs)@PANI composite fiber electrodes with excellent mechanical and electrochemical performance for all-solid-state symmetric supercapacitors. Journal of Colloid and Interface Science 513: 295–303.

9 Carbon Nanotubes and Graphene for Lithium-Ion Battery

Santwana Pati and Indu Elizabeth

CONTENTS

9.1 INTRODUCTION AND BACKGROUND

In the past decade global sales of smartphones have increased from 472 million units to 1535 million units [1]. The exponential increase might have been stagnant in the year 2020–21 due to the COVID pandemic, but it still shows an encouraging picture of how portable electronic devices are in demand more than ever. More and more consumers are getting added in the list of "Technologically Sound" who are capable of information acquisition and processing digitally. Innovations in portable electronic devices have majorly been driven by advancements in Li-ion battery (LIB) technology which has helped in reduced size, weight, and enhanced cycle life.

DOI: 10.1201/9781003231943-9

Another major area of breakthrough that has been catalyzed by LIB is the transportation sector. Electric vehicles are gradually replacing the noxious fumes of an internal combustion engine. Soon, everywhere the smell of gas stations will fade away into odorless charging stations. As more and more automobile industries are shifting toward electric vehicle production, the electrified future is much closer than we think. LIBs provide the ideal path to sustainable power and renewable transportation. LIBs offer high energy-to-weight ratios, low self-discharge rate, high open circuit voltage and a gradual loss of charge when not in use and low cost. Therefore, LIBs are a key part of the energy transition as the world swaps fossil fuel power for emissions free electrification. The Israeli firm Eviation is working on a prototype of an entirely electric aircraft which shows a carbon neutral future aviation [2].

Researchers all over the world have been working on developing novel and advanced LIBs that will not only enhance the efficiency of current smartphones and EVs but also allow other industries connected to batteries to be optimized such as new mining technology, innovative charging infrastructure and vehicle to grid applications.

Through this chapter, fundamental scientific and technology development of the LIB is presented. The progress in the research on CNT and graphene-based anode for LIB will be reviewed. The research work on CNTs will be classified based on various types of CNTs and their electrochemical performance. Similarly, Li-ion storage capability of graphene will be discussed focusing on the various factors affecting the electrochemical behavior of graphene. This chapter will also shed light on the future trend of LIB anode research based on CNT and graphene.

9.2 FUNDAMENTALS OF LIB

Batteries have been a well-known concept and a common household term. They can be categorized in two types: primary and secondary. The primary batteries are for one-time use and not rechargeable whereas the secondary batteries can be recharged. The focus of the research world are secondary batteries as they can be used for bulk storage of energy in the form of chemical energy [3, 4]. Some of the most common rechargeable batteries include Li-ion, Ni metal hydride, lead acid, etc. The battery properties depend on the active materials of which the anode and cathode are made and how the battery components are integrated. There are four distinct types of shapes for the rechargeable batteries: coin, cylindrical, prismatic and pouch as demonstrated in the Figure 9.1.

The Sony Corporation first commercialized the Li-ion battery in 1991 and since then it has become the most used rechargeable battery. Li-ion batteries offer higher gravimetric and volumetric energy density than other rechargeable batteries. They are also lightweight and show excellent cycling performance. All these advantages make Li-ion batteries a better option for portable electronic devices like mobile phones and laptops.

9.2.1 WORKING OF THE LIB

LIBs work on the principle of intercalation and deintercalation of Li ions. LIB consists of an anode, cathode, electrolyte, and a separator. Anode and cathode store the lithium. Electrolyte serves as the medium for the Li^+ ions movement. Meanwhile the electrical current flows outside through a device being powered (phone or laptop). The separator makes sure the current does not flow inside the cell. A very basic working of the LIB is shown in the Figure 9.2. Cathode is a lithium metal oxide, and the anode is composed of carbon generally. The electrolyte is the medium through which the Li ions move during charge and discharge. Electrolyte is typically a

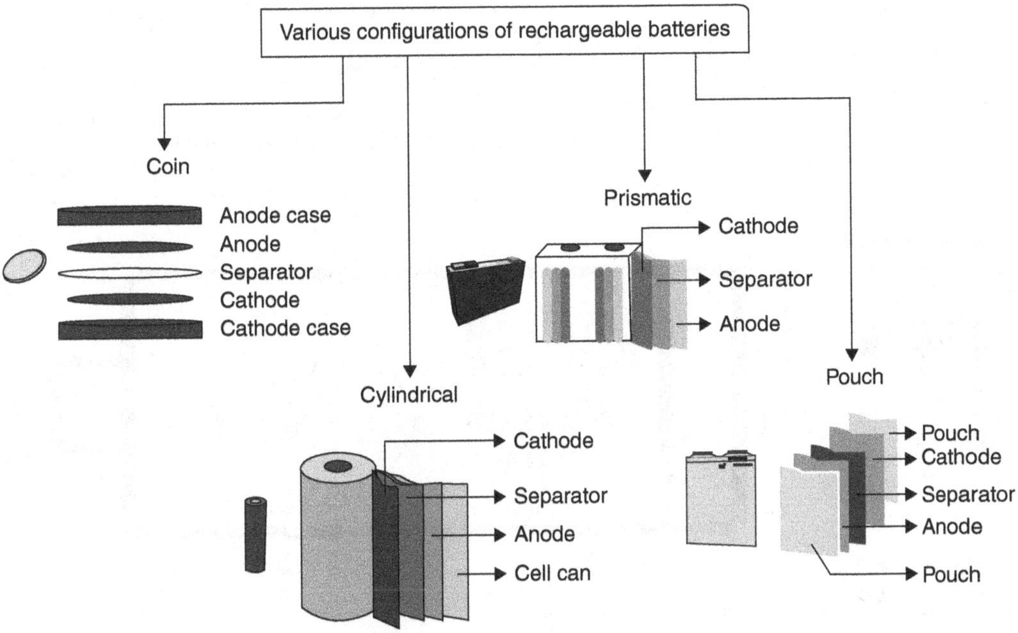

FIGURE 9.1 Various configurations of rechargeable batteries: Coin, Cylindrical, Prismatic and Pouch.

salt of lithium like $LiPF_6$ dissolved in an organic solvent. The following is the electrochemical reaction in the simplest manner:

$$\textbf{At Cathode, } LiMO_2 \leftrightarrow Li_{1-x}MO_2 + xLi^+ + xe^- \qquad \textbf{Eqn. (1)}$$

$$\textbf{At Anode, } xLi^+ + xe^- + 6C+ \leftrightarrow Li_xC_6 \qquad \textbf{Eqn. (2)}$$

LIBs provide power to millions everyday through laptops, mobile phones, electric cars etc. The various components of the battery along with the working principle have been discussed already. This chapter focuses on the anode and how carbon in various forms has been applied.

Anode is a very critical part of the LIB, and it decides the important characteristics of the battery. The active anode material occupies 10 weight percent of lithium rechargeable battery and releases electrons during the oxidation stage. Pure lithium metal itself, as an anode, provides the best capability, energy density and lightest weight. However, due to its highly reactive nature, it is usually unsafe and potentially explosive and flammable; irreversible dendrites can also be formed after many cycles. Graphite has the most fulfilling anode properties and has been widely used in commercial lithium batteries.

The following criteria must be met for new anode materials:

- High capacity and energy density
- Tremendous capacity retention
- Low irreversible capacity loss during the first cycle
- Good discharge voltage vs Li+, preferably between 0.3 and 0.5 V
- No co-intercalation of solvent molecules into the structure
- Good rate capability and performance at low temperature
- Low price and environmentally safe.

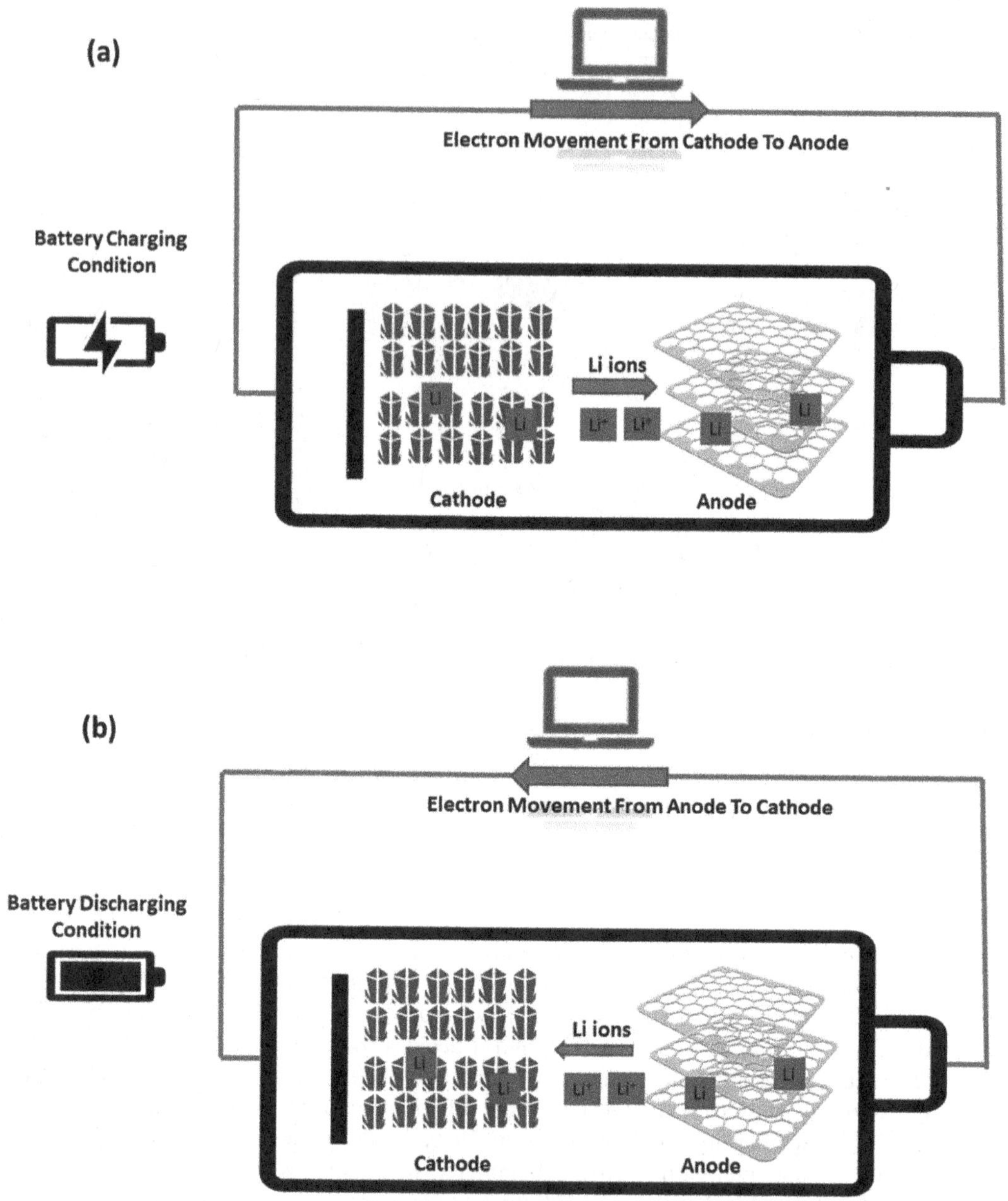

FIGURE 9.2 Lithium-ion battery operation during (a) charging and (b) discharging condition.

9.3 CARBON AS ANODE

Graphitic carbon is the dominant commercially available anode material. Therefore, modification of carbonaceous materials is the research focus. Scientists find scope for research to improve the electrode performance through the following properties: charge density, output voltage and cycle performance. Various forms of carbon including graphite, hard and soft carbon, carbon fibers, CNTs, and graphene have been investigated and their electrochemical properties have been analyzed. Carbon-based anodes offer various benefits like high performance, easy availability, low cost and no dendrite formation during recharge [5]. Graphite is one of the most conventional anode materials for batteries. The graphite anode is the commercial standard for the lithium-ion batteries. The commercial graphite anode is a layered intercalation material that allows for reversible lithium ion (de) insertion [3]. The average de-insertion potential is ~ 0.1 V, very close to metallic lithium. It is well known that Li intercalation reaction occurs only at the edge plane of graphite. Through the basal plane, intercalation is possible only at defect sites [4].

9.3.1 Determination of Theoretical Capacity

Faraday's first law of electrolysis states that mass of a substance altered at an electrolysis is directly proportional to the quantity of electricity transferred to that electrode [6]. Quantity of electricity refers to the quantity of electrical charge in coulombs. LiC_6 gives 6 carbon atoms with one electron through the circuit.

1 mole of electrons is equivalent to 6 carbon atoms (6 x 12 =72 gms)

1 faraday is equivalent to 9.648 x 104 C/mol.

Thus, 72 g gives 96485 C of electricity.

1g shall give 96485 /72 C/mol. = 1340 C/mol/g
= 1340 A-s/mol/g
= 0.3722 A-h/g
= 372 mAh/g

Therefore, as derived the theoretical capacity of graphite as an anode is 372 mAh/g. Several efforts have been done in investigating novel anode materials for high performance anode in LIBs. Some of them are CNFs (450 mAh/g), porous carbon (800–1100 mAh/g), Silicon (4200 mAh/g), Germanium (1600 mAh/g), Tin (994 mAh/g) and several others. But they all face major limitations like high volume expansion, poor electron transport, low coulombic efficiency, and cyclic capacity fading [3]. Nanostructured materials offer promising results as anode materials due to the following advantages:

- Smaller diffusion length
- Higher surface/volume ratio than bulk materials and enhanced surface area increases the capacity
- Rate of intercalation of Li-ions is higher hence the charging rate will be higher
- No cracking unlike bulk anode materials and volume contraction does not cause capacity loss in nanostructures
- Entanglement factor: enhanced electrical percolation and mechanical properties

9.4 CARBON NANOTUBES AS ANODE

CNTs are an allotrope of graphite in which sheets of graphite are rolled into a tube. With a high aspect ratio, CNT is almost a one-dimensional structure with a high length/diameter ratio. CNTs have stimulated the interest of the research community from a fundamental as well as application point of view. CNTs have been approved as the candidates for lithium-ion battery anode owing to its unique morphology, high conductivity, low density, high resilience and high tensile strength.

Standalone CNTs offer a capacity enhanced to almost 1000 mAh/g with significant chemical treatments to SWCNTs. But another major advantage of using CNTs is how they can act as a support matrix. CNTs form composites with various metals like tin, aluminum and magnesium and can be used as anode as the metals have high theoretical capacity.

9.4.1 Mechanism of Li-Ion Storage

Metals employ alloy formation with lithium ions when used as anode whereas carbon uses the intercalation technique. Li ions are inserted into the interlayers of graphite and the interlayer distance increases from 3.35 Å to 3.5 Å. This small expansion helps maintain the structural integrity of graphite and enhances the surface conductivity. Compared to graphite and fullerene, CNTs have the potential to accommodate more guest species in the interstitial sites. The number of Li ions is varied due to the different morphologies of CNTs.

9.4.2 Types of CNTs and Their Electrochemical Performance

CNTs are classified in several ways and the electrochemical properties as anode have been studied accordingly. Figure 9.3 shows detailed classification of the CNTs according to structure, morphology, preparation technique, etc. The whole section is ordered based on this manner and the electrochemical results are discussed accordingly.

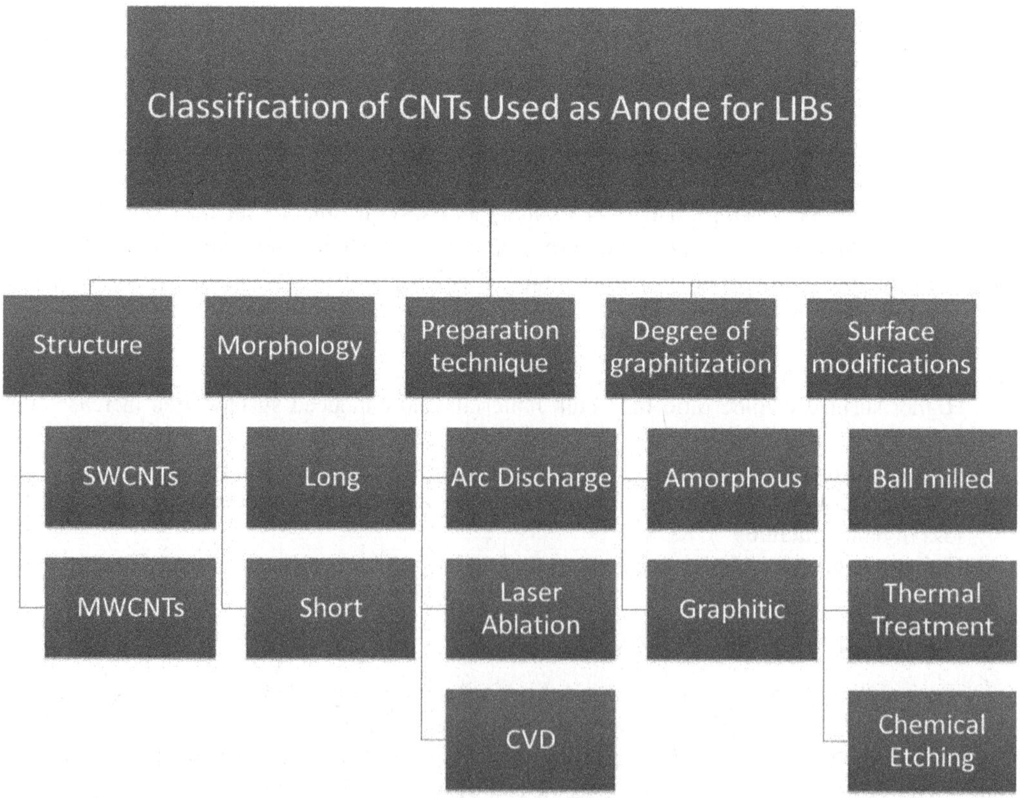

FIGURE 9.3 Classification of CNTs to be used as anode for LIBs.

9.4.2.1 Structure

Li-ion intercalation and deintercalation in the CNTs are completely dependent on the structure of the CNTs. SWCNT is a single sheet of graphene rolled in the shape of a cylinder in the dimensions a diameter of 1–2 nm and length of several micrometers. Similarly, MWCNTs are formed in the shape of several graphene sheets rolled into closed concentric cylinders.

9.4.2.1.1 SWCNTs as Anode

Electrochemical performance depends on the ability of the anode structure to form stable sites for Li-ion sites. In 1996, Thess et al. [7] showed an interesting structure that bundles of SWCNTs formed from 2-D triangular lattices arising from the Van der Waals attractions between nanotube sidewalls that can enhance the lithium battery capacity. In addition, there are calculations proposing a curvature-induced lithium condensation inside the core of the nanotubes [4]. Theoretically, SWCNTs are estimated to provide a capacity as high as 1116 mAh/g. However, due to the high surface area and formation of the SEI layer (solid electrolyte layer), the capacity practically is in the range of 400–600 mAh/g. Yang et al. reported capacities of 170 mAh/g and 266 mAh/g for two different samples for unetched SWCNTs [8], Wang et al. reported a capacity of 340 mAh/g [9] and Kawasaki et al. reported a capacity of 641 mAh/g for metallic CNTs [10]. Landi et al. presented a very interesting report with reversible lithium-ion capacity of 520 mAh/g for high purity SWCNTs [11]. The results show that the structural integrity and carbonaceous purity of individual SWCNTs is maintained during cycling, while the lithium insertion is accommodated by bundle channel expansion.

9.4.2.1.2 MWCNTs as Anode

CNTs do not exhibit any kind of expansion/contraction of structure and pulverization problem so they can sustain their capacity for long cycles as demonstrated in Figure 9.4. Lahiri et al. [12] fabricated a MWCNT-based anode grown on the current collector which shows good specific capacity, even at high current rate and excellent stability (nil capacity degradation) over 50 cycles.

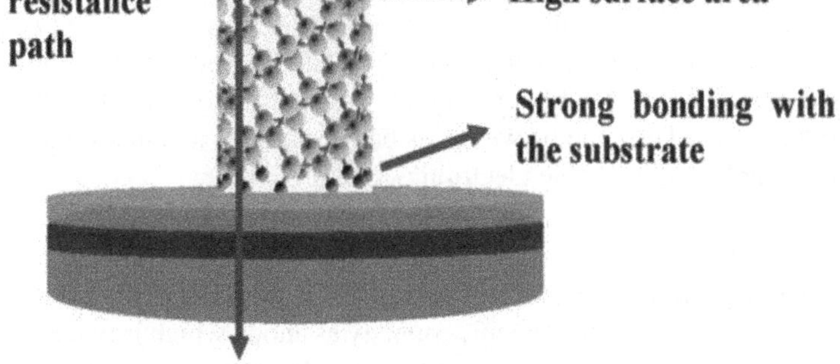

FIGURE 9.4 A schematic showing the advantages of the CNT based anode structure and the benefits it offers. Reprinted with permission from [12]. Copyright {2010} American Chemical Society.

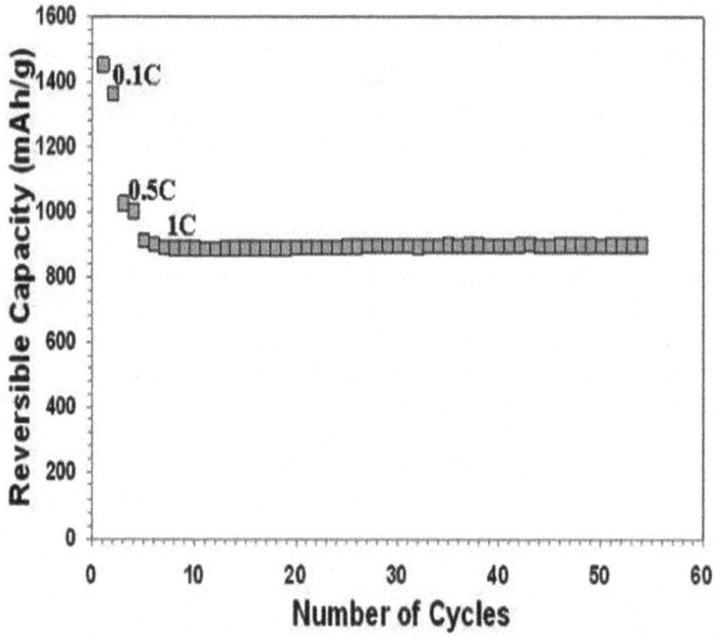

FIGURE 9.5 Exceptional stability of the reversible capacity (900 mAh/ g) of the MWNT-on-Cu anode in the long-run, at 1 C rate. Except for the first two cycles, virtually no capacity degradation was observed for this anode structure, in 50 cycles Reprinted with permission from [12]. Copyright {2010} American Chemical Society.

Capacity as high as 900 mAh/g was achieved using this anode with excellent capacity retention as shown in Figure 9.5 which is a 140% boost as compared to the theoretical capacity of graphite.

Paul et al. [13] demonstrated anodes composed of freestanding, binder-free and hierarchical multiwalled carbon nanotube (MWCNT) foam fabricated through a microwave plasma assisted chemical vapor deposition. The MWCNT-based electrode showed structural and chemical stability throughout the lithiation and delithiation procedure with a capacity of 790 mAh/g. As compared to random CNTs, vertically aligned CNTs (VA-CNTs) have a regular pore structure and larger surface area thereby exhibit better charge transport capacity. This makes them excellent electrode materials as compared to regular non-aligned CNTs [14, 15]. Wang et al. [16] reported a high reversible lithium storage capacity of 950 mAh/g using vertically aligned MWCNTs owing to the compartment structures. Another interesting report by Jing Ren et al. [17] shows a stretchable wire-shaped lithium-ion battery with aligned MWCNT-based anode as shown in Figure 9.6. The batteries are wire shaped, flexible and lightweight with very good capacity retention even after 1000 bending cycles. They can be woven as battery textiles and hence have a huge scope of applications for portable and wearable electronics.

9.4.2.2 Morphology

Yang et al. worked on short and long CNTs as anode for the LIBs [18]. Short CNTs fabricated using the CVD technique showed the specific capacity as anode which is twice that of long CNTs. The charge transfer resistance is lower in short CNTs and hence higher anode capacity is obtained. Similarly, Kang et al. reported their work on varied lengths CNTs and their electrochemical performance [19]. Small CNTs showed a specific capacity of 476 mAh/g and long CNTs showed

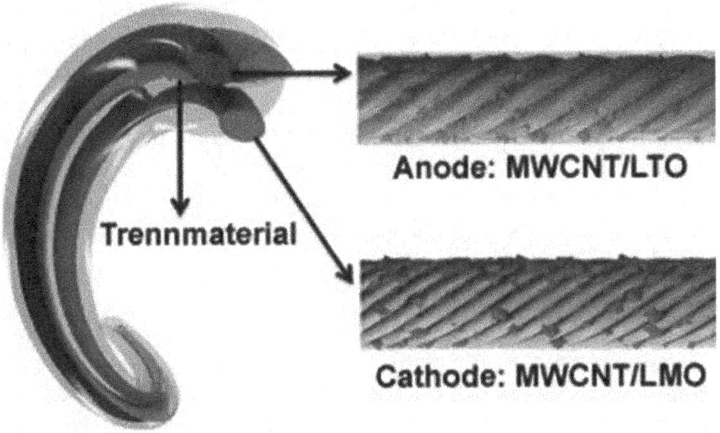

FIGURE 9.6 Wearable LIB using MWCNT anode. Reprinted with permission from [17]. Copyright {2014} John Wiley and Sons.

271 mAh/g. Long CNTs have complicated bunch structure. Hence the Li ions won't be able to diffuse deeply. Thereby most of the volume remains unused and results in lower capacity.

9.4.2.3 Preparation Technique

CNTs are prepared by three distinct techniques which are quite well known in nanoscience research. arc discharge, laser ablation and chemical vapor deposition (CVD) are used to prepare CNTs. Research articles suggest that CNTs produced by the various techniques tend to be quite different from each other.

Wang et al. [9] conducted a systematic examination of the electrochemical properties of MWCNTs produced by CVD technique. The CNTs were bundled with each other and showed a reversible capacity of 340 mAh/g. The lower-than-expected capacity could be due to the structural defects. Similarly, Frackowiak et al. [20] obtained MWCNTs using the CVD technique. MWCNTs were hypothetically assumed to provide space for lithium intercalation owing to the central canal inside the one-dimensional structure. Theoretically, in the catalytic MWCNTs, lithium can be stored in various places: through intercalation into the graphitic layers inside the tubes, through insertion into the canals, on to the pores or available structural defects on the surface of the CNT. Different types of bonding, between lithium and carbon can be expected theoretically.

There have been several reports on arc-discharge CNTs being used as anode for Li-ion batteries. CNTs prepared using the arc discharge technique are usually straight, well graphitized, and free of metal catalyst. They show limited electrochemical properties which might be due to lesser intercalation sites than CVD produced CNTs [8]. However, the electrochemical characteristics of the arc-produced CNTs are significantly improved by oxidation and thermal treatment.

Laser ablation is a sophisticated technique that uses the laser power for thermally stripping carbon atoms off from carbon bearing compounds [21]. This technique has been applied in different variations of the types of lasers and process parameters to obtain high purity SWCNTs. Gao et al. [22] reported their work on production of SWCNTs by laser ablating a graphite target using a 1064 nm Nd:YAG laser. The bundles of the SWCNTs were randomly oriented and showed a high irreversible capacity and large voltage hysteresis. After purification, the SWCNTs demonstrated a capacity as high as 650 mAh/g which is almost twice the capacity of graphite anodes, and the capacity was further enhanced to 1000 mAh/g after ball milling. Freestanding, laser-produced SWCNT anodes with Ni and Ti contacts, showed enhanced reversible capacity of almost 1250 mAh/g [23]. A detailed analysis of such reports suggests that laser ablation produces high purity CNT in which the inner core spaces of the SWCNTs are usually not accessible for intercalation because they have a closed structure and Li cannot intercalate easily. Purification processing cuts them into shorter segments and hence boosts the Li intake capacity. Empirical calculations conclude that chiral SWCNTs species may become more active to Li intercalation and storage with titanium contacts due to proper alignment of work function with the conduction states [23–25].

9.4.2.4 Degree of Graphitization

Degree of graphitization plays a significant role in the final applications of the CNTs. Amorphous CNTs show different properties from crystalline CNTs. Amorphous CNTs consist of walls of graphene sheets that show short-distance order and long-distance disorder. However, they are easy to synthesize and more economically viable than crystalline CNTs. And they have higher mechanical strength and higher surface area leading to better electrochemical performance [26–28]. Zhao et al. presented the electrochemical performance analysis of amorphous CNTs that showed a high capacity of 530 mAh/g and a good cycle life that retained 93% of the first cycle capacity [29]. Similarly, Wu et al. conducted a comparative study between slightly graphitized and well graphitized CNTs and obtained that slightly graphitized CNTs showed a high capacity of 640 mAh/g and the electrochemical capacity of the well graphitized CNTs was limited to 282 mAh/g [30]. The large surface area and the presence of abundant pores in amorphous CNTs lead to better contact between the electrode and electrolyte and enhanced ion transport. There is an immense scope of research in amorphous CNTs as anode owing to its low cost and simple synthesis technique and easy dopability with heteroatoms leading to enhanced electrochemical capacity.

9.4.2.5 Surface Modifications

Surface modifications of the CNTs can be done by ball milling, thermal treatment or chemical etching. Ball milling exposes more of the basal plane edges of the graphene sheets thereby increasing the active surface area. Reversible capacity of pristine SWCNTs increased to 1000 mA/g (more than twice) just by ball milling. Ball milling breaks the tubes and exposes the edges of the graphene layers, and enhanced surface area allows more Li ion intercalation [24].

Heat treatment results in removal of the amorphous carbon and thereby enhances the crystalline perfection of the CNT. This may result in improved performance of the CNTs as an anode for LIB. Yoon et al. [31] conducted experimental studies and concluded that heat-treated CNT films showed high capacity, good cycle stability and rate performance.

The chemical etching process involves reaction with strong acid such as nitric acid that results in large amounts of defects [32, 33]. Defective CNTs facilitate the insertion of several Li ions inside the tubular structure and thereby increase the lithium capacity and enhance the electrochemical performance of the LIB. This hypothesis has been confirmed by various reports with experimental results showing better anode capacity for etched CNTs as compared to pure CNTs. Eom et al. [34] reported a considerable increase in the capacity of LIB by using etched MWCNTs (681 mAh/g) over pure MWCNTs (351 mAh/g).

9.4.3 CNT-Based Nanocomposites as Anode

Different metals and metal oxides like Sn, SnO_2, TiO_2, Si, etc. have been used as anode materials for LIB [35–37]. They show a much higher theoretical anode capacity than carbonaceous materials. Despite the high Li intercalation capacities, these metals develop cracks during the cycle and hence lead to breakdown of the electrode. The electrode undergoes a huge volume change during the lithiation and de-lithiation process. Therefore, the electrode faces a quick capacity fading and hence cannot be used as a successful anode material for LIB. CNTs with their tubular structure and high resilience can be combined with these metals to provide structural stability by accommodating the large volume changes during the cycles. This hypothesis has been put to practical applications and various metal-CNT nanocomposites have been fabricated and applied as anode for LIB and resulted in enhanced cycle performance.

Silicon is the most attractive metal to be used as anode for LIB due to its soaring high theoretical capacity of 4200 mAh/g. However, the cracking of the electrode during the cycling process due to large volumetric changes leads to failure of silicon as an anode material after a few cycles. Si/MWCNT nanocomposites used as anode show a capacity as high as 1770 mAh/g with a negligible capacity loss over many cycles [34]. The CNT can accommodate the large volume expansion of the silicon layer and thereby control the overall volume changes of the electrode. There have been many other reports of using Si/MWCNTs of various architectures as anode for enhanced electrochemical performance [38–41].

Carbon nanotube networks provide the conduction pathways and voids for the silicon expansion during the electrochemical cycling session. Therefore, a composite of silicon with CNT shows better electrochemical performance than raw silicon anodes. Figure 9.7 demonstrates this significant development in the discharge capacity due to addition of CNT in the composite of silicon to be used as anode [42].

Another commonly used metal as anode for LIB is tin owing to its high theoretical capacity of 992 mAh/g which is almost thrice the capacity of graphite as anode. However, a pure Sn electrode faces 300% volume expansion during the cycling process and hence leads to electrode cracking. Similar to the case of silicon, MWCNTs can act as a buffer zone to prevent the breakdown of the Sn electrode during the volume changes and hence improve the overall electrochemical cycle

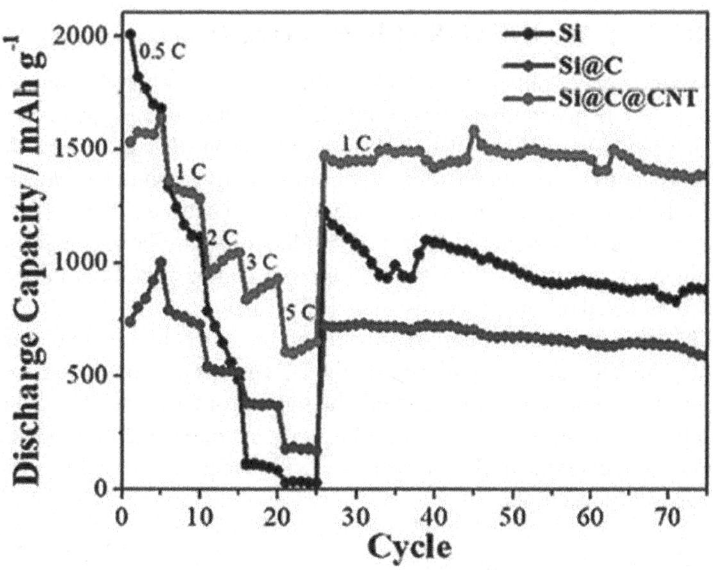

FIGURE 9.7 Rate performance of the pure Si, Si@C, and Si@C@CNT electrodes. Reprinted from [42] with permission from Elsevier.

performance. There have been various reports on Sn/MWCNTs as well as Sn/SnO_2/MWCNT nanocomposites fabricated using different techniques and used as anode material for LIB for better electrochemical performance [43–46].

Based on the previous studies, it was observed that CNTs are a promising material as an effective anode material for LIB applications. Various types of CNTs, with different structure, morphology, degree of graphitization, preparation technique, surface modifications have been discussed in this section in view of their electrochemical properties. In the end, metal-CNT nanocomposites as anode are reviewed with some examples.

It is interesting to note that the various components of LIB don't contribute to energy efficiency and hence lead to degradation of the overall energy density of the system. Moreover, the binder in the electrodes decreases the available specific area of the active materials thereby weakening the active lithium-ion transport. The binders also hamper the working temperature range due to fluctuations with temperature. Therefore binder-free, electrical conductor-free and current collector-free configurations can help achieve a higher energy density in LIBs. As a result, the development of a flexible, lightweight, binder-free and current collector-free electrode configuration to increase the energy density of LIBs has drawn the attention of some researchers [46].

The freestanding design of electrodes has several advantages over the conventional electrode design. First of all, the dead weight of the electrode significantly decreases due to removal of the binders and current collector. Secondly, it becomes easy to handle and hence flexible designs are also achievable especially for the lightweight and wearable electronic gadgets. At last, the problems due to the metal collectors, like corrosion, manufacturing issues, etc. are easily avoided in the case of freestanding electrodes.

Large flexible batteries can be easily placed in hollow spaces of the automobile body of future hybrid and electric vehicles. Needless to stress that high power and high energy density are expected. The structural and electrochemical properties of the electrode play an essential role in deciding the overall battery performance. Hence, the development of flexible electrodes, free-standing with high energy and power density, with long cycle life and safety becomes vital.

9.5 GRAPHENE AS ANODE FOR LI-ION BATTERY

In the most used anode material for Li-ion battery, the graphitic carbon, Li-ions are stored reversibly between its graphene layers up to a maximum lithiated state of LiC_6 at a low electrochemical potential. The theoretical possibility of reducing the lithiation time and improving the rate capability due to the lower transportation distance as compared to graphite has triggered the research for exploring graphene and its variants as an anode for Li-ion batteries. Owing to the large surface area, high aspect ratio, and outstanding electrical conductivity, graphene is expected to show high reversible lithium storage capacity since the lithium ions can be stored not only on both sides of graphene but also on its edges and defective sites [47–49].

Various studies have been carried out to understand the correlation between the electrochemical behavior of graphene and its synthesis methods, number of constituent layers, degree of ordering, specific surface area, its physical and chemical properties etc. These factors are found to have significant effects on the Li ion storage capacity, its rate capability, potential profile, and irreversible loss during cycling. In this section, we will discuss these features that influence the Li ion storage capability of graphene.

9.5.1 Methods of Graphene Preparation

Top-down and bottom-up approaches are followed for the synthesis of graphene. The top-down approach mainly involves preparation of graphene from bulk graphite through exfoliation methods either mechanically or chemically. The bottom-up approach involves synthesis of graphene from chemical precursors through techniques like chemical vapor deposition (CVD).

In CVD technique, graphene growth takes place through the decomposition of hydrocarbons on metal catalysts. The type of hydrocarbon precursors and the catalyst play a significant role in controlling the growth mechanism as well as number of layers in the grown graphene. By optimization of the growth parameters like temperature, time, gas flow, precursors, etc., it is possible to grow defect free single or multilayers of graphene using the CVD technique [50]. But due to the high cost, controlled experimental conditions, and low yield, CVD-grown graphene is not preferred as active material for electrodes in Li-ion batteries.

Reduce graphene oxide (rGO) can be prepared from bulk graphite by exfoliating graphene layers through oxidation process followed by reduction process. The most used method for the preparation of GO is the Hummers method. The electrical and chemical properties of the rGO mainly depends on the GO precursor and the reduction techniques used. So, it is important to optimize the synthesis route for obtaining high quality rGO for efficient energy storage in Li-ion batteries [51].

9.5.2 Number of Layers in Graphene

The number of layers in graphene determines its physical, structural, and electronic characteristics. The Li ion storage capacity of graphene is also influenced by the number of layers. The mechanism of Li ion storage also varies depending on the number of layers. In multi-layer graphene, Li intercalates between the graphene layers and forms Li-intercalated Li-C compounds as in graphite. In these staged intercalation compounds, the number of constituent graphene layers between the intercalated Li layer is denoted by the stage index N. The mechanism of Li ion storage in few-layers graphene (number of layers between 2 and 10), is still not very clear. Lee et al. [52], through DFT calculations, showed that the Li ions prefer to intercalate into the outermost interlayers compared to the inner ones. It was observed that for 6-layer graphene, the outermost gallery was filled first followed by the inner ones. But for 7-layer graphene, it was found that both the interior and exterior interlayers were getting filled simultaneously. This shows that the number of layers in the graphene determines the intercalation mechanism.

In bi-layer graphene, Li ions intercalate between the two layers like bulk graphite, but the staging mechanism is different as the intercalation in bi-layer graphene is only restricted to two dimensions. Research has found that the Li-ion diffusion in bi-layer graphene is faster compared to bulk graphite. In single-layer graphene, Li ions are chemically or physically adsorbed on one side or both sides of graphene. DFT calculations by Garay-Tapia et al. [53] have shown that in low Li-ion concentration, Li ions can form strong ionic bonds with C, while at higher concentration it is more of covalent bonding with carbon atoms.

9.6 INTRINSIC AND EXTRINSIC DEFECTS IN THE GRAPHENE

Single or multi-vacancy and Stone-Wales defects are the most common intrinsic defects observed in the graphene structures. Vacancy defects are mainly caused when the C atoms in the carbon hexagon go missing. Stones-Wales defect occurs when two carbon hexagons rearrange leading to the formation of non-hexagonal rings without the removal of any C atom. These defects can lead to increased Li-storage capacity due to higher charge transfer from Li ion to graphene. The intrinsic defects present in the graphene can act as preferential sites for Li-ion accumulation, therefore the specific capacity of graphene can be tuned by varying the density of defects in the graphene [54].

Extrinsic defects mainly come from the presence of foreign atoms like oxygen or other functional groups attached to the carbon hexagonal rings. The DFT simulations by Stournara and Shenoy [55] showed that Li-ion storage capacity and the lithiation potential can be varied depending upon the oxygen concentration. The presence of pores in graphene also enhances Li ion storage. There are also many other defects like edge effect, pores, stacking order degree of graphitization, etc., which also affect the Li storage capacity of graphene [56].

9.7 METAL-GRAPHENE COMPOSITE ANODES FOR LI-ION BATTERIES

Metals, due to their high specific capacity, are being widely used as anode for Li-ion batteries. But they experience high volume expansion during lithiation and de-lithiation due to which the cycling stability of metals are limited. Composites of metals with carbon materials help to reduce the stress due to the volume change and improve its cyclability and specific capacity.

In the work by Mingbo Ma et al. [57] three-dimensional (3D) lamellar SiOC-graphene composite was synthesized through hydrothermal reaction and electrostatic self-assembly process, in which SiOC powders encapsulated by amorphous carbon were homogeneously dispersed in graphene sheets. The composite anode material exhibited excellent rate capability and high specific capacity of 676 mAh g^{-1} at 200 mA g^{-1} current density. A full cell was also demonstrated with this anode and $LiFePO_4$ as cathode which demonstrated a stable voltage platform and good cyclability for 200 cycles (See Figure 9.8).

Transition metal phosphide Ni_5P_4 wrapped in fluffy graphene was synthesized through chemical vapor deposition and tested as anode material for Li-ion batteries [58]. The porous structure of Ni_5P_4 contributed to greater Li-ion diffusion, whereas the graphene provided a stable solid electrolyte interface and an electron-conducting layer. The composite anode material showed a high reversible capacity of 739 $mAhg^{-1}$ at a current density of 500 mAg^{-1} for 300 cycles.

Though graphene is a promising anode material for Li-ion batteries, it has many disadvantages. The irreversible Li-ion loss especially during the first electrochemical cycle, higher production cost, structural stability on repeated cycling and poor thermal stability as compared to graphitic carbon has been identified as some of the major concerns in using graphene-based anodes. Also, the charge/discharge potential profiles of graphene-based anodes are higher than the average potential exhibited by graphitic carbon due to the different Li intercalation /deintercalation

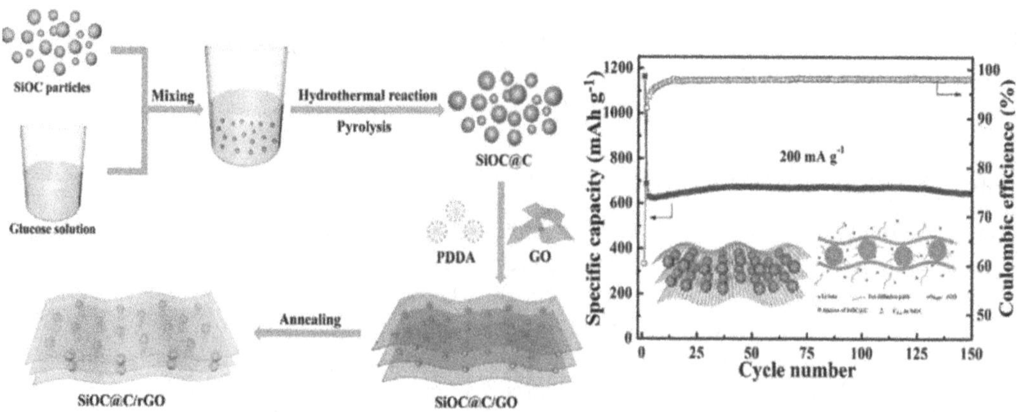

FIGURE 9.8 Fabrication of 3D lamellar SiOC-graphene composite. Reprinted from [57] with permission from Elsevier.

mechanism in graphene. As a result of this, the overall voltage of Li-ion full cell will be lower as compared graphitic carbon [59].

9.8 CONCLUDING REMARKS AND FUTURE SCOPE

The recent award of Nobel prize in 2019 to John B. Goodenough, M. Stanley Whittingham, and Akira Yoshino for their pioneering work on LIB reinforces the important role that LIBs play in current day technology. LIB is a ripe area for innovations and research groups around the world are tirelessly working to maximize the potential with future applications such as advanced robotics, drones, satellites, miniature electrical devices, and fully electric aircrafts.

This chapter discusses in detail the application of CNTs and graphene as anode for LIBs. Researchers have worked on various types of CNTs that were showcased categorically. CNT-metal nanocomposites show the greatest potential to revolutionize batteries. Similarly, graphene has been intensively investigated as a potential anode material. Based on the discussion, it can be concluded that graphene makes significant improvement to cycle performance of anode and shows higher potential that graphite anodes. An interesting hybrid structure of graphene-CNT carpets that was applied as anodes and cathodes in binder-free lithium-ion capacitors, producing stable devices with high energy densities (120 Wh/kg) was reported by Salvatierra et al. in 2017 [60].

However, there is still a long way to go for commercial applications of graphene and CNTs as anode material as currently the cycling stability, low reversible capacity, and cost of production are a major hinderance. The lack of reproducibility of nanomaterials is a major challenge to achieve lab-to-industrial fabrication. Thus, more efforts should be devoted to developing high rate and stable cycling anodes that are commercially viable using CNTs, graphene and nanocomposites based on them.

REFERENCES

1. Statista (2021) Global smartphone shipments forecast from 2010 to 2022(in million units). www.statista.com/statistics/263441/global-smartphone-shipments-forecast/
2. Puiu Tibi (2020) *The Future is Bright for Lithium-Ion Batteries*. Found. Lindau Nobel Laureate Meet.
3. Goriparti S, Miele E, De Angelis F, Di Fabrizio E, Proietti Zaccaria R, Capiglia C (2014) Review on recent progress of nanostructured anode materials for Li-ion batteries. J Power Sources 257:421–443.
4. Zhao M, Xia Y, Liu X, Tan Z, Huang B, Li F, Ji Y, Song C (2005) Curvature-induced condensation of lithium confined inside single-walled carbon nanotubes: First-principles calculations. Phys Lett A 340:434–439.
5. Wu YP, Rahm E, Holze R (2003) Carbon anode materials for lithium ion batteries. J Power Sources 114:228–236. https://doi.org/10.1016/S0378-7753(02)00596-7
6. Ehl RG, Ihde AJ (1954) Faraday's electrochemical laws and the determination of equivalent weights. J Chem Educ 31:226.
7. Andreas T, Roland L, Pavel N, Hongjie D, Pierre P, Jerome R, Chunhui X, Hee LY, Gon KS, G. RA, T. CD, E. SG, David T, E. FJ, E. SR (1996) Crystalline Ropes of Metallic Carbon Nanotubes. Science (80-) 273:483–487. https://doi.org/10.1126/science.273.5274.483
8. Yang S, Song H, Chen X, Okotrub A V, Bulusheva LG (2007) Electrochemical performance of arc-produced carbon nanotubes as anode material for lithium-ion batteries. Electrochim Acta 52:5286–5293. https://doi.org/10.1016/j.electacta.2007.02.049
9. Wang GX, Ahn J, Yao J, Lindsay M, Liu HK, Dou SX (2003) Preparation and characterization of carbon nanotubes for energy storage. J Power Sources 119–121:16–23. https://doi.org/10.1016/S0378-7753(03)00117-4

10. Kawasaki S, Hara T, Iwai Y, Suzuki Y (2008) Metallic and semiconducting single-walled carbon nanotubes as the anode material of Li ion secondary battery. Mater Lett 62:2917–2920.
11. Landi BJ, Ganter MJ, Schauerman CM, Cress CD, Raffaelle RP (2008) Lithium Ion Capacity of Single Wall Carbon Nanotube Paper Electrodes. J Phys Chem C 112:7509–7515. https://doi.org/10.1021/jp710921k
12. Lahiri I, Oh S-W, Hwang JY, Cho S, Sun Y-K, Banerjee R, Choi W (2010) High capacity and excellent stability of lithium ion battery anode using interface-controlled binder-free multiwall carbon nanotubes grown on copper. ACS Nano 4:3440–3446.
13. Paul R, Etacheri V, Pol VG, Hu J, Fisher TS (2016) Highly porous three-dimensional carbon nanotube foam as a freestanding anode for a lithium-ion battery. RSC Adv 6:79734–79744.
14. Bulusheva LG, Arkhipov VE, Fedorovskaya EO, Zhang S, Kurenya AG, Kanygin MA, Asanov IP, Tsygankova AR, Chen X, Song H (2016) Fabrication of free-standing aligned multiwalled carbon nanotube array for Li-ion batteries. J Power Sources 311:42–48.
15. Lu W, Goering A, Qu L, Dai L (2012) Lithium-ion batteries based on vertically-aligned carbon nanotube electrodes and ionic liquid electrolytes. Phys Chem Chem Phys 14:12099–12104.
16. Wang GX, Yao J, Liu HK, Dou SX, Ahn J-H (2006) Growth and lithium storage properties of vertically aligned carbon nanotubes. Met Mater Int 12:413–416.
17. Ren J, Zhang Y, Bai W, Chen X, Zhang Z, Fang X, Weng W, Wang Y, Peng H (2014) Elastic and wearable wire-shaped lithium-ion battery with high electrochemical performance. Angew Chemie 126:7998–8003.
18. Yang S, Huo J, Song H, Chen X (2008) A comparative study of electrochemical properties of two kinds of carbon nanotubes as anode materials for lithium ion batteries. Electrochim Acta 53:2238–2244. https://doi.org/10.1016/j.electacta.2007.09.040.
19. Kang C, Lee H-J (2018) Morphological control of three-dimensional carbon nanotube anode for high-capacity lithium-ion battery. Jpn J Appl Phys 57:05GC05.
20. Frackowiak E, Gautier S, Gaucher H, Bonnamy S, Beguin F (1999) Electrochemical storage of lithium in multiwalled carbon nanotubes. Carbon N Y 37:61–69. https://doi.org/10.1016/S0008-6223(98)00187-0.
21. Lam C, James JT, McCluskey R, Arepalli S, Hunter RL (2006) A Review of Carbon Nanotube Toxicity and Assessment of Potential Occupational and Environmental Health Risks. Crit Rev Toxicol 36:189–217. https://doi.org/10.1080/10408440600570233
22. Gao B, Kleinhammes A, Tang XP, Bower C, Fleming L, Wu Y, Zhou O (1999) Electrochemical intercalation of single-walled carbon nanotubes with lithium. Chem Phys Lett 307:153–157.
23. DiLeo RA, Castiglia A, Ganter MJ, Rogers RE, Cress CD, Raffaelle RP, Landi BJ (2010) Enhanced capacity and rate capability of carbon nanotube based anodes with titanium contacts for lithium ion batteries. ACS Nano 4:6121–6131.
24. Simon GK, Goswami T (2011) Improving anodes for lithium ion batteries. Metall Mater Trans A 42:231–238.
25. Arepalli S (2004) Laser ablation process for single-walled carbon nanotube production. J Nanosci Nanotechnol 4:317–325.
26. Tan KH, Ahmad R, Leo BF, Yew MC, Ang BC, Johan MR (2012) Physico-chemical studies of amorphous carbon nanotubes synthesized at low temperature. Mater Res Bull 47:1849–1854. https://doi.org/10.1016/j.materresbull.2012.04.073
27. Xu L, Zhang Y, Zhang X, Huang Y, Tan X, Huang C, Mei X, Niu F, Meng C, Cheng G (2014) Designed synthesis of tunable amorphous carbon nanotubes (a-CNTs) by a novel route and their oxidation resistance properties. Bull Mater Sci 37:1397–1402.
28. Nai J, Zhao X, Yuan H, Tao X, Guo L (2021) Amorphous carbon-based materials as platform for advanced high-performance anodes in lithium secondary batteries. Nano Res 1–14.
29. Zhao T, Liu Y, Li T, Zhao X (2010) Electrochemical performance of amorphous carbon nanotube as anode materials for lithium ion battery. J Nanosci Nanotechnol 10:3873–3877.
30. Wu GT, Wang CS, Zhang XB, Yang HS, Qi ZF, He PM, Li WZ (1999) Structure and lithium insertion properties of carbon nanotubes. J Electrochem Soc 146:1696.

31. Yoon S, Lee S, Kim S, Park K-W, Cho D, Jeong Y (2015) Carbon nanotube film anodes for flexible lithium ion batteries. J Power Sources 279:495–501. https://doi.org/10.1016/j.jpowsour.2015.01.013
32. Xiong Z, Yun YS, Jin H-J (2013) Applications of Carbon Nanotubes for Lithium Ion Battery Anodes. Mater. 6.
33. Sehrawat P, Julien C, Islam SS (2016) Carbon nanotubes in Li-ion batteries: A review. Mater Sci Eng B 213:12–40.
34. Eom JY, Park JW, Kwon H-S, Rajendran S (2006) Electrochemical insertion of lithium into multiwalled carbon nanotube/silicon composites produced by ballmilling. J Electrochem Soc 153:A1678.
35. Liu X-M, Huang Z dong, Oh S woon, Zhang B, Ma P-C, Yuen MMF, Kim J-K (2012) Carbon nanotube (CNT)-based composites as electrode material for rechargeable Li-ion batteries: A review. Compos Sci Technol 72:121–144.
36. Roselin LS, Juang R-S, Hsieh C-T, Sagadevan S, Umar A, Selvin R, Hegazy HH (2019) Recent advances and perspectives of carbon-based nanostructures as anode materials for Li-ion batteries. Materials (Basel) 12:1229.
37. Elizabeth I, Singh BP, Gopukumar S (2019) Electrochemical performance of Sb2S3/CNT free-standing flexible anode for Li-ion batteries. J Mater Sci 54:7110–7118.
38. Kim T, Mo YH, Nahm KS, Oh SM (2006) Carbon nanotubes (CNTs) as a buffer layer in silicon/CNTs composite electrodes for lithium secondary batteries. J Power Sources 162:1275–1281.
39. Na Y-S, Yoo H, Kim T-H, Choi J, Lee WI, Choi S, Park D-W (2015) Electrochemical performance of Si-multiwall carbon nanotube nanocomposite anode synthesized by thermal plasma. Thin Solid Films 587:14–19.
40. Tocoglu U, Cevher O, Guler MO, Akbulut H (2014) Coaxial silicon/multi-walled carbon nanotube nanocomposite anodes for long cycle life lithium-ion batteries. Appl Surf Sci 305:402–411.
41. Cetinkaya T, Tocoglu U, Cevher O, Guler MO, Akbulut H (2014) Electrochemical performance of silicon/MWCNT composite electrodes for lithium ion batteries. Acta Phys Pol A 125:285–287.
42. Zhu X, Choi SH, Tao R, Jia X, Lu Y (2019) Building high-rate silicon anodes based on hierarchical Si@C@CNT nanocomposite. J Alloys Compd 791:1105–1113.
43. Alaf M, Gultekin D, Akbulut H (2013) Electrochemical properties of free-standing Sn/SnO2/multi-walled carbon nano tube anode papers for Li-ion batteries. Appl Surf Sci 275:244–251.
44. Korusenko PM, Nesov SN, Bolotov V V, Povoroznyuk SN, Sten'kin YA, Pushkarev AI, Fedorovskaya EO, Smirnov DA (2019) Structure and electrochemical characterization of SnOx/Sn@MWCNT composites formed by pulsed ion beam irradiation. J Alloys Compd 793 :723–731.
45. Uysal M, Cetinkaya T, Alp A, Akbulut H (2014) Production of Sn/MWCNT nanocomposite anodes by pulse electrodeposition for Li-ion batteries. Appl Surf Sci 290:6–12.
46. Elizabeth I, Mathur RB, Maheshwari PH, Singh BP, Gopukumar S (2015) Development of SnO2/multiwalled carbon nanotube paper as free standing anode for lithium ion batteries (LIB). Electrochim Acta 176:735–742.
47. Dahn JR, Zheng T, Liu Y, Xue JS (1995) Mechanisms for lithium insertion in carbonaceous materials. Science (80-) 270:590–593.
48. Sonia FJ, Jangid MK, Ananthoju B, Aslam M, Johari P, Mukhopadhyay A (2017) Understanding the Li-storage in few layers graphene with respect to bulk graphite: experimental, analytical and computational study. J Mater Chem A 5:8662–8679.
49. Elizabeth I, Singh BP, Bijoy TK, Reddy VR, Karthikeyan G, Singh VN, Dhakate SR, Murugan P, Gopukumar S (2017) In-situ conversion of multiwalled carbon nanotubes to graphene nanosheets: an increasing capacity anode for Li Ion batteries. Electrochim Acta 231:255–263.
50. Mattevi C, Kim H, Chhowalla M (2011) A review of chemical vapour deposition of graphene on copper. J Mater Chem 21:3324–3334.
51. Zhao J, Pei S, Ren W, Gao L, Cheng H-M (2010) Efficient preparation of large-area graphene oxide sheets for transparent conductive films. ACS Nano 4:5245–5252.
52. Lee E, Persson KA (2012) Li absorption and intercalation in single layer graphene and few layer graphene by first principles. Nano Lett 12:4624–4628.

53. Garay-Tapia AM, Romero AH, Barone V (2012) Lithium adsorption on graphene: from isolated adatoms to metallic sheets. J Chem Theory Comput 8:1064–1071.
54. Banhart F, Kotakoski J , Krasheninnikov A V (2011) Structural defects in graphene. ACS Nano 5:26–41.
55. Stournara ME, Shenoy VB (2011) Enhanced Li capacity at high lithiation potentials in graphene oxide. J Power Sources 196:5697–5703.
56. Mukhopadhyay A, Guo F, Tokranov A, Xiao X, Hurt RH, Sheldon BW (2013) Engineering of Graphene Layer Orientation to Attain High Rate Capability and Anisotropic Properties in Li-Ion Battery Electrodes. Adv Funct Mater 23:2397–2404.
57. Ma M, Wang H, Xiong L, Huang S, Li X, Du X (2022) Self-assembled homogeneous SiOC@ C/graphene with three-dimensional lamellar structure enabling improved capacity and rate performances for lithium ion storage. Carbon N Y 186:273–281.
58. Zhang C, Park G, Lee B-J, Xia L, Miao H, Yuan J, Yu J-S (2021) Self-Templated Formation of Fluffy Graphene-Wrapped Ni5P4 Hollow Spheres for Li-Ion Battery Anodes with High Cycling Stability. ACS Appl Mater Interfaces 13:23714–23723.
59. Sonia FJ, Aslam M, Mukhopadhyay A (2020) Understanding the processing-structure-performance relationship of graphene and its variants as anode material for Li-ion batteries: A critical review. Carbon N Y 156:130–165.
60. Salvatierra RV, Zakhidov D, Sha J, Kim ND, Lee S-K, Raji A-RO, Zhao N, Tour JM (2017) Graphene carbon nanotube carpets grown using binary catalysts for high-performance lithium-ion capacitors. ACS Nano 11:2724–2733.

10 Carbon Nanotubes and Graphene-Based Thermoelectric Materials

A Futuristic Approach for Energy Harvesting

Kriti Tyagi, Ajay K. Verma, Bhasker Gahtori and Sanjay R. Dhakate

CONTENTS

10.1 INTRODUCTION

More than one-third of the current world-wide greenhouse gas emission can be accounted to the pedigree of electricity production such as natural gas, oil, coal, etc. It becomes indispensable to augment the standard of living by seeking for more reliable and cleaner form of energy generation [1].

Owing to the vast development of transportation and industry sectors, the world-wide energy demand has escalated dramatically. The overall energy consumption of all the countries is

DOI: 10.1201/9781003231943-10

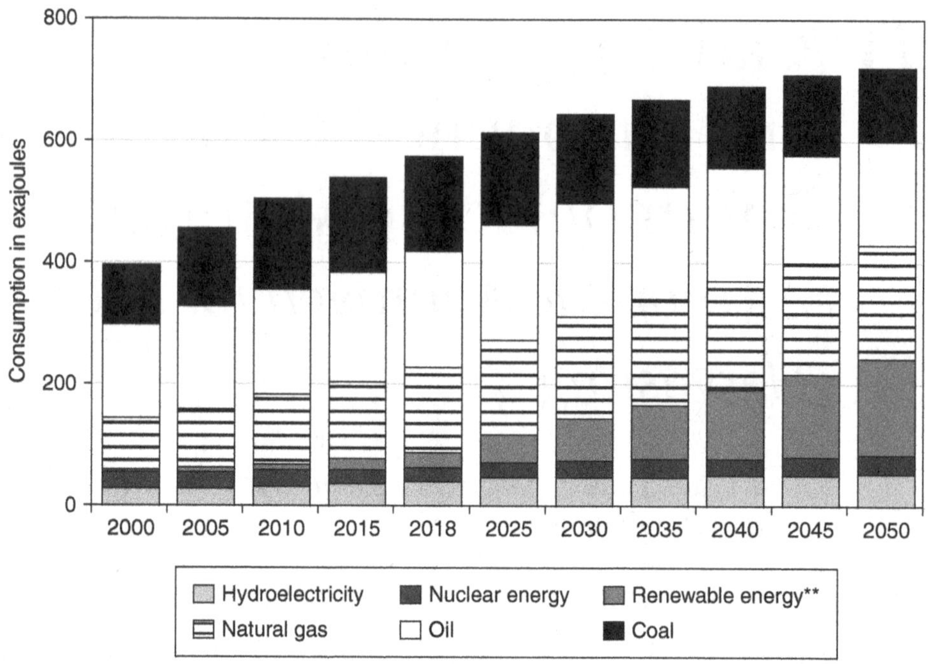

FIGURE 10.1 The projected global energy consumption. Copyright Statista 2022.

expected to rise significantly by the year 2050 (Figure 10.1). Also, the area and growth of a country's population marks a significant contribution toward the energy demand.

There are several reasons for the energy crisis situation [2]:

1. With the current consumption model relying substantially on the exploitation of non-renewable energy sources, the reserves of coal, oil, gas, uranium, etc. are fast depleting.
2. The unceasingly increasing world population poses a threat, as a result of which, the universal energy requirement is expected to escalate by at least 50% by the year 2030.
3. The infrastructure needs to be updated so as to maximize the energy production.
4. The continuous use of prevailing traditional energy sources leads to, in addition to several other effects, an increase in global warming. Thus, energy crisis and environmental crisis are inter-related.
5. The substantial contribution to ongoing energy crisis is a consequence of the problem of dispensable use of energy resources, particularly, electricity and fuel.

In order to provide a solution to the rapidly enhancing energy crisis problem we need to explore renewable sources of energy. Among various other advantages over conventional fossil fuels, these sources of energy don't contribute toward greenhouse gas emission and are known as alternative energy solutions. These sustainable, clean, and green energy solutions comprise of elements such as solar, hydropower, wind, biomass, geothermal, heat, etc. Another measure to prevent an energy crisis lies in minimal consumption of energy by providing modern and improved energy infrastructure in the form of smart cities.

In our day-to-day life, and also in the industrial sector, a large amount of the heat generated is unavoidably wasted into the environment. The foremost contribution to unused heat arises from automotive units, wherein only about 20% the total energy is converted into beneficial energy

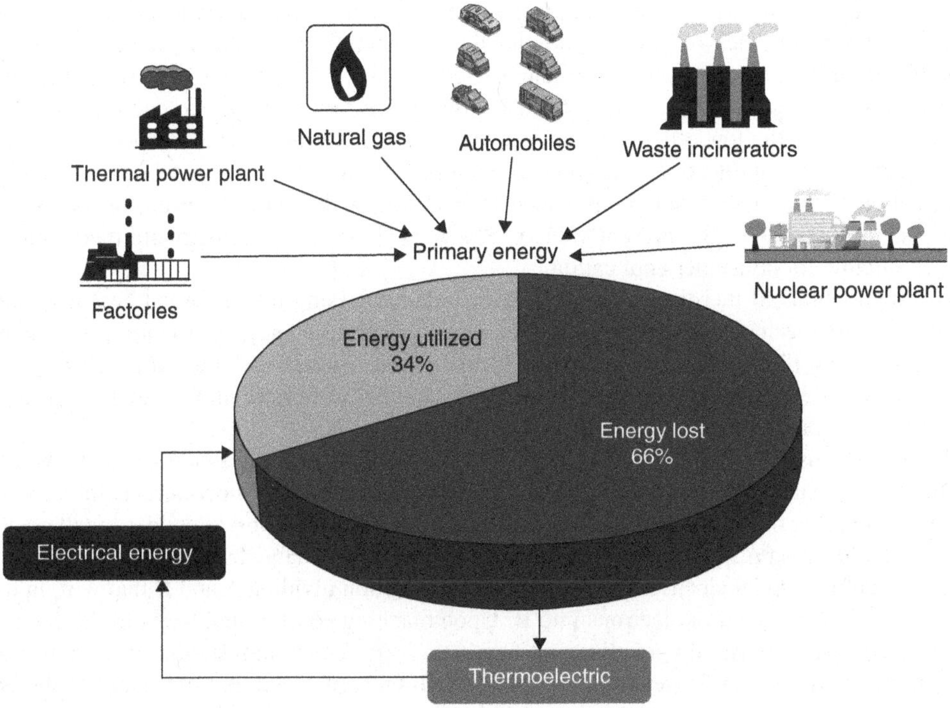

FIGURE 10.2 Sectors resulting in large amount of waste heat energy.

and the rest is discarded as heat. Both automotive and aeronautics depict the best examples of low efficiency conversion though they incur high energy usage. Almost 75% of the total energy generated during combustion vanishes in the engine coolant or exhaust/turbine as heat energy. To be accurate, around 60% or more of the energy created is seldom used, since a major part of it gets dissipated as waste heat.

Figure 10.2 depicts different sectors generating an immense amount of waste heat, in addition to, the efficiently used energy from non-renewable sources of energy. The wasted heat energy has the possibility to be converted to valuable electrical energy employing thermoelectric (TE) materials. Thus, thermoelectric materials and devices are thought of as viable solutions to the neglectance of waste heat energy revival by efficiently transforming it to a useful form of energy.

10.2 THERMOELECTRIC AND RELATED EFFECTS

Sustainable development can be brought about when important issues pertaining to clean energy generation and its conservation are addressed [3–5]. The focus needs to be shifted from major dependence on non-renewable fossil fuels to the development of alternative power generation methods that are eco-friendly, efficient, and cost–effective [4, 6, 7]. The huge increase in renewable energy generation is leading to rapid progression of thermoelectric energy exchange modules [8–10]. Since more than 60% of the aggregate energy generated is lost as waste heat [11], thermoelectric modules with the potential to directly transform waste heat ejected from various sources into electrical energy [12, 13] employing Seebeck effect, can prove to be a viable solution to the increasing energy requirements, while simultaneously addressing issues related to global warming [14].

The thermoelectric energy generation technology has many advantages, including scalability, solid-state operation, lack of moving parts leading to maintenance free operation, environment friendly [15] and reliable operation over long period of time [4, 16]. Thermoelectric devices have a varied range of application ranging from energy generation, cooling to autonomous power systems. However, the low efficiency and expensiveness of thermoelectric devices restricts their use for commercial applications. In order to contest with the prevailing power production and refrigeration techniques, novel thermoelectric materials with thermoelectric conversion efficiency about three times the present value (~5–7%) [13] are required to enable thermoelectric devices suitable for consumer application.

The applications of thermoelectric devices include harnessing the discarded heat from automobile exhaust mechanism that leads to escalation of fuel efficiency, distributed power generation, preserving biological specimens through thermoelectric refrigeration, infrared detectors, improvement in cooling techniques for computers, laptops, and improving the quality of life by using thermoelectric devices in cooking stoves, etc. [4, 17, 18].

When two dissimilar metals, A and B, having dissimilar work functions and chemical potentials are connected together and both maintained at the same temperature T, initially $\mu_A \neq \mu_B$, a small number of electrons will start to diffuse from A into B, consequently, a small positive charge is build-up on A and, at the same time, a small negative charge builds-up on B. This leads to generation of a trivial electric field right at the intersection dividing A and B that will, in turn, oppose further movement of electrons into B. Upon reaching equilibrium, the both the chemical potentials in must essentially equalize and hence $\mu_A = \mu_B$. The manifestation of a temperature gradient results in voltage development between both ends of metal, but, inversely, if the ends of A and B are at same temperatures, there will be absence of voltage difference (Figure 10.3).

10.2.1 Seebeck Effect

Thomas Johann Seebeck observed the first of the thermoelectric effects in 1821 [20]. The Seebeck effect states that when two electrically conducting dissimilar materials are linked in a closed circuit, maintaining temperature gradient at the two junctions, electric current tends to flow in the closed circuit. A voltage difference (Δ V) will be developed between the two materials when the circuit is open (Figure 10.4), due to diffusion of high energy carriers from hot end across to the cold end of the conductor. It is observed that the voltage difference varies directly with the

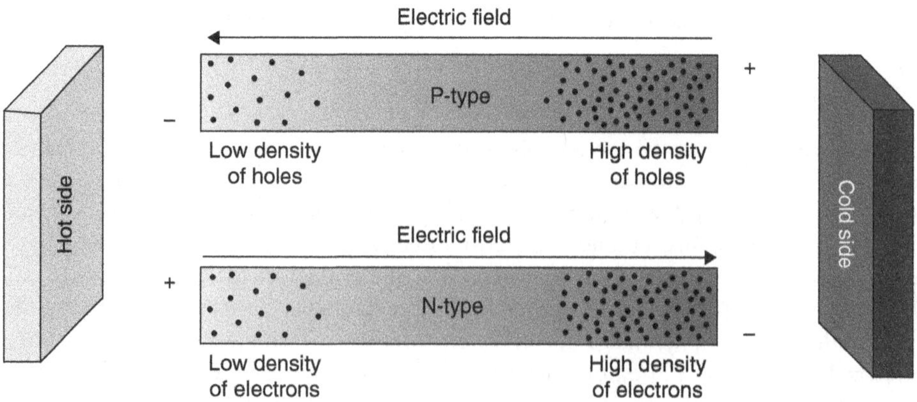

FIGURE 10.3 A schematic depiction of TE effect. Reprinted with permission from [19]. Copyright 2019 @ from IntechOpen.

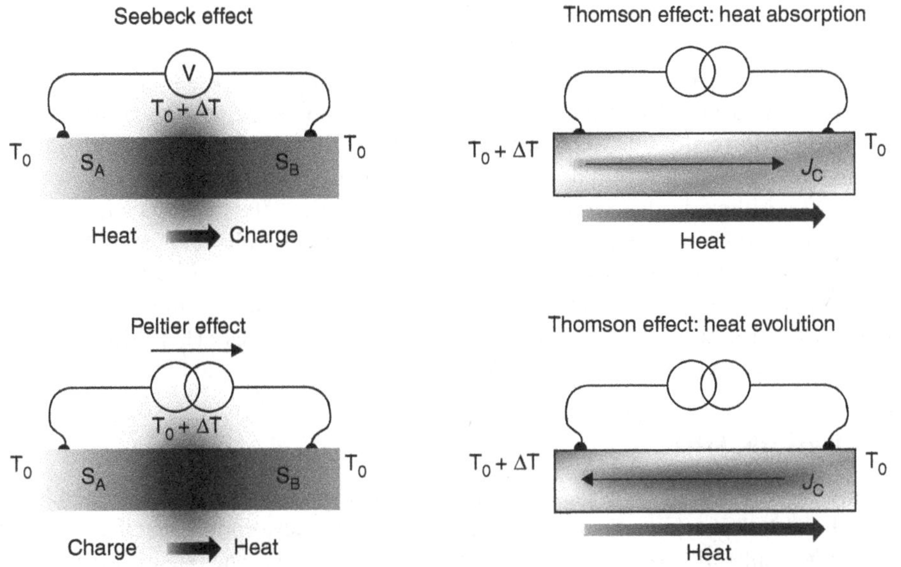

FIGURE 10.4 Schematic of a Seebeck, Peltier and Thomson effect. Reprinted with permission from [21]. Copyright 2020 @ from American Physical Society.

temperature gradient between the two junctions. The fraction of voltage difference to the temperature gradient is termed as Seebeck coefficient (*S*), alternatively called thermopower.

$$Q = \tau I \Delta T \tag{10.1}$$

where, *d*V is the developed voltage, *d*T is applied temperature difference, *S* is Seebeck coefficient. The Seebeck coefficient, $S = \frac{dV}{dT}$, is a measure of amount of the thermoelectric voltage generated as a result of the temperature gradient maintained in the material. The Seebeck coefficient fluctuates with temperature and relies strongly on the configuration of the conductor. On application of temperature gradient to the material having electrons as majority charge carriers, voltage and Seebeck coefficient are both negative. Such materials are defined as n-type thermoelectric materials. Similarly, if holes are the majority charge carriers, Seebeck coefficient has a positive value and material is considered as p-type thermoelectric material.

10.2.2 Peltier Effect

A related phenomenon was studied by Jean Charles Anthanase Peltier [22], a few years after Seebeck's observation. Peltier effect is opposite of Seebeck effect as it deals with conversion of electricity into heat transfer. If an electric current is passed through the intersection of two dissimilar materials, one of the junctions becomes hot while the other becomes cold. The difference in Fermi energy of the two materials causes this effect. Heat absorption and rejection sites change on changing the path of current flow. This effect is termed as Peltier effect (Figure 10.4).

The heat generated or absorbed is directly proportional to the applied current.

$$Q \,\alpha\, I \tag{10.2}$$

$$Q = \Pi . I \tag{10.3}$$

where, Q is defined as heat absorbed or generated, I is the applied current, and Π denotes Peltier coefficient measured in volts. Peltier coefficient is thus defined as amount of heat liberated or absorbed at the intersection of two different materials while one coulomb magnitude of charge crosses the junction. The functioning principle of the thermoelectric refrigerators is centered on this effect [23]. Although, the Seebeck effect and the Peltier effect both occur at the junction between two dissimilar materials they are not merely interfacial phenomenon but depend upon the bulk characteristics of the different materials involved. In a conductor, the electric current is carried by electrons. These electrons retain varied energies in dissimilar materials. When current travels through a material to another, the electron energy changes. This difference in energy manifests as cooling or heating at the intersection.

10.2.3 Thomson Effect

The third thermoelectric effect was anticipated and then successfully demonstrated by William Thomson (Lord Kelvin) in the year 1851. It is related to the absorption or release of heat along a conductor on application of temperature gradient (Figure 10.4). The rate of heat unconfined or absorbed varies directly to the temperature difference and also to the current density.

$$Q = \tau I \Delta T \tag{10.4}$$

where, τ is Thomson coefficient.

It was later established by Thomson, that the Seebeck effect and the Peltier effect are thermodynamically reversible and the connection between Seebeck and Peltier coefficients is given by:

$$\Pi = S . T \tag{10.5}$$

where, Π, S are Peltier and Seebeck coefficient respectively and T is the temperature in Kelvin.

10.3 REQUIREMENTS FOR FABRICATING A TE DEVICE

Thermoelectric devices have the capability to convert thermal energy to electrical energy on application of a temperature difference. A thermoelectric module comprises of n- and p-type legs joined electrically in series combination and thermally in parallel (Figure 10.5). The advantage of a thermoelectric device is that it includes no moving parts, is portable, environment friendly, makes no noise, etc. The efficiency of a thermoelectric device is shown:

$$\eta = \eta_c \frac{\sqrt{1+ZT}-1}{\sqrt{1+ZT}+\frac{T_C}{T_H}} \tag{10.6}$$

is the Carnot's efficiency, $\Delta T = T_H - T_C$, T_H is the temperature at the hot end, T_C is the temperature of the cold end, ZT is the efficiency of the material.

$$\eta_C = \frac{\Delta T}{T_H} \tag{10.7}$$

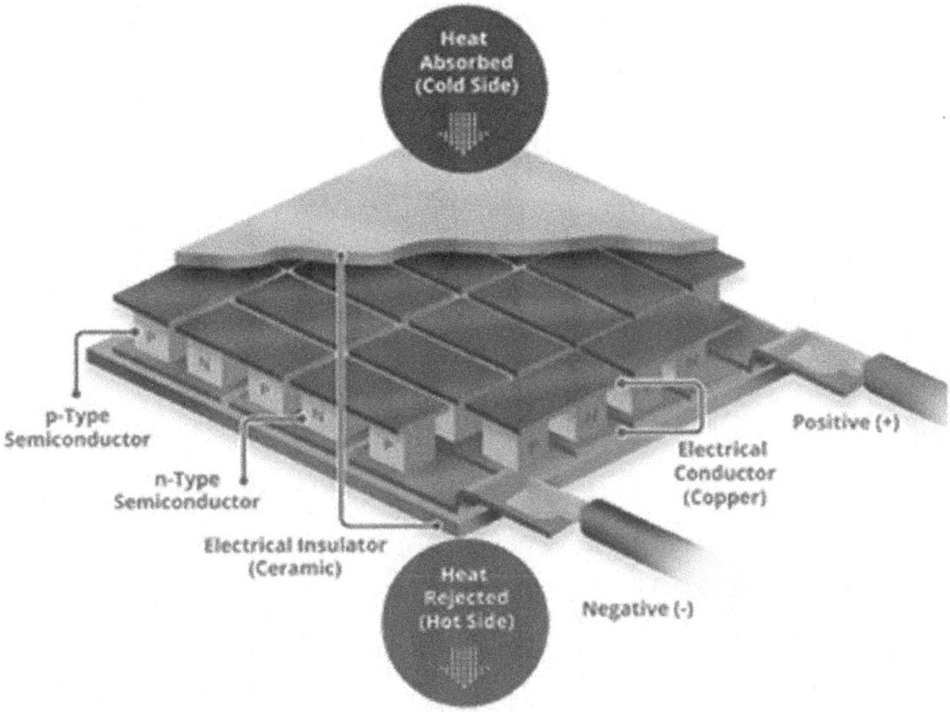

FIGURE 10.5 Depiction of an actual thermoelectric device. Reprinted with permission from [25]. Copyright 2019 @ from MDPI.

Thermoelectric efficiency was first derived by Edmund Altenkirch in 1911 [24]. He proposed the theory indicating coefficient of performance in a thermoelectric cooler and the efficiency value of thermoelectric generator.

The thermoelectric figure-of-merit (ZT) is a material property and is described as:

$$ZT = \frac{S^2 \sigma T}{\kappa} \tag{10.8}$$

$$\kappa = \kappa_e + \kappa_l \tag{10.9}$$

where, S, σ, are the Seebeck coefficient and electrical conductivity, respectively, κ_e and κ_l are electronic and lattice thermal conductivity, respectively, T is absolute temperature. The term $S^2\sigma$ is known as power factor. The Seebeck coefficient S measures the entropy in the system expressed generally in μV/K. The value of the Seebeck coefficient for a conductor is not absolute but relative in manner. The electrical conductivity of a thermoelectric material is reported to lie in between 10^2–10^3 S/m [26]. The electrical conductivity of semiconductor material depends on the carrier concentration as well as mean-free path taken by the charge carriers. Efforts are being made to minimize the lattice part of thermal conductivity by scattering phonons over a wide range of wavelengths using different approaches.

From equation 10.8, it can be inferred that in an attempt to obtain high ZT value, the material must possess a high Seebeck coefficient and high electrical conductivity so as to permit electricity conduction. This would result in a potential difference through the sample. Since, all these are inter-dependent parameters, it is highly difficult to decrease κ with simultaneous decrease in σ.

These parameters are mutually dependent on each other owing to Wiedmann-Frenz law:

$$\kappa_e = L\sigma T \tag{10.10}$$

where, L denotes Lorentz number, given by

$$L = \left(\frac{\Pi^2}{3}\right)\left(\frac{k_B}{e}\right)^2 \tag{10.11}$$

where, k_B symbolizes Boltzmann constant, *e* is the electronic charge. Wiedmann-Frenz law applies to the electronic contribution to thermal conductivity, which amounts to merely 1/3 of the total contribution to electrical conductivity [27].

The dependence of *S* on carrier concentration (Pisarenko relation) is expressed as:

$$S = \frac{8\pi^2 k_B^2}{3eh^2} m^* T(\frac{\pi}{3n})^{\frac{2}{3}} \tag{10.12}$$

where, k_B, *e*, *h*, m^*, *n* denote Boltzmann constant, electronic charge, Planck's constant, carrier effective mass and carrier concentration, respectively.

The electrical conductivity dependence on *n* is given by:

$$\sigma = ne\mu \tag{10.13}$$

where, μ is carrier mobility. It can be inferred increasing the carrier concentration of a material, causes S to decrease which leads to an increase in σ. Thus, overall effect on Z is minimized. Also, a large value of effective mass produces large *S* but tends to degrade carrier mobility given by:

$$\mu = e\tau / m^* \tag{10.14}$$

where, τ is scattering time.

Figure 10.6 illustrates the three main parameters w.r.t. carrier concentration [27]. Since σ and S differ in a reciprocal manner, allowing an enhancement in ZT is difficult. Additionally, σ and *S* are inversely linked, so it becomes difficult to enhance the power factor beyond a certain optimum value. A ZT value of 2.5–3 is required to enable widespread usage of thermoelectric device for power generation and refrigeration applications. Metals showcase a very small value of the Seebeck coefficient hence, nearly all thermoelectric materials fall in the category of semiconductors [28].

10.4 DIFFERENT TYPES OF THERMOELECTRIC MATERIALS

The low efficiency of TE devices limits their application when compared with other commercially available devices. However, the important role of TE materials in recovery of automotive or industrial waste heat [30–37] utilization of solar thermal energy [10, 38–44] and application of solid-state refrigeration [45–47], cannot be denied. The ZT for TE materials from 1950 till today has increased stepwise. The ZT value of most of the state-of-the-art TE materials, for e.g. Bi_2Te_3 [48–50], SiGe [51–53] and PbTe [54–57], lie less than or near unity. Several techniques are employed to decouple the inter-dependent parameters that constitute the figure-of-merit.

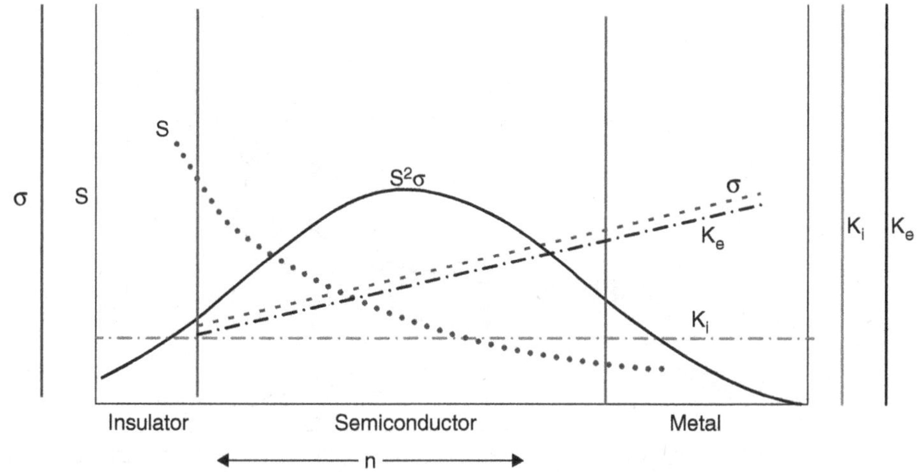

FIGURE 10.6 Variation of Seebeck coefficient (S), electrical conductivity (σ), power factor ($S^2\sigma$) and thermal conductivity (κ) with carrier concentration. Reprinted with permission from [29]. Copyright 2018 @ from Institute of Physics.

Approaches such as band gap engineering, controlled disorder and nanostructuring were applied to design new materials and novel methods of synthesis resulting in a significant increase of ZT.

10.4.1 Inorganic Thermoelectric Materials

New materials are being intensively searched for their suitability for thermoelectric applications. Figure 10.7 illustrates the ZT of various thermoelectric materials alongwith the year illustrating important milestones. The thermoelectric materials are further categorized into various generations depending upon the value of ZT. The materials having average ZT value ~1 are termed first generation, materials having ZT between 1 and 1.7 fall under second generation and the recently studied materials with the ZT value > 1.8 are specified as third generation materials. Also, according to their operating temperature range they are classified as low-, mid- and high-temperature thermoelectric materials. The current state-of-the-art thermoelectric materials are divided into various categories like skutterudites, half-heuslers, zintl phase, clathrates, oxides, SiGe alloys, chalcogenides, etc. Over the years, continuous research has been conducted in an attempt to synthesize high efficiency thermoelectric materials. Figure 10.8 shows the timeline of prominent thermoelectric materials since the year 1950. The potential thermoelectric compounds are characterized by complex crystal structure that helps to achieve low κ and complex Fermi surface required for obtaining high values of power factor. Slack, in 1994, predicted the impact of loosely bonded atom in crystal cage of skutterudites leading toward low lattice thermal conductivity [59]. This further ignited a deep search for novel filled skutterudites and other compounds bearing crystalline cages. This led to evolution of clatharates compounds having similar crystalline cage structure [60]. Other compounds with complex structure sans the crystalline cage geometry, e.g., zintl and complex chalcogenides, were also researched upon [61, 62]. Some other compounds with intrinsically low κ_L and sufficiently large Fermi surface were also studied [63, 64].

One of the prominent challenges for devices produced employing bulk synthesized inorganic materials is reduced flexibility. In order to address the issue, inorganic materials can be prepared as thin films deposited on various substrates, e.g., flexible organic substrates, CNTs, inorganic

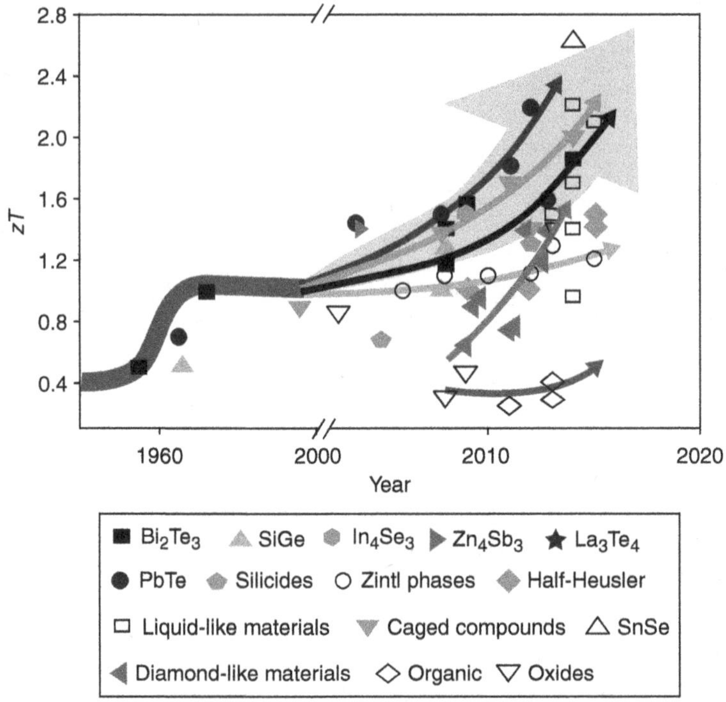

FIGURE 10.7 The evolution of ZT for various thermoelectric materials. Reprinted with permission from [58]. Copyright 2019 @ from Science Press.

thin-film materials, 2D materials. However, the research on these favorable choices for flexible TEGs is still in its premature stage.

10.4.2 Organic Thermoelectric Materials

The inability to use thermoelectricity in practical applications is mostly due to the high cost and toxicity of materials. Since organic TE materials have the potential to be used in a range of specialized applications, comprising flexible thermoelectrics in wearable clothing, watches, biomedical, electronics, automobiles, and space missions; they have a great deal of appeal in this direction. This is because these materials have inherent benefits like mechanical flexibility, less toxicity, light weight, and low material cost [66–69]. As it was stated in prior sections, the current commercialized thermoelectric materials have strong thermoelectric performance in low, mid, and high temperature regime respectivily. Despite their great performance, they cannot be used often owing to various intrinsic limitations including high cost, brittleness, toxicity, and material shortage, among others. Since organic materials have unique advantages such light weight, mechanical flexibility, cheap cost, great reinforcing, and ample material availability, they are being investigated as a solution to these problems [66, 70, 71]. Despite the several advantages, the commercialization of organic thermoelectric generators has been hampered by their poor TE performances (TEGs). To improve performance of organic material is a tedious work because they have complex structure, and their performance depends on several parameters [72]. Furthermore, there are scarcely any theoretical models that can direct us toward creation of high performance, organic TE materials. The fundamental investigations into organic TE materials are not yet well developed, so there are lot of research opportunities available for exploration of organic TE materials.

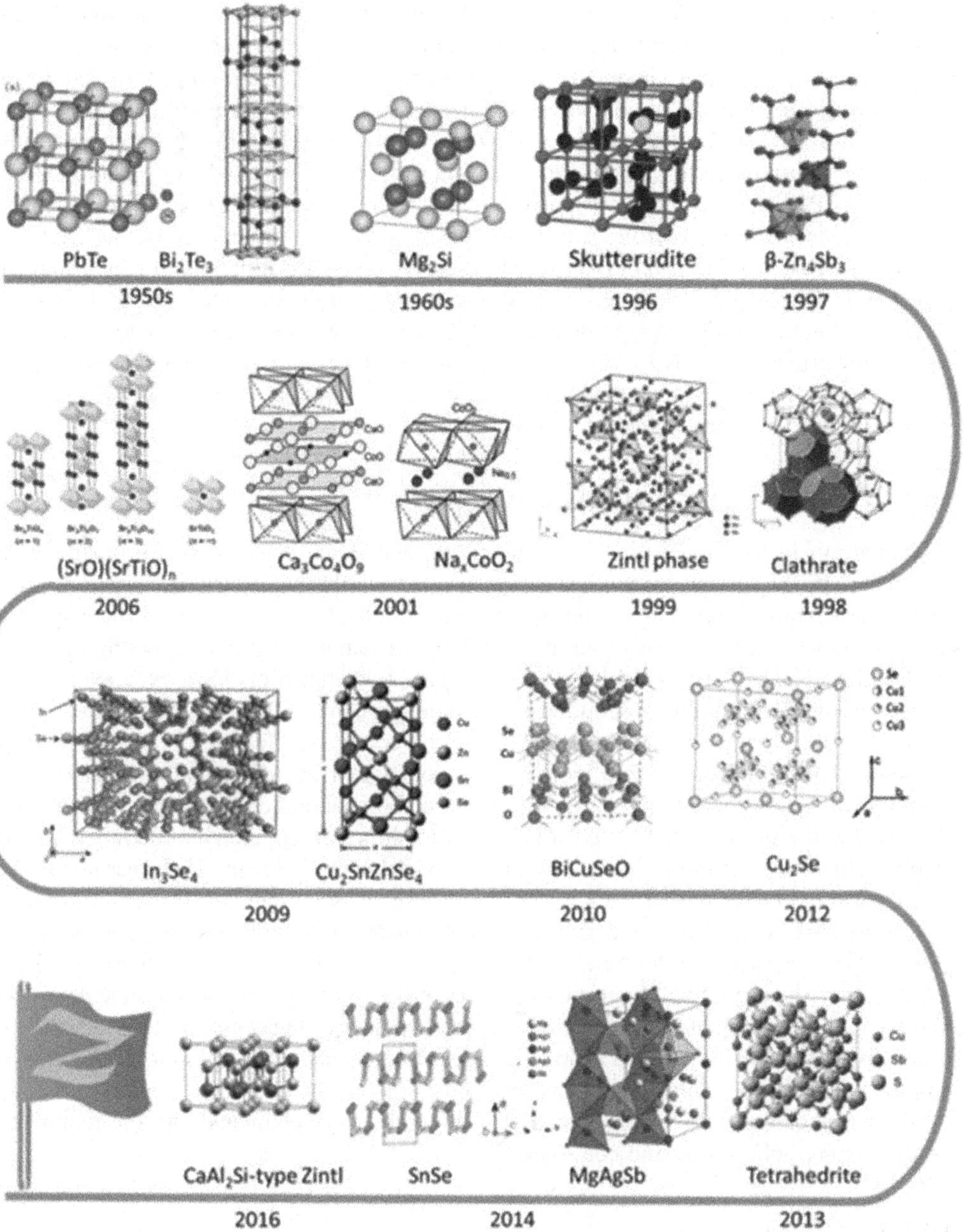

FIGURE 10.8 The timeline showcasing selected thermoelectric materials. Reprinted from [65], Copyright (2017), with permission from Elsevier.

The best suitable organic materials required for the construction of flexible thermoelectric devices are conducting polymers. These offer several benefits, including being lightweight, flexible, geometrically adaptable, and most significantly able to be scaled up for commercial purpose, Organic materials, in comparison to inorganic thermoelectric materials, have an innately lower thermal conductivity, which makes organic materials more suitable for use in thermoelectric applications. [66, 73].

PEDOT [poly(3,4-ethylenedioxythiophene)] is considered as the most prosperous conducting polymer [74], due to the fact that it has a low density, high conductivity after being doped with the proper dopants, is environmentally safe and stable, and is simple to synthesize [75]. Its insolubility in water and other common solvents is a cause for caution when utilizing PEDOT, which restricts its uses. The solution lies in its emulsification with PSS forming an aqueous solution. For flexible thermoelectric applications, the most often studied conducting polymers include polyaniline (PANI), polyacetylenes, poly(3,4-ethylenedioxythiophene) (PEDOT), polycarbazole, polythiophenes, and polypyrrole. Among the potential flexible thermoelectric materials studied are PEDOT:Tos (Tosylate) and PEDOT:PSS (Polystyrene Sulfonate), both of which show p-type conduction. Due to their poor electron affinity and potential for interaction with oxygen and water found in the air, their n-type thermoelectric counterparts are challenging to synthesize. The organic materials' utility is limited, nevertheless, by the structural disorder that prevents both electrical and thermal conductivity. Furthermore, these material-based thermoelectric devices showcase figure-of-merit too low to be used in commercial applications.

10.5 CARBON-BASED THERMOELECTRICS

In the thermoelectric arena, thus far, inorganic semiconductors are exhaustively studied due to degeneracy in their nature which leads to commendable thermoelectric performance [76]. The heat source required for maintaining temperature gradient can be in any form/shape and are scarcely planar, implying that material flexibility is a required attribute in configuring thermoelectric device. Though thermoelectric devices based on thin films have been attempted, such devices are unable to maintain required temperature gradient in the cross-plane direction. These devices are thus rendered insufficient for energy harvesting applications [77]. Also, majority of elements constituting degenerate inorganic semiconductor are toxic and expensive rare-earth metals. Low-dimensional forms of carbon such as carbon nanotubes, graphene, etc., with remarkable conducting properties, have been predicted to showcase appreciable thermoelectric power factors. This confirms the idea that low-dimensional materials having high thermoelectric performance as compared to their bulk forms, as proposed by Dresselhaus [78–81].

Carbon-based compounds, like graphene, carbon nanotubes, fullerene, etc., possess distinctive physical properties like high carrier mobility, outstanding electrical conductivity, good thermal stability and thus, qualifies themselves as composite additives in the thermoelectric matrix. The addition of such materials results in the improvement of thermoelectric performance of the nanocomposites. Owing to their distinctive electrical, mechanical, opto-electronic and optical properties, carbon derivatives have become most sought out material as composite additives. Additionally, carbon falls in the category of most abundant elements with the majority of its isotopes being light weight and non-toxic.

10.5.1 GRAPHENE

An important requirement toward improvement of ZT is enhancement in the thermal resistance of the interfacial region with little or no compromise of the electrical properties. As carbon-based compounds are remarkable conductors, they qualify themselves as suitable materials for accomplishing the effect. In particular, graphene showcases various fascinating properties like high surface area (~2600 m^2/g), extraordinary electrical conductivity (~6000 S/cm), remarkable carrier mobility (~ 20,000 cm^2/Vs) and exceptional mechanical properties (mechanical stiffness ~ 130 GPa) and has a two-dimensional single-layer structure with sp^2-hybridization. Also, graphene in the form of nanoribbons exhibited an increased Seebeck coefficient of ~4000 μV/K, as a result

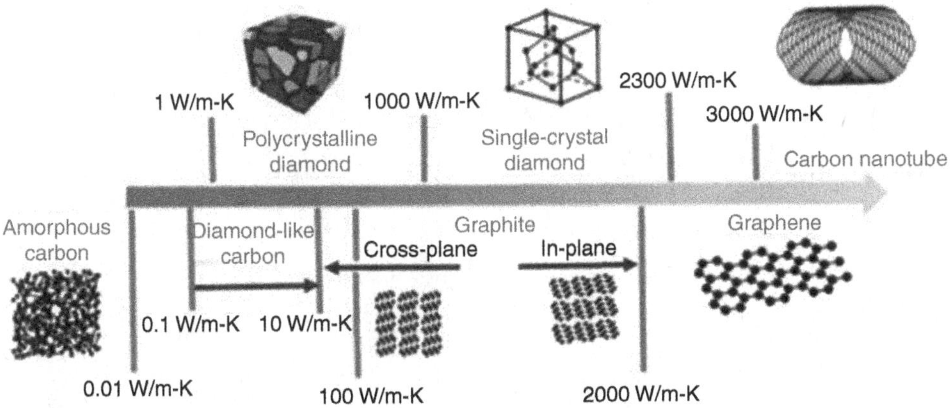

FIGURE 10.9 Thermal conductivity distribution of carbon-based compounds. Reprinted with permission from [92]. Copyright 2018 @ from Advances in Condensed Matter Physics.

of band structure modification [82]. These intriguing features of graphene qualifies them as potential candidates for synthesizing different functional nanocomposites for unrealized applications in fuel cells, conductive polymers, bio sensors, super-capacitors, etc. A notable drawback of graphene as TE material is high value of κ_L (Figure 10.9). However, several strategies for reducing κ_L of graphene are existent, e.g., introduction of vacancies [83, 84], dislocations, nanoholes [85, 86], isotopes [87]. Graphene in the form of nanoribbons exhibits low κ_L because of the dual effect of phonon edge localization [88–90] and boundary scattering [91].

By using carbon additives with different compounds TE performance is expected to be better because of the following reasons: (a) realizing an increased density of states (DOS) in the vicinity of Fermi level, and (b) the defects that arise at the interface between carbon derivative and the thermoelectric compound further aids in lattice scattering. Another advantage of using carbon additives lies in the fact that these compounds are rare-earth free, cost-efficient, and also non-toxic in nature. Also, these are considered to be attractive materials for device fabrication because of their flexible nature. Because of such remarkable properties these additives are widely researched materials for wearable thermoelectric devices [93–97].

Since graphene and carbon nanotubes possess anisotropic structures, electrons and phonons showcase different transport behavior in different tube/layer alignment. Graphene, having higher carrier mobility when incorporated into a thermoelectric material possessing lower thermal conductivity, is predicted to result in achieving improved thermoelectric performance [98, 99].

Further, an improvement in thermoelectric properties of Bi_2Te_3/graphene composites have been observed on varying the material dimensions to nanometer scale [100]. With graphene being a flexible material, its devices are globally researched as wearable thermoelectrics and also in smart textiles owing to excellent mechanical properties and high value of electrical conductivity (Figure 10.10).

10.5.2 Carbon Nanotube

10.5.2.1 CNT Doping

In the preliminary study on improvement of thermoelectric properties of CNTs, argon plasma treatment was adopted on pure CNTs [101]. Argon irradiation introduced structural defects in CNT walls and the defects aggregated with extended irradiation of argon. The introduction of

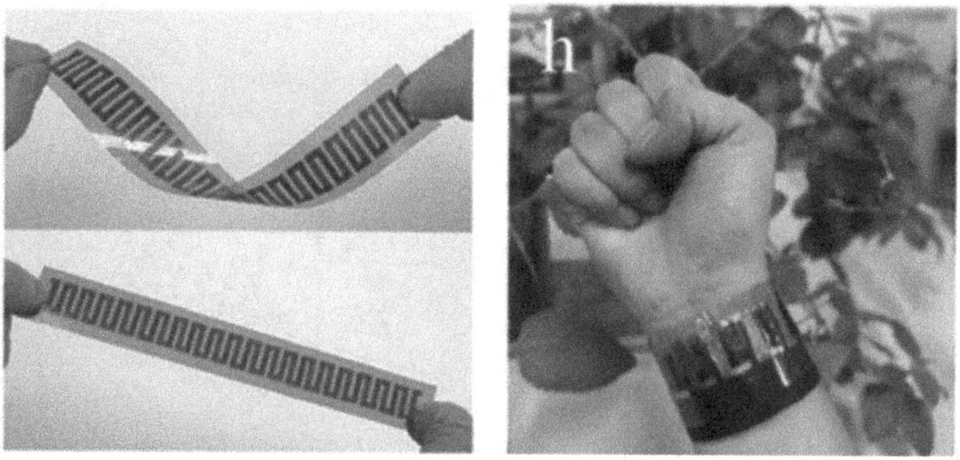

FIGURE 10.10 (a) Images of a graphene-based flexible device [94]; (b) Prospective usage of a wearable TE device. Reprinted with permission from [96]. Copyright 2019 @ from Wiley.

defects led to the enhancement of semiconducting properties of CNTs. The Seebeck coefficient reported an increase at the operating temperature while a degradation in electrical conductivity was observed when CNTs were subjected to argon irradiation. Metallic SWCNTs have also been studied as potential candidates for thermoelectric applications. The fine tuning between Fermi energy and electrolyte gating as observed in metallic SWCNTs solves the interdependence of S and σ, thus leading to exhibition of improved thermoelectric performance.

10.5.2.2 CNT Composites

With a notion of exploiting the benefits of individual components, composites of CNTs with various polymers were focused on [102–104]. Studies focusing on the effects of doping carbon nanotubes (CNT) in different class of TE alloys have been described by different researchers. In addition to exhibiting high electrical conductivity as a result of large mobility, single-walled carbon nanotubes (SWCNTs) showcase remarkable thermoelectric properties. The de-Broglie wavelength of semiconducting SWCNTs is greater than the diameter (L) (Figure 10.11), thus suggesting that s-SWCNTs with smaller diameter qualifies as good TE material. Over the span of years, inorganic materials exhibiting remarkable thermoelectric performance have been utilized to develop composites in conjunction with CNT [105–109]. Apart from high temperature application studies, mid temperature as well as room temperature studies showcasing enhanced thermoelectric performance for inorganic material/CNT composites have been reported [110]. CNTs were dispersed in Cu_2Se matrix to obtain composite material with exceptional thermoelectric properties at the operating temperature.

With the fast emergence of conducting polymers, PEDOT:PSS has proved to be the primary choice among other widely used polymers [112]. The junction structure as shown in Figure 10.12 depicts the potential to showcase exceptional thermoelectric transport properties. An array of connected electron channels are created at the interface. Owing to exceptional transmission characteristics of CNT, the σ values of the resultant composite film is as high as ~40,000 S/m. Also, since there is a mismatch in the vibrational spectra of CNT and PEDOT:PSS, the phonon propagation at the nodal sites is obstructed, this results in prevention of enhancement of κ and decrease in the S.

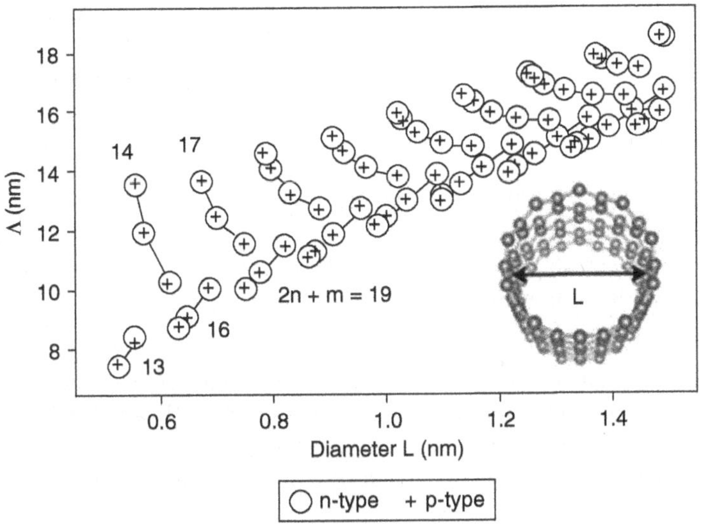

FIGURE 10.11 de-Broglie wavelength Λ depicted as a function of s-SWNT diameter. Reprinted with permission from [111]. Copyright 2019 @ from MDPI.

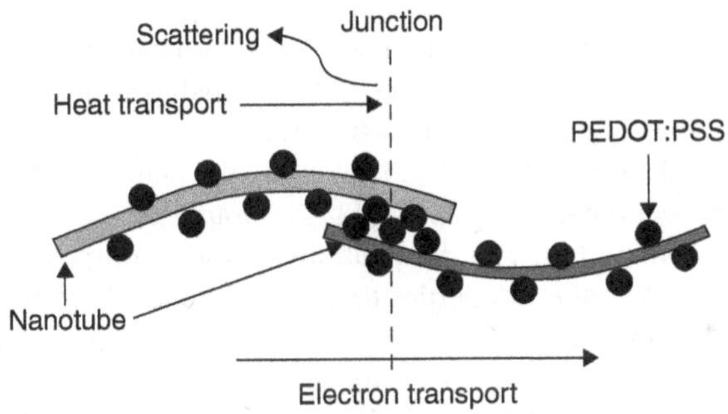

FIGURE 10.12 Junctions formed in the composites by coating of PEDOT:PSS particles on CNTs. Reprinted with permission from [113]. Copyright 2018 @ from MDPI.

CNTs also play an important role in the arrangement of conducting polymer chains [114]. In case of CNT-PANI composites, PANI chains pack densely onto the CNT surface owing to strong π–π interaction present for PANI and CNTs (Figure 10.13). The good crystallinity in these polymers can be accounted for efficient thermoelectric properties in the resulting composite.

10.5.3 Hybrid Composites

Composites of organic materials with conductive materials are developed with an objective to improve the thermoelectric properties of organic materials. In this course, carbon derivative-based elastic TE materials have emerged as competent candidates for useful applications. As suggested theoretically, these are proposed to be favorable thermoelectric materials due to the presence of low dimensionality [115–118]. CNTs are known to enhance the tensile strength of

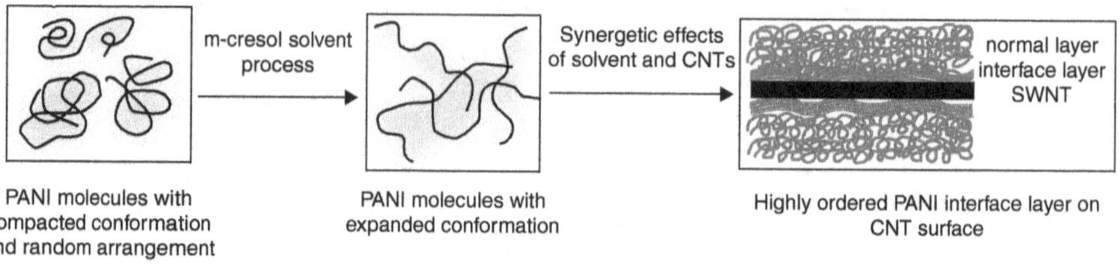

FIGURE 10.13 Highly ordered PANI interface layer present on CNT surface. Reprinted with permission from [113]. Copyright 2018 @ from MDPI.

the material they are amalgamated with. The electrical conductivity of the polymers is largely enhanced by adding a meager amount of CNT while preserving low κ. CNT fillers when doped with non-conducting poly(vinyl acetate) (PVAc) matrix results in dramatic enhancement of electrical conductivity with insignificant changes to the intrinsically low κ inherent to polymer matrix.

Thermal conductivity value of CNTs networks is flexibily tuned from 0.1 $Wm^{-1}K^{-1}$ to metal like conductivity [119]. Crystallographic defects reduce thermal conductivity by phonon scattering while larger size defects, like Stone Wales defects, scatter phonons in wide range frequencies, thus, resulting in enhanced reduction in thermal conductivity [120]. In addition to TE preferable electronic and thermal properties, CNTs also exhibits very high mechanical strength [121]. Because of prominently large aspect ratio CNTs exhibits small percolation threshold, which give dual advantage of cost-efficiency and flexibility in nature.

Graphene strongly conducts electricity and heat along the plane. A Single graphene layer is ~ 100 times stronger than steel for same thickness [122]. Graphene possesses high values of σ and hence a high value of κ, which limits its applications in some composites, however, in organic composites it compensates the other properties to maintain good TE properties.

Addition of reduced graphene oxide (rGO) to organic material results in enhancement of transport properties [123]. Incorporation of a meager amount of rGO strengthens the phonon scattering mechanism at grain boundaries and also increases mobility and charge carrier concentration in certain classes of thermoelectric compounds, especially, skutterudites, chalcogenides, metal oxide-based composites, etc. Graphene and rGO mainly work as stimulators of the inherent performance of TE materials, since appreciable growth in ZT value on graphene or rGO addition was witnessed primarily for alloys based on TE materials possessing low preliminary ZT. Significant enhancement in ZT was not observed for already processed thermoelectric materials possessing high electrical conductivity.

Addition of CNTs and graphene to organic materials results in improved mechanical properties and enhanced conductivity. A deep insight into alignment and orientation of these materials will lead to improved performance. Despite the slightly higher thermal conductivities of CNTs and graphene they are used for enhancement of the TE properties of different materials by composite approach. Both graphene and CNT are used in organic and inorganic materials for better improvement in TE performance.

The polymer-inorganic TE composites also lie in the category of hybrid thermoelectric composites. In this case, a polymer matrix having inert nature acts as a host for the thermoelectric filler material. It also allows the thermoelectric material to take charge of transport properties. As the polymer matrix is highly disordered on different length scales it provides the key advantage of lower thermal conductivity.

10.6 THE FUTURE PROSPECTIVE OF ORGANIC TE MATERIALS

The TE properties of the organic materials are dependent on its chemical structure and micro-structural morphology. Techniques like doping, post-treatment, alignment and crystallinity, etc. can prove to be effective in enhancing the ZT values, although the technological processes need to be optimized. In addition, since n-type polymers are highly unstable in air, the majority of conducting polymers and resultant TE composites possess p-type conduction. The conducting polymers like poly[K_x(Ni-ett)] have shown appreciable performance as an n-type polymer, however, its insoluble nature restricts its applications. This would considerably affect the progress toward growth of wearable TE devices. Therefore, there is an urgent requirement to focus future research and development toward stabilization of n-type polymers and also resultant TE composites.

The ZT values of corresponding p- and n-type leg are not exactly equal; this demands an optimization and reform of the geometrical cross-sectional area of both legs. Also, the human skin does not possess smooth texture, thus adding to the complexity in enhancement of conversion efficiency. Therefore, efforts need to be focused toward optimizing parameters like device dimension, structure, geometry, arrangement, etc. For high-temperature TE applications, there is no substitute for inorganic materials, which faces the major hurdle of poor mechanical flexibility. Researchers have been involved in the use of inorganic materials as various forms of thin films like employing flexible substrates for depositing inorganic thin films, CNT-derived thin film, complex inorganic material thin film, 2D material-based thin films etc., however, the research in this direction is still in its infancy.

To conclude, no standards have yet been established to measure the TE properties of prevalent organic TE materials. To promote development, a standard technology needs to be established for characterization of organic materials and resultant wearable TEGs.

10.7 CONCLUSIONS

Carbon in different forms can prove to be favorable candidates for futuristic development of low-cost and lightweight polymer-based composites for energy harvesting employing thermoelectrics. High performance material synthesis seems feasible by engaging organic materials for thermoelectric generation. CNTs and graphene being one-dimensional semiconductors have the potential to showcase improved thermoelectric performance as compared to bulk as well as 2D materials. Encouraging results obtained in recent years indicates the capability of carbon-based TE materials to be used in TEGs for various emerging and sophisticated applications.

REFERENCES

1. Charisiou, N.D., et al., *Effect of active metal supported on SiO2 for selective hydrogen production from the glycerol steam reforming reaction*. Bioresources, 2016. **11**(4): p. 10173–10189.
2. Zhang, H., et al., *A novel phase-change cement composite for thermal energy storage: Fabrication, thermal and mechanical properties*. Applied energy, 2016. **170**: p. 130–139.
3. Petroleum, B., *BP Statistical Review of World Energy. June 2013. London: British Petroleum*. 2013.
4. Rowe, D.M., *CRC Handbook of Thermoelectrics*. 1995: CRC press.
5. Sootsman, J.R., D.Y. Chung, and M.G. Kanatzidis, *New and old concepts in thermoelectric materials*. Angewandte Chemie International Edition, 2009. **48**(46): p. 8616–8639.
6. Afshar, O., et al., *A review of thermodynamics and heat transfer in solar refrigeration system*, in *Renewable and Sustainable Energy Reviews*. 2012. p. 5639–5648.

7. Xi, H., L. Luo, and G. Fraisse, *Development and applications of solar-based thermoelectric technologies*. Renewable and Sustainable Energy Reviews, 2007. **11**(5): p. 923–936.
8. Crabtree, G.W. and N.S. Lewis, *Solar energy conversion*. Physics today, 2007. **60**(3): p. 37–42.
9. Rowley, J.C., *Geothermal energy development*. Physics Today, 2008. **30**(1): p. 36–45.
10. Tritt, T.M., H. Böttner, and L. Chen, *Thermoelectrics: Direct Solar Thermal Energy Conversion*. MRS Bulletin, 2008. **33**(04): p. 366–368.
11. Zhang, X. and L.-D. Zhao, *Thermoelectric materials: Energy conversion between heat and electricity*. Journal of Materiomics, 2015. **1**(2): p. 92–105.
12. Shakouri, A., *Recent developments in semiconductor thermoelectric physics and materials*. Materials Research, 2011. **41**(1): p. 399.
13. Vining, C.B., *An inconvenient truth about thermoelectrics*. Nature Materials, 2009. **8**(2): p. 83–85.
14. Ismail, B.I. and W.H. Ahmed, *Thermoelectric power generation using waste-heat energy as an alternative green technology*. Recent Patents on Electrical & Electronic Engineering (Formerly Recent Patents on Electrical Engineering), 2009. **2**(1): p. 27–39.
15. Weidenkaff, A., et al. *Development of Thermoelectric Oxides for Renewable Energy Conversion Technologies*. 2008. Pergamon.
16. Zebarjadi, M., et al., *Perspectives on thermoelectrics: from fundamentals to device applications*. Energy & Environmental Science, 2012. **5**(1): p. 5147–5162.
17. Goldsmid, H., *Thermoelectric Refrigeration*. 1964. New York: Plenum Press.
18. MacDonald, D.K.C., *Thermoelectricity: An Introduction to the Principles*. 2006: Courier Corporation.
19. Tzounis, L., *Organic thermoelectrics and thermoelectric generators (TEGs)*. Advanced thermoelectric materials for energy harvesting applications, 2019: p. 7.
20. Seebeck, T.J., *Ueber die magnetische Polarisation der Metalle und Erze durch Temperaturdifferenz*. Annalen der Physik, 1826. **82**(3): p. 253–286.
21. Morrison, K. and F.K. Dejene, *Thermal Imaging of the Thomson Effect*. Physics, 2020. **13**: p. 137.
22. Peltier, J.C.A. *Nouvelles expériences sur la caloricité des courants électriques*. in *Annales de Chimie et de Physique*. 1834.
23. DiSalvo, F.J., *Thermoelectric Cooling and Power Generation*. Science, 1999. **285**(5428): p. 703–706.
24. Altenkirch, E., *Über den Nutzeffekt der Thermosäule*. Physikalische Zeitschrift, 1909. **10**: p. 560.
25. Kumar, P.M., et al., *The design of a thermoelectric generator and its medical applications*. Designs, 2019. **3**(2): p. 22.
26. Cadoff, I.B. and E. Miller, *Thermoelectric Materials and Devices: Lectures Presented During the Course on Thermoelectric Materials and Devices Sponsored by the Dept. of Metallurgical Engineering in Cooperation with the Office of Special Services to Business and Industry, New York University, New York, NY, June 1959 and 1960*. Vol. 2. 1960: Reinhold Pub. Corp.
27. Snyder, G.J. and E.S. Toberer, *Complex thermoelectric materials*. Nat Mater, 2008. **7**(2): p. 105–114.
28. Goldsmid, H.J., *The Thermoelectric and Related Effects*, in *Introduction to Thermoelectricity*. 2010, Springer. p. 1–6.
29. Masood, K.B., et al., *Odyssey of thermoelectric materials: foundation of the complex structure*. Journal of Physics Communications, 2018. **2**(6): p. 062001.
30. Wu, C., *Analysis of waste-heat thermoelectric power generators*. Applied Thermal Engineering, 1996. **16**(1): p. 63–69.
31. Rowe, D. and G. Min, *Evaluation of thermoelectric modules for power generation*. Journal of Power Sources, 1998. **73**(2): p. 193–198.
32. Rowe, D.M., *Thermoelectrics, an environmentally-friendly source of electrical power*. Renewable energy, 1999. **16**(1): p. 1251–1256.
33. Crane, D.T. and G.S. Jackson, *Optimization of cross flow heat exchangers for thermoelectric waste heat recovery*. Energy Conversion and Management, 2004. **45**(9): p. 1565–1582.
34. Yang, J. and T. Caillat, *Thermoelectric materials for space and automotive power generation*. MRS bulletin, 2006. **31**(3): p. 224–229.

35. Bell, L.E., *Cooling, heating, generating power, and recovering waste heat with thermoelectric systems*. Science, 2008. **321**(5895): p. 1457–1461.
36. Kyratsi, T. *Thermoelectric materials and applications on the recovery of waste heat energy*. in *Organized by the Hellenic Physical Society with the Cooperation of the Physics Departments of Greek Universities: 7th International Conference of the Balkan Physical Union*. 2010. AIP Publishing.
37. Niu, X., J. Yu, and S. Wang, *Experimental study on low-temperature waste heat thermoelectric generator*. Journal of Power Sources, 2009. **188**(2): p. 621–626.
38. Telkes, M., *Solar thermoelectric generators*. Journal of Applied Physics, 1954. **25**(6): p. 765–777.
39. Goldsmid, H., J. Giutronich, and M. Kaila, *Solar thermoelectric generation using bismuth telluride alloys*. Solar Energy, 1980. **24**(5): p. 435–440.
40. Omer, S. and D. Infield, *Design optimization of thermoelectric devices for solar power generation*. Solar Energy Materials and Solar Cells, 1998. **53**(1–2): p. 67–82.
41. Scherrer, H., et al., *Solar thermolectric generator based on skutterudites*. Journal of Power Sources, 2003. **115**(1): p. 141–148.
42. Zhang, Q.J., et al. *Recent development in nano and graded thermoelectric materials*. in *Materials Science Forum*. 2005. Trans Tech Publ.
43. Khattab, N. and E. El Shenawy, *Optimal operation of thermoelectric cooler driven by solar thermoelectric generator*. Energy Conversion and Management, 2006. **47**(4): p. 407–426.
44. Baxter, J., et al., *Nanoscale design to enable the revolution in renewable energy*. Energy & Environmental Science, 2009. **2**(6): p. 559–588.
45. Tritt, T.M., *Thermoelectric phenomena, materials, and applications*. Annual Review of Materials Research, 2011. **41**: p. 433–448.
46. Gorsse, S., et al., *Nanostructuration via solid state transformation as a strategy for improving the thermoelectric efficiency of PbTe alloys*. Acta Materialia, 2011. **59**(19): p. 7425–7437.
47. Chowdhury, I., et al., *On-chip cooling by superlattice-based thin-film thermoelectrics*. Nature Nanotechnology, 2009. **4**(4): p. 235–238.
48. Gordiakova, G. and S. Sinani, *The thermoelectric properties of bismuth telluride with alloying additives*. Soviet Physics-Technical Physics, 1958. **3**(5): p. 908–911.
49. Wright, D.A., *Thermoelectric Properties of Bismuth Telluride and its Alloys*. Nature, 1958. **181**(4612): p. 834–834.
50. Goldsmid, H.J., *The Electrical Conductivity and Thermoelectric Power of Bismuth Telluride*. Proceedings of the Physical Society, 1958. **71**(4): p. 633.
51. Steele, M. and F. Rosi, *Thermal Conductivity and Thermoelectric Power of Germanium-Silicon Alloys*. Journal of Applied Physics, 1958. **29**(11): p. 1517–1520.
52. Abrikosov, N.K., et al., *Thermoelectric properties of silicon-germanium- boron alloys*. Sov Phys Semiconductors, 1969. **2**(12): p. 1468–1473.
53. Rowe, D. and R. Bunce, *The thermoelectric properties of heavily doped hot-pressed germanium-silicon alloys*. Journal of Physics D: Applied Physics, 1969. **2**(11): p. 1497.
54. Wyrick, R. and H. Levinstein, *Thermoelectric voltage in lead telluride*. Physical Review, 1950. **78**(3): p. 304.
55. Putley, E., *Thermoelectric and galvanomagnetic effects in lead selenide and telluride*. Proceedings of the Physical Society. Section B, 1955. **68**(1): p. 35.
56. Kolomoets, N., T. Stavitskaia, and L. Stilbans, *An investigation of the thermoelectric properties of lead selenide and lead telluride*. Soviet Physics-Technical Physics, 1957. **2**(1): p. 59–66.
57. Gershtein, E., T. Stavitskaia, and L. Stilbans, *A study of the thermoelectric properties of lead telluride*. Soviet Physics-Technical Physics, 1957. **2**(11): p. 2302–2313.
58. Qi-Hao, Z., B. Sheng-Qiang, and C. Li-Dong, *Technologies and applications of thermoelectric devices: current status, challenges and prospects*. Journal of Inorganic Materials, 2019. **34**(3): p. 279.
59. Slack, G.A. and V.G. Tsoukala, *Some properties of semiconducting IrSb3*. Journal of Applied Physics, 1994. **76**(3): p. 1665–1671.

60. Nolas, G., et al., *Semiconducting Ge clathrates: Promising candidates for thermoelectric applications*. Applied Physics Letters, 1998. **73**(2): p. 178–180.
61. Kim, S.-J., et al., *Ba4In8Sb16: Thermoelectric properties of a new layered Zintl phase with infinite zigzag Sb chains and pentagonal tubes*. Chemistry of Materials, 1999. **11**(11): p. 3154–3159.
62. Brown, S.R., et al., *Yb14MnSb11: New high efficiency thermoelectric material for power generation*. Chemistry of Materials, 2006. **18**(7): p. 1873–1877.
63. Zhao, L.-D., et al., *Ultrahigh power factor and thermoelectric performance in hole-doped single-crystal SnSe*. Science, 2016. **351**(6269): p. 141–144.
64. Parker, D., X. Chen, and D.J. Singh, *High three-dimensional thermoelectric performance from low-dimensional bands*. Physical Review Letters, 2013. **110**(14): p. 146601.
65. Liu, W., et al., *New trends, strategies and opportunities in thermoelectric materials: a perspective*. Materials Today Physics, 2017. **1**: p. 50–60.
66. Chen, G., W. Xu, and D. Zhu, *Recent advances in organic polymer thermoelectric composites*. Journal of Materials Chemistry C, 2017. **5**(18): p. 4350–4360.
67. Peng, S., et al., *A review on organic polymer-based thermoelectric materials*. Journal of Polymers and the Environment, 2017. **25**(4): p. 1208–1218.
68. Zhang, Y., et al., *Recent advances in organic thermoelectric materials: Principle mechanisms and emerging carbon-based green energy materials*. Polymers, 2019. **11**(1): p. 167.
69. Zhang, Y. and S.-J. Park, *Flexible organic thermoelectric materials and devices for wearable green energy harvesting*. Polymers, 2019. **11**(5): p. 909.
70. Liang, L., et al., *Large-area, stretchable, super flexible and mechanically stable thermoelectric films of polymer/carbon nanotube composites*. Journal of Materials Chemistry C, 2016. **4**(3): p. 526–532.
71. Fan, Z., et al., *Polymer films with ultrahigh thermoelectric properties arising from significant seebeck coefficient enhancement by ion accumulation on surface*. Nano Energy, 2018. **51**: p. 481–488.
72. Venkatasubramanian, R., et al., *Thin-film thermoelectric devices with high room-temperature figures of merit*. Nature, 2001. **413**(6856): p. 597–602.
73. Toshima, N., N. Jiravanichanun, and H. Marutani, *Organic thermoelectric materials composed of conducting polymers and metal nanoparticles*. Journal of Electronic Materials, 2012. **41**(6): p. 1735–1742.
74. Li, J., et al., *Thermoelectric properties of flexible PEDOT: PSS/polypyrrole/paper nanocomposite films*. Materials, 2017. **10**(7): p. 780.
75. Li, Y., et al., *PEDOT-based thermoelectric nanocomposites–A mini-review*. Synthetic Metals, 2017. **226** : p. 119–128.
76. Beretta, D., et al., *Thermoelectrics: From history, a window to the future*. Materials Science and Engineering: R: Reports, 2019. **138**: p. 100501.
77. Kim, S.J., J.H. We, and B.J. Cho, *A wearable thermoelectric generator fabricated on a glass fabric*. Energy & Environmental Science, 2014. **7**(6): p. 1959–1965.
78. Hone, J., et al., *Thermoelectric power of single-walled carbon nanotubes*. Physical Review Letters, 1998. **80**(5): p. 1042.
79. Dragoman, D. and M. Dragoman, *Giant thermoelectric effect in graphene*. Applied Physics Letters, 2007. **91**(20): p. 203116.
80. Zuev, Y.M., W. Chang, and P. Kim, *Thermoelectric and magnetothermoelectric transport measurements of graphene*. Physical Review Letters, 2009. **102**(9): p. 096807.
81. Dresselhaus, M.S., et al., *New directions for low-dimensional thermoelectric materials*. Advanced Materials, 2007. **19**(8): p. 1043–1053.
82. Rakshit, M., D. Jana, and D. Banerjee, *General strategies to improve thermoelectric performance with an emphasis on tin and germanium chalcogenides as thermoelectric materials*. Journal of Materials Chemistry A, 2022. **10**(13): p. 6872–6926.
83. Wang, Y., S. Chen, and X. Ruan, *Tunable thermal rectification in graphene nanoribbons through defect engineering: a molecular dynamics study*. Applied Physics Letters, 2012. **100**(16): p. 163101.
84. Haskins, J., et al., *Control of thermal and electronic transport in defect-engineered graphene nanoribbons*. ACS Nano, 2011. **5**(5): p. 3779–3787.

85. Feng, T. and X. Ruan, *Ultra-low thermal conductivity in graphene nanomesh*. Carbon, 2016. **101**: p. 107–113.
86. Yu, J.-K., et al., *Reduction of thermal conductivity in phononic nanomesh structures*. Nature Nanotechnology, 2010. **5**(10): p. 718–721.
87. Chen, S., et al., *Thermal conductivity of isotopically modified graphene*. Nature Materials, 2012. **11**(3): p. 203–207.
88. Wang, Y., et al., *Phonon lateral confinement enables thermal rectification in asymmetric single-material nanostructures*. Nano Letters, 2014. **14**(2): p. 592–596.
89. Wang, Y., B. Qiu, and X. Ruan, *Edge effect on thermal transport in graphene nanoribbons: A phonon localization mechanism beyond edge roughness scattering*. Applied Physics Letters, 2012. **101**(1): p. 013101.
90. Wang, Y., et al., *Two-dimensional thermal transport in graphene: a review of numerical modeling studies*. Nanoscale and Microscale Thermophysical Engineering, 2014. **18**(2): p. 155–182.
91. Zou, J., et al., *Metal grid/conducting polymer hybrid transparent electrode for inverted polymer solar cells*. Applied Physics Letters, 2010. **96**(20): p. 96.
92. Chakraborty, P., et al., *Carbon-based materials for thermoelectrics*. Advances in Condensed Matter Physics, 2018. **2018**.
93. Tu, N.D.K., J.A. Lim, and H. Kim, *A mechanistic study on the carrier properties of nitrogen-doped graphene derivatives using thermoelectric effect*. Carbon, 2017. **117**: p. 447–453.
94. Juntunen, T., et al., *Inkjet printed large-area flexible few-layer graphene thermoelectrics*. Advanced Functional Materials, 2018. **28**(22): p. 1800480.
95. Hsieh, Y.-Y., et al., *High thermoelectric power-factor composites based on flexible three-dimensional graphene and polyaniline*. Nanoscale, 2019. **11**(14): p. 6552–6560.
96. Wu, B., et al., *High-performance flexible thermoelectric devices based on all-inorganic hybrid films for harvesting low-grade heat*. Advanced Functional Materials, 2019. **29**(25): p. 1900304.
97. Novak, T.G., et al., *Complementary n-type and p-type graphene films for high power factor thermoelectric generators*. Advanced Functional Materials, 2020. **30**(28): p. 2001760.
98. Liang, B., et al., *Fabrication and thermoelectric properties of graphene/composite materials*. Journal of Nanomaterials, 2013.
99. Agarwal, K., et al., *Nanoscale thermoelectric properties of Bi2Te3–Graphene nanocomposites: Conducting atomic force, scanning thermal and kelvin probe microscopy studies*. Journal of Alloys and Compounds, 2016. **681**: p. 394–401.
100. Li, S., et al., *Graphene quantum dots embedded in Bi2Te3 nanosheets to enhance thermoelectric performance*. ACS Applied Materials & Interfaces, 2017. 9(4): p. 3677–3685.
101. Zhao, W., et al., *Flexible carbon nanotube papers with improved thermoelectric properties*. Energy & Environmental Science, 2012. **5**(1): p. 5364–5369.
102. Qu, S., et al., *A novel hydrophilic pyridinium salt polymer/SWCNTs composite film for high thermoelectric performance*. Polymer, 2018. **136**: p. 149–156.
103. Fan, W., C.-Y. Guo, and G. Chen, *Flexible films of poly (3, 4-ethylenedioxythiophene)/carbon nanotube thermoelectric composites prepared by dynamic 3-phase interfacial electropolymerization and subsequent physical mixing*. Journal of Materials Chemistry A, 2018. **6**(26): p. 12275–12280.
104. Gao, C. and G. Chen, *A new strategy to construct thermoelectric composites of SWCNTs and poly-Schiff bases with 1, 4-diazabuta-1, 3-diene structures acting as bidentate-chelating units*. Journal of Materials Chemistry A, 2016. **4**(29): p. 11299–11306.
105. Zhao, W., et al., *N-type carbon nanotubes/silver telluride nanohybrid buckypaper with a high-thermoelectric figure of merit*. ACS Applied Materials & Interfaces, 2014. **6**(7): p. 4940–4946.
106. Khasimsaheb, B., et al., *The effect of carbon nanotubes (CNT) on thermoelectric properties of lead telluride (PbTe) nanocubes*. Current Applied Physics, 2017. **17**(2): p. 306–313.
107. Li, C., et al., *Fabrication of flexible SWCNTs-Te composite films for improving thermoelectric properties*. Journal of Alloys and Compounds, 2017. **723** : p. 642–648.
108. Gao, W., et al., *Enhanced thermoelectric properties of CNT dispersed and Na-doped Bi2Ba2Co2Oy composites*. Ceramics International, 2017. **43**(7): p. 5723–5727.

109. Nunna, R., et al., *Ultrahigh thermoelectric performance in Cu 2 Se-based hybrid materials with highly dispersed molecular CNTs*. Energy & Environmental Science, 2017. **10**(9): p. 1928–1935.
110. Lei, J., et al., *Enhancement of thermoelectric figure of merit by the insertion of multi-walled carbon nanotubes in α-MgAgSb*. Applied Physics Letters, 2018. **113**(8): p. 083901.
111. Hung, N.T., A.R.T. Nugraha, and R. Saito, *Thermoelectric properties of carbon nanotubes*. Energies, 2019. **12**(23): p. 4561.
112. JIANG, F., C. LIU, and J. XU, *The evolution of organic thermoelectric material based on conducting poly (3, 4-ethylenedioxythiophene)*. Chinese Science Bulletin, 2017. **62**(19): p. 2063–2076.
113. Wang, X., H. Wang, and B. Liu, *Carbon nanotube-based organic thermoelectric materials for energy harvesting*. Polymers, 2018. **10**(11): p. 1196.
114. Choi, J., et al., *Enhanced thermopower in flexible tellurium nanowire films doped using single-walled carbon nanotubes with a rationally designed work function*. Carbon, 2015. **94**: p. 577–584.
115. Meng, C., C. Liu, and S. Fan, *A promising approach to enhanced thermoelectric properties using carbon nanotube networks*. Advanced Materials, 2010. **22**(4): p. 535–539.
116. Choi, Y., et al., *Effect of the carbon nanotube type on the thermoelectric properties of CNT/Nafion nanocomposites*. Organic Electronics, 2011. **12**(12): p. 2120–2125.
117. Kim, G.H., D.H. Hwang, and S.I. Woo, *Thermoelectric properties of nanocomposite thin films prepared with poly (3, 4-ethylenedioxythiophene) poly (styrenesulfonate) and graphene*. Physical Chemistry Chemical Physics, 2012. **14**(10): p. 3530–3536.
118. Xiang, J. and L.T. Drzal, *Templated growth of polyaniline on exfoliated graphene nanoplatelets (GNP) and its thermoelectric properties*. Polymer, 2012. **53**(19): p. 4202–4210.
119. Kumanek, B. and D. Janas, *Thermal conductivity of carbon nanotube networks: A review*. Journal of Materials Science, 2019. **54**(10): p. 7397–7427.
120. Mingo, N., et al., *Phonon transmission through defects in carbon nanotubes from first principles*. Physical Review B, 2008. **77**(3): p. 033418.
121. Yu, M.-F., et al., *Strength and breaking mechanism of multiwalled carbon nanotubes under tensile load*. Science, 2000. **287**(5453): p. 637–640.
122. Lee, C., et al., *Measurement of the elastic properties and intrinsic strength of monolayer graphene*. Science, 2008. **321**(5887): p. 385–388.
123. Okhay, O. and A. Tkach, *Impact of Graphene or Reduced Graphene Oxide on Performance of Thermoelectric Composites*. Carbon, 2021. **7**(2): p. 37.

11 Carbon Nanotubes and Graphene for Thermal Management

Jeevan Jyoti and Surya Kant Tripathi

CONTENTS

DOI: 10.1201/9781003231943-11

11.1 INTRODUCTION

The electronic industries are moving toward the devices which are more efficient, compact in size, more powerful, automatically accompanied by enhanced energy and power density, which possess new challenging tasks for thermal management materials (TMM). The electronic devices have led to significant demand of innovative thermal management (TM) technology to enhance the system performance and consistency by eliminating high heat flux produced within electronic devices. There are many well-known thermal problems such as performance degradation, equipment failure, security risks, thermal dissipation, etc. Thermal dissipation is a more critical issue in electronic devices that affect the performance, reliability and durability of electronic devices. Thermal dissipation has attained a crucial consideration of researchers in the development of electronic technology which is used in electronic chips, light-emitting diodes (LED), energy storage and conversion systems [1, 2]. Increasing demands of miniaturization of electronic and high power devices require efficient TMMs. It is the ability to control the working temperature of electronic devices such as integrated circuits using temperature monitoring and device cooling. TMMs are significantly demanded to avoid the performance deterioration and efficiency of electronic devices. TMMs not only control the noise but also have the ability to control the temperature level of the system. The main challenges of TM systems are: harsh environments, reducing the production cost, reduced form factors, design and manufacturing for the advanced technologies and materials, reliability, stringent standards, increasing the consumer's needs and demands, etc.

The compact design of electronic devices, as well as thermal dissipation, is quite a challenging task. The efficient TM and thermal interface materials (TIM) are important requirements of modern electronic technology [3]. Small-scale TM systems are necessary to manage the thermal dissipation in electronic devices such as smartphones, smartwatches, laptops, integrated circuits (IC), etc. The high thermal conductive materials (TCM) are in essential demand and play a significant role in the reliability, service life and operational performances of electronic devices [4, 5]. The heat exchanger and temperature maintenance play a vital role in identifying the overall size, cost, efficiency, and quality of the TMMs. The basic modifications of the system deal in design concern the shape, variation in structure, surface area enhancement, etc. Various materials have been used as the TMMs. Conventional materials such as copper, nickel, silver, steel and aluminum were used for heat transfer, but these materials suffered from issues like heavy-weight, continuous maintenance, corrosion and manufacturability [6]. Researchers are continuously exploring new materials to improve the heat transfer rate of the thermal system. For the improvement of the thermal system and to avoid mechanical constraints, various nanofillers are used to overcome the problem of the metallic heat exchangers. Polymer composites have been significant in addressing the heavy weight in the TMMs. Generally, polymers have a low value of thermal conductivity (TC) in the range of 0.1 to 0.5 W/mK, which is quite far away from the industrial requirements [7]. The lower value of TC has limited the applications of TMMs.

Conductive fillers reinforcement in the polymer matrix is an important method to enhance the thermal conductivity of polymer composites. Various metals, carbon and ceramic-based nanofillers are used as reinforcement materials in the polymer. The commonly used thermally conducting fillers include metal fillers (Cu, Al, Ag) [8, 9], ceramic fillers (such as aluminum nitride, aluminum oxide, boron nitride) [10, 11] and carbon fillers (graphene, carbon nanotube, graphite, etc.) [12–16]. These fillers are extensively used for enhancing the thermal properties of polymer composites. Metal-based fillers have the highest TC, but they are associated with weight and cost issues. On the other hand, ceramic fillers exhibit the low value of TC. Carbon-based fillers have high value TC, low thermal expansion, and corrosion resistance. The next-generation electronic devices need the composite materials which fulfill the requirements of low density, flexibility, light in weight, economically oriented heat conduction as well as ultrahigh TC. Polymer composite-based TMs have been extensively used as packing materials in electronic devices because they possess lightweight, excellent chemical resistance, easy processing and are cheapest.

Carbon nanofillers such as carbon nanotubes (CNTs) and graphene have attracted significant attention both in experimental and theoretical work due to their high TC, stiffness, unique electron mobility, etc. [17, 18]. Thermal properties are considered to be a significant factor for the applications of carbon nanofillers in TM at the nanoscale. The value of TC at room temperature ranges from 0.01 W/mK in amorphous carbons to thousand W/mK in CNTs and graphene. The room temperature TC of CNTs is recorded as ~3,000 to 3,500 W/mK [19, 20]. Experimental calculations show that the TC of graphene is in the range of 3,000 to 5,000 W/mK at room temperature [21]. Based on these outstanding properties, CNTs and graphene showed significant potential in the solar cell, secondary batteries, super capacitors and heat sinks. Most of the heat in carbon nanofillers is carried out by lattice vibrations, i.e., acoustic phonons. For the past few decades, polymer composites are generally being used as a heat sink and TIM. Fillers have excellent potential for improving heat transport properties. The properties of polymer composites directly depend upon the dispersion of fillers within the polymer matrix. The TC of carbon fillers depends upon the hybridization of filler, structural disorder, dimension, thickness, and grain size. The TC of the material depends on the mobility of electrons, photons, and phonons. In the carbon nanofillers, phonons are considered as the thermal energy carriers and the contribution of the electron is negligible [21, 22]. The lattice vibrations of the atoms or molecules describe the contribution of phonon frequencies with high accuracy. Graphene and CNTs have been extensively used in various TMMs. In this chapter, we will discuss the mechanisms, theoretical calculations, and applications of CNTs and graphene in TMMs.

11.2 PHONON AND THERMAL TRANSPORT IN SOLID MATERIALS

Thermal conductivity is the ability of intrinsic material to transport thermal energy. The thermal transfer mode is obeyed by Fourier's law of heat conduction.

$$q = -\lambda \nabla t \tag{11.1}$$

Where the symbol λ is TC of the materials (W/mK), ∇t represents the temperature gradient, and q is heat flux. The negative sign in equation (11.1) exhibits the flow of heat from higher temperature to lower temperature. The total TC is the sum of lattice, phonon and electron components ($\lambda_{total} = \lambda_{lattice} + \lambda_{phonon} + \lambda_{electron}$). The thermal resistance is usually applied to determine the efficiency of thermal transport in the electronic devices and is calculated as:

$$R = \frac{\Delta T}{q} \tag{11.2}$$

Where ΔT is the temperature difference between the two regions with units' centigrade/ Kelvin. The thermal resistance is the reciprocal of the TC and the unit is Kelvin-meters/Watt (Km/W). According to the basic perspective of quantum theory, TC is defined as the ability to transport thermal energy to the surrounding atoms mainly by collision [23]. According to the second law of thermodynamics, heat transfer occurs from a high temperature to a lower temperature region.

In solid materials, electrons, holes, and phonons act as a carrier for heat transportation (atomic lattice vibrations). The contribution of phonons in the field of TC is dominated in semiconductors and insulators. While in metals the electronic contributions are more important than the phonon contribution. The basic mechanism of thermal conduction in the polymers is through phonons. The TC of the polymers can be calculated by the Debye equation:

$$k = \frac{1}{3}Cvl \tag{11.3}$$

Where the symbol 'C' represents the specific heat, v is the average velocity of the phonon and l is the mean free path of the phonon. The mean free path length is the average distance of the phonon which will travel without scattering. The scattering of the phonon can be caused by the scattering of other phonons, impurities, defects, as well as boundaries [24]. The following equation represents the concept of phonon dispersion and the relationship between the lattice vibration frequency ω and wave vector k. The group velocity v is given by:

$$v = \frac{d\omega_k}{dk} \tag{11.4}$$

The phonon dispersion plays an important role to study the transport of thermal energy within the materials. The scattering of phonons through the anharmonicity of lattice, with normal and Umklapp process in the crystal lattice. In next section, we will explain the mechanism of the phonon-phonon interaction in crystalline and amorphous materials.

11.3 MECHANISM OF HEAT CONDUCTION

11.3.1 Crystalline Materials

In this section, we are focusing on the mechanism behavior of the TC. It is important to understand the mechanism of the TC within amorphous as well as crystalline materials. Generally, crystalline materials have a high value of TC. In crystalline materials, atoms are arranged regularly. The common example of crystalline materials such as CNTs, graphene, diamond, silicon, boron nitride, metals, glass ceramics, etc, shows the higher TC as compared to the amorphous materials [25–27]. In perfect crystalline structures, the heat gets a transfer from the higher temperature (Hot surface) to lower temperature (Cold surface) through the lattice vibrations with a similar frequency (v =E/h, where h is the Planck constant). Toberer et al. [28] reported the TC and underlying transportation properties of crystalline materials. Figure 11.1(a) shows the mechanism of TC in the crystalline structure. The mechanism of TC followed crystalline materials are given below [29, 30]:

- Heat energy is transmitted to the surface atoms of the specimens.
- Atoms present on the surface gain vibrational energy.
- This vibrational energy transfers the thermal energy in the form of a wave with the adjacent atoms with the same speed.
- Thermal energy is distributed with the help of phonon to the entire crystal.
- When the vibrational energy reaches the opposite surface of the sample, the heat conduction or radiation gets completed.
- It is partially transferred by radiation or conduction to the surrounding area.

Figure 11.1(a) indicates heat source contact with the crystalline lattices which show that the first layer of the atoms conducts the heat in the form of phonon vibrations. The crystalline lattices are densely packed atoms and there exists strong chemical bonding between them. The vibration of the first layer atoms rapidly spreads to the adjacent atom in the form of the wave, and the adjacent atoms pass the vibrations to the next atoms, which exhibit quick heat transfer within the crystalline materials. CNTs and graphene possess ideal crystalline structures in which all the atoms are arranged in the hexagonal lattice structures. Each carbon atom is covalently bonded to the other three carbon atoms.

When the carbon atoms of these fillers come in contact with the heat source, it vibrates, and the vibrations will rapidly pass to the other adjacent atoms. The researchers believe that the heat transfer in CNTs and graphene is due to the phonon waves and some researchers have proved this assumption [31–33].

The crystal structure is an important factor of TC. Figure 11.1(b) shows the phonon vibration in terms of the temperature gradient. The vibrations of atoms in the material are described by the

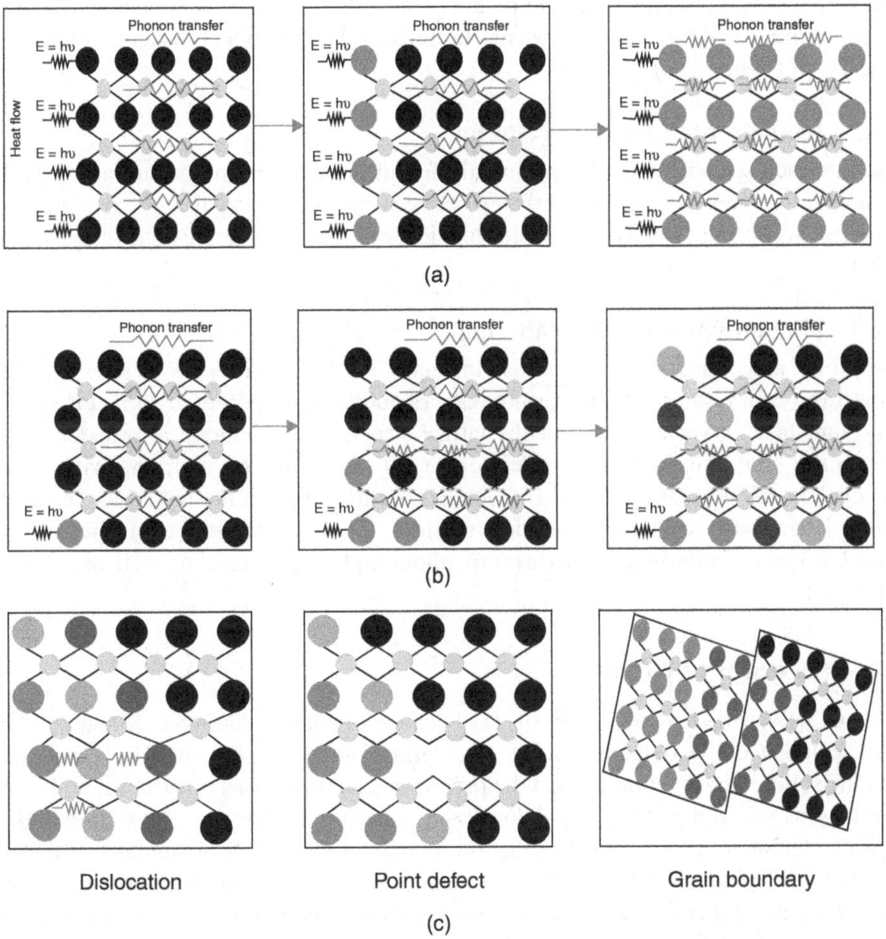

FIGURE 11.1 (a) mechanism of thermal conductivity in crystalline materials, (b) Schematic representation of temperature gradient in crystalline materials and (c) phonon scattering within the crystalline materials, in presence of defects.

Fourier law. In crystalline materials, the presence of a defect, discontinuity, and interface in crystalline material, significantly decrease the TC. The defects present in the crystalline structures are due to point defects, vacancies, isotopic impurities, grain boundaries and dissociation as shown in Figure 11.1(c). Any defect present in the crystalline materials affects the phonon scattering, i.e., phonons (wave) will no longer move as a single wave inside the crystal. Phonon scattering plays a significant role at the atomic level in understanding the concept of thermal transport in imperfect crystalline materials [34]. The phonon scattering shows essentially the three mechanisms:

11.3.1.1 Phonon-Defect (PD) Scattering

In the imperfect crystal point defects, grain boundary, edge dislocation can be simply imagined due to the phonon-defect scattering. Any type of defect present in the crystalline structure acts as the barrier for phonon movement of the surrounding atoms, leading to some scattering. The atoms of the lattice are given to raise the scattering of phonons such as reflection, diffraction, and refraction of phonon, which have a negative impact on the phonon transportation. Lattice defects are mainly the point defects, impurities, dislocation, vacancy defects etc. The collisions of a phonon with crystal boundary or defect do not change the energy of the phonon. The preference of defects within the lattices makes the shorter mean free path, which leads to a lower value of intrinsic TC of crystal.

11.3.1.2 Phonon-Interface (PI) Scattering

Phonon-interface scattering can be commonly found in polymer composite materials. In composite materials, there are many interfaces between the fillers and the polymer matrix which exhibit more scattering. This type of scattering depends upon the dimensions of the samples and geometry.

11.3.1.3 Phonon-Phonon (PP) Scattering

In crystalline and amorphous materials, intrinsic TC arises from PP scattering. This type of scattering occurs due to the anharmonicity of interatomic potential [35]. PP scattering is an important fact, in which atoms of the lattice structure will vibrate to transfer the thermal energy. In the harmonic oscillation (ideal case), all the atoms are vibrating with an identical frequency. Various normal modes or frequencies are available in the materials i.e., atoms present in the lattice will exhibit anharmonic vibrations. These numerous frequencies will describe how the Umklapp scattering (U-process) or phonon-phonon scattering will occur within the material [36].

11.3.2 Amorphous Materials

Amorphous solids have a low value of TC because of their disordered structures. Allen and Feldman calculated the dynamic lattice calculations and analyzed the vibrational modes in amorphous solids. These vibrations can be propagated as propagons, diffusions, and locons [37, 38]. Propagons are transmitting and delocalizing the thermal carriers with a preferably identifiable wave vector in a high wavelength (low frequency) range. It is similar to the traditional representation of phonon, which extends throughout the system and transmits thermal energy. Diffusions are delocalized modes and contribute to heat conduction in a diffusive manner rather than the propagation of energy in terms of phonons. Propagons can be defined in terms of wavelength and wave vectors. Locons are low wavelength (high frequency) modes that are localized and are non-propagating modes. The contribution of TC in locons is negligible [38]. Locons are incapable of transporting thermal energy in harmonic solids [39].

11.3.3 Polymer and Composites Materials

For the past few decades, polymer composites are generally being used as a heat sink and thermal interface materials (TIM). Generally, polymers have a low value of TC; the mean free path of phonon is extremely small due to scattering with other phonons. In the polymer material, the phonon scattering phenomenon is different as compared to crystalline material. The internal impurities, lattice defects significantly influence the phonon scattering and degenerated phonons transport affects the efficiency. In the polymers, the vibrational frequency of phonons mismatches interfaces between the atoms as shown in Figure 11.2(a). Various research groups have carried out the reinforcement of conductive fillers with the polymer matrix in favor of improving the properties of the materials. Fillers have excellent potential for improving heat transport properties. The properties of polymer composites directly depend upon the dispersion of fillers within the polymer matrix. The most significant and commonly used part of passive TM is TIM. The conducting fillers provide a continuous conducting network between the fillers and polymer matrix as shown in Figure 11.2(b). These conducting networks have minimized the phonon scattering problem.

There are three types of TIMs phase change materials (PCM), solid-state heat spreaders, curing and non-curing thermal pastes. The basic function of the TIMs is used to fill the grooves and voids formed by the imperfect surface finishing between the two contact surfaces. The carbon nanofillers have been used to fill the voids between the surfaces and the conduction pathway across the interface.

The polymer-based composites are good thermal interface adhesives, which improve the interactions between the two contact surfaces. The polymer composites offer excellent TC and improve the mechanical integrity of the electronic packaging [40]. Low thermal contact resistance and high TC are essential features of TIMs. Carbon nanofillers such as graphene and carbon nanotubes are being widely used as reinforcement in polymer composites due to their excellent thermal properties [41]. The high TC alone cannot assure good TIM performance. The performance of TIMs generally depends upon the concentration of filler, orientation, dimension, shape, dispersion of fillers within the polymer, which enhances the thermal contacts among the mating surfaces [42]. The addition of nanofillers within the polymer matrix overcomes the problem of

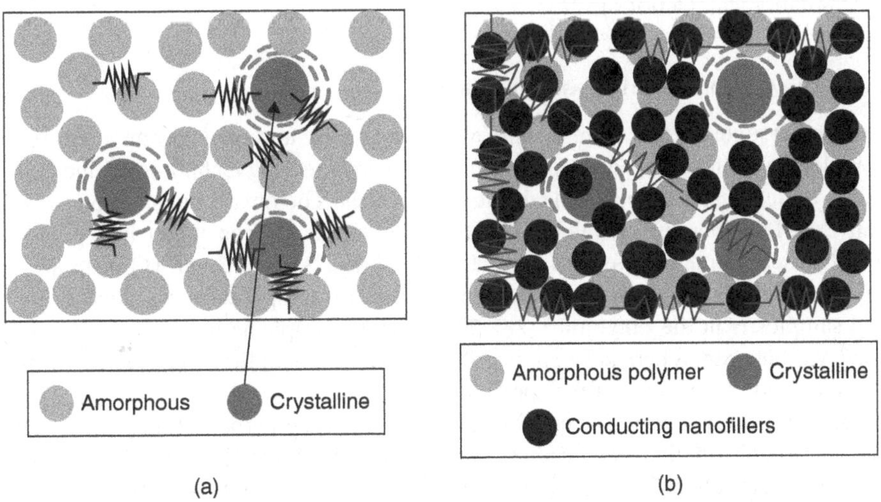

FIGURE 11.2 Mechanism of TC measurement in (a) amorphous polymer and (b) carbon nanofillers reinforcement in polymer composites.

the low value of TC of the polymer. CNTs and graphene are considered the significant materials for TC. The addition of a low concentration of CNTs and graphene significantly enhances the TC of polymer composites [43].

11.4 THERMAL CONDUCTIVITY OF CARBON NANOFILLERS

In recent years, carbon nanofillers have attracted significant research attention of the scientific communities [44]. Carbon nanofillers possess a low bulk density as compared to metallic fillers. The conductivity of nanofillers depends on the π–bonding structure and defect-free structure. CNTs and graphene nanofillers have been considered to be more popular due to their outstanding mechanical, thermal and electrical properties. These nanofillers act as excellent nanomaterials for absorbing the heat dissipation of electronic devices.

11.4.1 Thermal Conductivity of CNTs

CNTs were firstly observed in 1991 by Iijima and these have gained significant research attention. Various groups have reported excellent mechanical, physical, optical, thermal, and electrical properties of CNTs. The value of TC of CNTs and graphene is reported up to 3,000 to 3500 W/mK [45] and 3,000 to 5,000 W/mK [46], respectively. The value of TC for CNTs in the longitudinal axis is much higher as compared to the transverse direction. The quantitative and direct measurement of thermal transport properties of discrete CNTs remains difficult because of nanoscale experimental measurements [47]. So, it is very difficult to obtain the high value of TC experimentally. The theoretical calculations and simulations have been used to find out the value of TC [27, 48, 49]. The transportation of the thermal energy in CNTs is supposed to be via a phonon conduction mechanism. The conduction of phonon in nanotubes is affected by various parameters such as boundary scattering surface, mean free path of the phonons, the number of phonon active modes and inelastic Umklapp scattering [50]. The TC of CNTs depends on the diameter and length of the tube, atomic arrangement, morphology, structural defects and presence of the impurities [51]. The theoretical and experimental calculation of finite TC represents K_T as of individual CNTs is shown in Figure 11.3(a) which shows that the equivalent length of the nanotube is large in all the simulation calculations [52].

11.4.2 Thermal Conductivity of Graphene

Graphene is a 2D material that forms a lattice of hexagonally arranged carbon atoms. Graphene has been widely used in various electronic applications in emerging technologies due to its excellent thermal, optical, electronic and mechanical properties. The hexagonal structure of graphene acts as an excellent thermal conductor due to the intrinsic phonon mean free path. Thus, it possesses significant potential for use in TM applications. The value of TC of graphene is in the range of 5000 W/mK at room temperatures. At room temperature, in-plane TC of freely suspended samples is in the range of 3,000–5000 W/mK. Graphene flakes are outstanding heat conductors. Various researchers have studied that graphene has unlimited potential for heat conduction and depends on the size of the specimen.

A monolayer of graphene is sp_2 hybridization in which the carbon atoms are arranged in a honeycomb structure with the nearest neighbor distance of 1.42 Å [53]. To understand the concept of heat transport in graphene, lattice vibration mode plays a significant role. The hexagonal lattice structure of the graphene with the unit cell consists of two carbon atoms and the lattice constant a=2.46 Å [54]. Figure 11.3 (b) shows that the unit cell of graphene is marked with dashed lines,

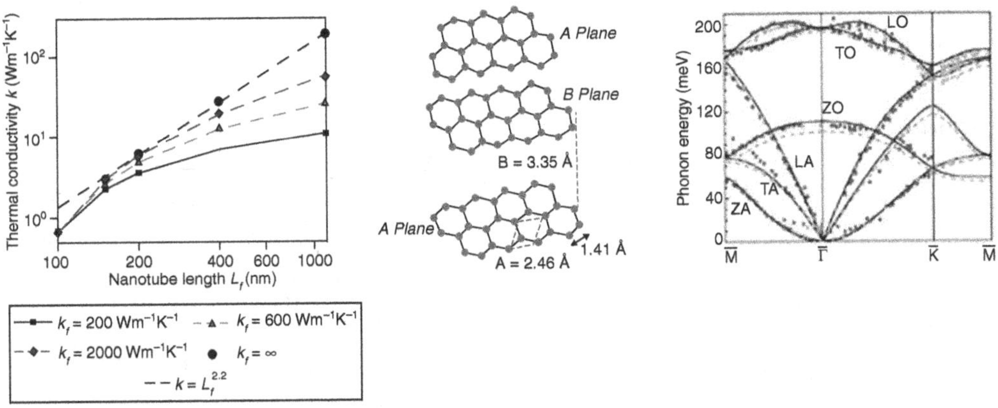

(a) (b) (c)

FIGURE 11.3 TC of CNTs versus length of CNTs L_T for thermal conductivity (K_T) = 2000 W/mK (diamonds), K_T = 600 W/mK (triangles), K_T = 200 W/mK (diamonds), and $K_T = \infty$ (circles). Reprinted permission from [52] Copyright@ 2012 Applied Physics Letter, (b) Atomic arrangement in graphene sheet. In the bottom sheet, the dashed line represents the unit cell. The areal density of the graphene atoms is 3.82 x 10^{15} cm^{-2} and (c) dispersion of phonon along $\bar{M}$to $\bar{M}$ crystallographic axis. The presence of linear in-plane acoustic, TA mode (transverse acoustic TA and longitudinal acoustic LA) as well as out-of-plane acoustic (ZA) modes with quadratic dispersion is shown 3(c). Phonon dispersion of graphite using the inelastic neutron scattering (green dots), HREELS (red dots), inelastic X-ray scattering green dots and graphene are shown by gray-dashed and solid lines. Reprinted with permission from [55], copyright @2016 IOP Publishing.

which contain N=2 carbon atoms. Figure 11.3(c) clearly shows the formation of three acoustic (A) and 3 N-3=3 (3x2–3=3) optical (O) phonon modes. Longitudinal modes (L) represent the atomic vibration along the direction of wave propagation. On the other hand, transverse modes (T) correspond to in-plane vibrations perpendicular to the propagation direction. The out-of-plane atomic vibrations also occur in graphene, known as flexural (Z) phonons. Longitudinal optical (LO), longitudinal acoustic (LA), transverse optical (TO), transverse acoustic (TA) phonons correspond to the vibration within the graphene. The figure shows the phonon dispersion of the graphite and graphene. Figure 11.3 (c) shows the comparison curve of the phonon dispersion of graphene and graphite. Phonon dispersion curves of graphite are measured by high resolution electron energy loss spectroscopy (HREELS) marked as red dots, inelastic X-ray scattering marked as green and inelastic neutron scattering as open circles compared to density functional theory (DFT) calculations of graphene.

The allotropes of carbon fillers have occupied the excellent value of TC. TC of various allotropes of carbon span an outstanding large range of the 5-order magnitude from the lowest of 0.01 W/mK in amorphous at room temperature as shown in Figure 11.4(a). The TC of graphene is 2000–5000 W/mK.

11.5 THERMAL PROPERTIES OF CARBON NANOFILLERS REINFORCED POLYMER COMPOSITES

Due to the outstanding values of TC, carbon nanofillers behave as excellent materials which improve the TC of the polymer composites. Thermal analysis is considered to be vital in

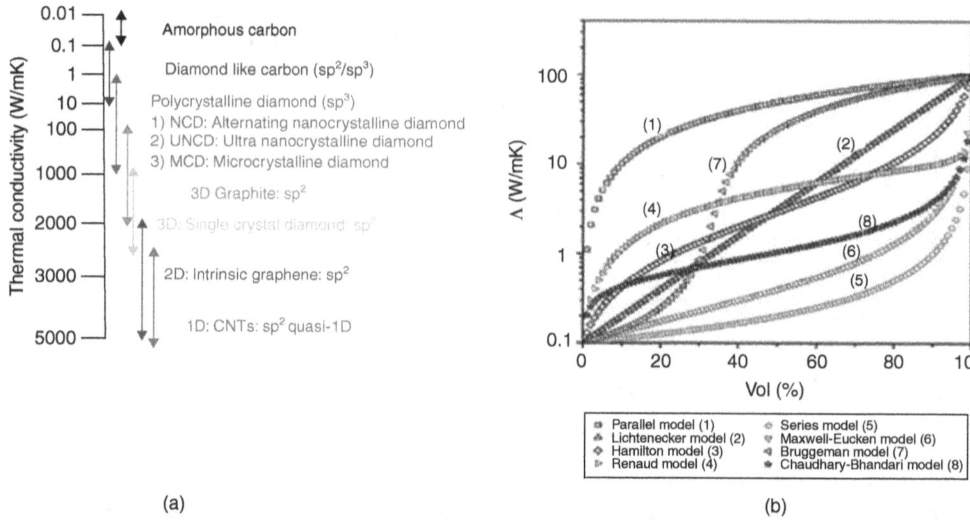

FIGURE 11.4 (a) TC values of the various carbon nanofillers based on the literatures, (b) Comparison study of thermal conductivity of the various theoretical models as a function of volume fraction. Reprinted with permission from [56], Copyright@2019 from Elsevier.

engineering structures, numerous research groups have explored the TC of nanocomposites. These nanomaterials and their structures have initiated one of the hot research topics in various emerging technologies and are being explored by the scientific community because of their extensively improved thermal and mechanical properties [57, 58]. TC of polymer and filler reinforcement polymer composites can be separated into two parts according to the fabrication techniques. Aggregates embedded of the polymer matrix, such as crystallization, orientation, micro-scale ordered structures, cross-linking, and intermolecular interaction, all have significant effects on the TC values of polymers. Various other parameters such as aspect ratio, size of nanofillers, volume fraction, interfacial interaction, intrinsic TC, defects present in fillers, dispersion of fillers within matrix, etc. has also affected the effectiveness of TC polymer composites. The filler concentration significantly affects the thermal properties of the composites. At the higher concentration of fillers, it may start agglomerates. The effects of filler agglomeration are not clear, they may serve the crack initiation sites or mechanical defects, but in some cases the value of TC has been improved. The percolation theory helps to understand the effect of filler loading on the energy transfer within the polymer composites. The relationship between the TC, volume fraction and percolation threshold can be calculated by a scaling law [59]:

$$K^{*} \alpha\, K\left(f - f_c\right)^t \tag{11.5}$$

Where K* is the TC of composites, K is TC of filler, f is volume fraction, f_c is the percolation threshold, and t is the percolation exponent.

The carbon nanofiller reinforcement in polymer composites materials overcomes the problem of the metallic-based heat exchangers. The design and manufacturing of polymer/polymer composites-based heat exchangers have been prompted because of their simple processing, easy to handle, excellent mechanical properties, resistance to corrosion as compared to other ceramics and metal-based composites. The TC of polymer composites plays an important role in TMMs.

The addition of suitable nanofiller reinforcement in the polymer matrix is to improve the TC of polymer composites. The external conductions by humidity, sample aging, operating temperature or heat flux also affect the TC of the polymer composites. The temperature has significantly affected the value of TC of polymer composites. The value of TC increases as the temperature increases because the rise in the temperature promotes the thermal motion of molecules.

11.6 THEORETICAL MODELING OF TC AND THERMAL EXPANSION

When the thermal energy is moved within the material, it undergoes expansion because the atoms of the lattices vibrate with the large amplitude. The change in temperature within the material is "∂T" and "∂L" is the change in length, then the fraction increase in the length per unit change in temperature, then coefficient of linear expansion (α) is given as:

$$\alpha = \frac{1}{L}\left(\frac{\partial L}{\partial T}\right) \tag{11.6}$$

The value of thermal expansion is not limited in one dimension; they are volumetric. Thus, the volumetric coefficient of thermal expansion β is given as:

$$\beta = \frac{1}{V}\left(\frac{\partial V}{\partial T}\right) \tag{11.7}$$

In the case of the homogenous dispersion of carbon nanofiller reinforced composites, the relation between the linear and volumetric coefficient of thermal expansion is represented as

$$\beta = 3\alpha \tag{11.8}$$

11.6.1 Coefficient of Thermal Expansion (CTE) Models

In the early 19th century, researchers had significant interest in the theoretical modeling of the thermal properties of composites. Many models have been proposed in the literature, which depend upon the shape, concentration of fillers. The rule of mixture is the simplest approach to analyse the coefficient of thermal expansion for the homogenous dispersion of filler within the composites [60]. The rule of mixture is the first order approximation and given as:

$$\gamma_C = \gamma_f \varnothing + \gamma_m (1 - \varnothing) \tag{11.9}$$

Where γ_m, γ_f *and* γ_C are the CTE of matrix, filler, and composites, respectively and $\varnothing$ is the volume fraction of the fillers and calculated in the term of weight fraction w is:

$$\varnothing = \frac{w}{w + (1 - w)\frac{d_f}{d_m}} \tag{11.10}$$

Where d_f and d_m are the density of filler and matrix.

Turner et al. [61] reported the sophisticated model of the composites. The equation can be written as:

$$\gamma_C = \frac{\varnothing \frac{M_f \gamma_f}{d_f} + (1-\varnothing) M_m \gamma_m / d_m}{\varnothing \frac{M_f}{d_f} + (1-\varnothing) M_m / d_m} \tag{11.11}$$

Where M_m and M_f are the bulk modulus of matrix and the filler, respectively. Schapery et al. [62] reported the lower and upper limits of CTE of composites.

$$\text{Lower bound } \gamma_C^- = \alpha_m + \frac{M_f}{M_C^+}\left[\frac{(M_m - M_C^+)(\gamma_f - \gamma_m)}{(M_m - M_f)}\right] \tag{11.12}$$

$$\text{Upper bound } \gamma_C^+ = \alpha_m + \frac{M_f}{M_C^-}\left[\frac{(M_m - M_C^-)(\gamma_f - \gamma_m)}{(M_m - M_f)}\right] \tag{11.13}$$

Where the symbols + and – stand for the upper and lower limit, respectively, and M_C^+ and M_C^- are the upper and lower limits of bulk modulus of the composites and are provided by Hashin and Shitrikman [63]:

$$M_C^+ = M_f + \frac{\varnothing}{\frac{1}{M_m - M_f} + \left(\frac{3\varnothing}{3M_f + 4\mu_f}\right)} \tag{11.14}$$

$$M_C^- = M_m + \frac{\varnothing}{\frac{1}{M_f - M_m} + \left(\frac{3(1-\varnothing)}{3M_m + 4\mu_m}\right)} \tag{11.15}$$

Where μ_m and μ_f are the value of shear modulus of matrix and filler, respectively, the value of μ and M can be calculated from Young's modulus E and Poisson ratio ν:

$$\text{M} = \frac{E}{2(1+\nu)} \text{ and } \mu = \frac{E}{(2-3\upsilon)} \tag{11.16}$$

11.6.2 Modeling of TC

Various theoretical and empirical models have been proposed to calculate the TC of composites in two phase mixtures [64–66]. The two-phase polymer composites are the simplest to measure the TC. The arrangement of material in either series or parallel with respect to heat flow gives the lower or upper bound of effective TC. For the upper and lower bound is describe as follow:

$$k_c = \varnothing k_f + (1-\varnothing) k_m \tag{11.17}$$

And for the series conduction model

$$\frac{1}{k_C} = \frac{\varnothing}{k_f} + \frac{(1-\varnothing)}{k_m} \tag{11.18}$$

The series and parallel model are not successfully modeled because the TC of material totally depends upon the conductivity of the phases, sizes, shapes and spatial distribution of filler within the matrix.

Geometrical model (Lichtenecker model) a well-known model to determine the value of TC of the composites is given as:

$$k_c = k_m^{1-\varnothing} + k_f^{\varnothing} \tag{11.19}$$

Taking the natural logarithmic on both sides of the equation (11.19) is:

$$\log k_c = (1-\varnothing) logk_m + \varnothing logk_f \tag{11.20}$$

Renaud, Krischer and Chaudhary-Bhandari models explain TC by various series and parallel combinations [67]:

$$k_c = f\left((1-\varnothing)k_m + \varnothing k_f\right) + (1-f)\left(\frac{1-\varnothing}{k_m} + \frac{\varnothing}{k_f}\right)^{-1} \tag{11.21}$$

$$k_c = \left(\frac{1-f}{(1-\varnothing)k_m + \varnothing k_f} + f\left(\frac{1-\varnothing}{k_m} + \frac{\varnothing}{k_f}\right)\right)^{-1} \tag{11.22}$$

$$k_c - \left((1-\varnothing)k_m + \varnothing k_f\right)^f \left(\frac{1-\varnothing}{k_m} + \frac{\varnothing}{k_f}\right)^{(f-1)} \tag{11.23}$$

Where f is the adjustable parameter ranging from 0 to 1, is indicating the contribution of series and parallel structures.

The modification of the Maxwell model using the Hamiltonian equations and adjustable parameter k is introduced.

$$k_c = k_m \frac{(k-1)k_m + k_f - (k-1)(k_m - k_f)\varnothing}{(k-1)k_m + k_f + (k_m - k_f)\varnothing} \tag{11.24}$$

Nielsen et al. [68] reported the empirical model to predict the TC of polymer composites:

$$k_c = k_m \frac{1 + A\beta v_f}{1 - \beta v_f \varphi}, \text{ Where } \beta = \frac{\frac{k_f}{k_m} - 1}{\frac{k_f}{k_m} + A}, \varphi = 1 + \frac{1 - v_m}{v_m^2} v_f \text{ and } A = E_c - 1 \tag{11.25}$$

Here A is the aspect ratio, v_m is the maximum volume fraction of the fillers, E_c is the Einstein constant which is related to the shape and orientation of particles.

The Halpin-Tsai (HT) model is a semi empirical model. This model incorporates two parts, one for the longitudinal and other for the transverse thermal conductivities are given as [69]:

$$k_c = \frac{3}{8}\left(1-2\left(\frac{l}{d}\right)*\left[\frac{\frac{k_f}{k_m}-\frac{d}{4t}}{\frac{k_f}{k_m}+\frac{l}{4t}}\right]\varnothing_f\right)*\left(1-\left[\frac{\frac{k_f}{k_m}-\frac{d}{4t}}{\frac{k_f}{k_m}-\frac{d}{2t}}\right]\varnothing_f\right)*k_f$$
$$+\frac{5}{8}\left(1-\left[2\frac{\frac{k_f}{k_m}-\frac{d}{4t}}{\frac{k_f}{k_m}-\frac{d}{2t}}\right]\varnothing_f\right)*k_f*\left(1-\left[\frac{\frac{k_f}{k_m}-\frac{d}{4t}}{\frac{k_f}{k_m}+\frac{d}{2t}}\right]\varnothing_f\right)*k_f \quad (11.26)$$

Where d, t and l are the diameter, thickness, and length of CNT filler, k_m and k_f are the TC of matrix and fillers, respectively. $\varnothing_f$ is the volume fraction of the CNTs. Figure 11.4(b) shows the comparison study of all theoretical models. The graph showed that the TC enhanced with the increase in the concentration of fillers.

The TC of polymer composites depends on good thermal interface adhesives, which improve the interactions between the mating surfaces. The polymer composites offer excellent TC and improve the mechanical strength of the electronic packaging [40]. Low thermal contact resistance and high TC are essential features of TIMs. Carbon nanofillers such as graphene and carbon nanotubes are being widely used as reinforcement agents in polymer composites due to their excellent thermal properties as shown in Figure 11.4(b) [41]. The high TC alone cannot assure good TIM performance. The performance of TIMs generally depends upon the concentration of filler, orientation, dimension, shape, dispersion of fillers within the polymer, which enhances the thermal contacts among the matting surfaces.

TC of the enhancement factor (TCEF) is calculated from the following expression [70]:

$$\text{TCH} = \frac{\textit{thermal conductivity enhancement percentage}}{\textit{Filler volume \%}} \quad (11.27)$$

In industrial applications, the ideal TIM requires the high value of TC with the combination of easy processing, high flexibility, electrical insulations, etc.

11.7 METHODS OF TC MEASUREMENTS IN CARBON NANOFILLERS

11.7.1 Optothermal Raman

Balandin et al. [46] reported that the Optothermal Raman technique which was used to measure the TC of suspended single-layer graphene with a spot size of 0.5–1.0 μm as shown in Figure 11.5(a). They found that the value of TC was as high as 4800–5300 W/mK. In this technique, a laser was used as the source of heat and focused on the center of a specimen [71, 72]. The temperature gradient is used to measure the TC through the Raman spectrum. The peak shift in the Raman spectrum exhibits a linear relationship with temperature change. Other results were analyzed using the

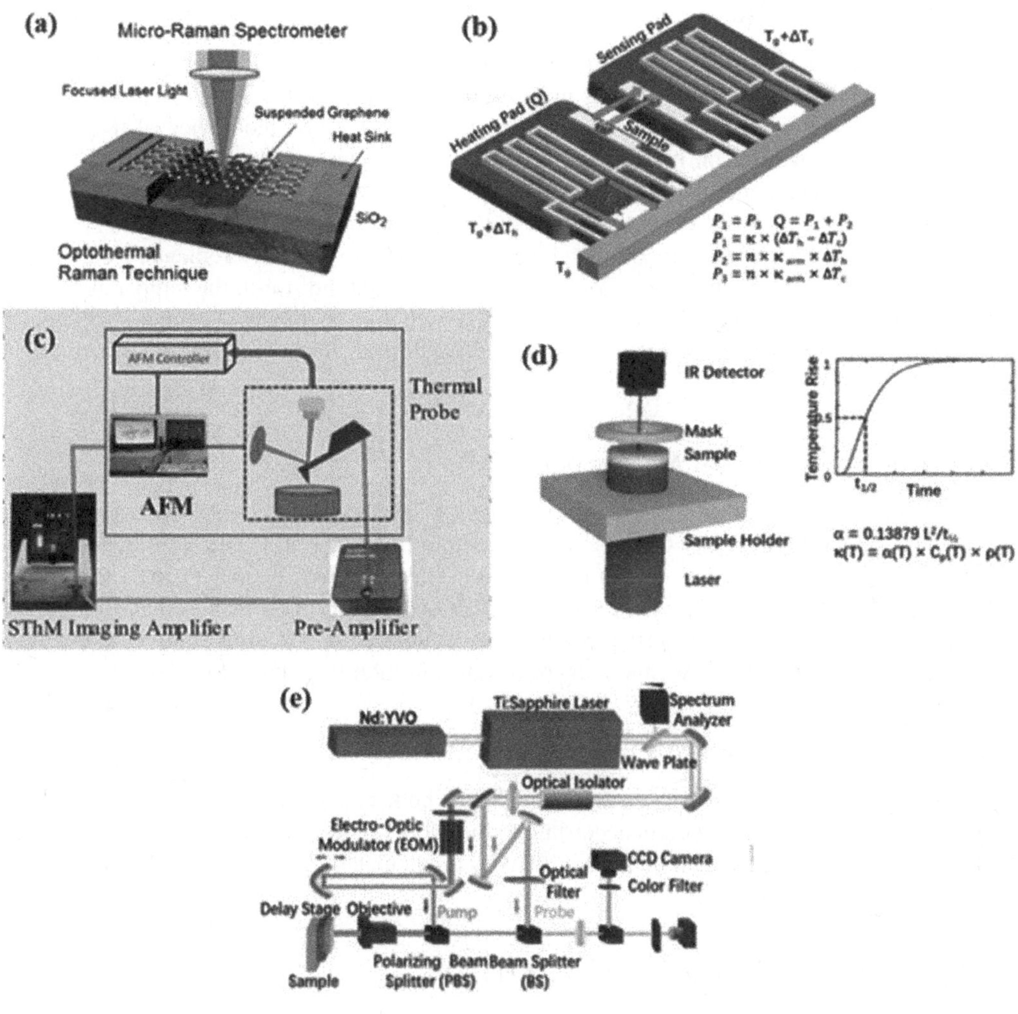

FIGURE 11.5 Experimental set up of various measurements techniques (a) Optothermal Raman technique (b) suspended pad technique, (c) Scanning thermal microscopy Reprinted with permission from [85], copyright@2021 from Elsevier (d) laser flash technique and (e) Time-domain thermoreflectance (TDRT), where orange and blue lines represent the probe laser beam and pump respectively Reprinted with permission from [77], copyright@2018 from Elsevier.

difference between the transmitted laser power through the graphene flakes and empty holes [73] or the assumption of absorption of laser power of graphene flake [74].

This technique is based optically and non-destructive technique, it has also been used to analyze the TC of the various 2D materials such as molybdenum disulfide (MoS_2) [75], tungsten disulfide WS_2 [76], hexagonal boron nitride (h-BN) [74], etc. But the experimental error in this technique is found to be very high. The accuracy of this technique depends upon various factors such as laser spot size, interfacial thermal conductance of contact samples, the optical absorption of the samples, heat dissipation to the surrounding [77]. Various research groups have focused to reduce the effect of heat loss on the surrounding environment. Additionally, improved accuracy in this technique is also required.

11.7.2 Suspended Pad

In 2001, Kim et al. invented the suspended technique to measure the thermal properties of the nanoscale materials [78]. This method is most useful for analyzing the TC of carbon nanofillers (carbon nanotube, nanowires, graphene, etc). Figure 11.5(b) shows the device of this method, which consists of two silicon nitride membranes, with a patterned platinum (Pt) coil on each pad. The micro heaters and thermometers are used for providing the heat and they recorded the temperature of the two membranes, respectively [77, 79]. The pads are suspended in six silicon nitride arms coated with Pt for thermal and electrical conductivity measurements. Heat transfer within the samples can be measured from the heating pad and the rise in the temperature of the sensor. This technique is providing consistent results with high accuracy. The main challenges of this technique are in the sample transferring process and time consuming, and the device manufacturing is also difficult [80].

11.7.3 Scanning Thermal Microscopy (STM)

This technique is based on atomic force microscopy (AFM), a scanning probe microscope which gives temperature mapping based on the tip as shown in Figure 11.5(c). The scanning tip is used to analyse the local temperature within the samples. This technique is used to analyse the grain boundaries of 2D materials, local hot spots and heat dissipation [81–83]. In 2000, this method was firstly used to measure the thermal conductivity of the CNTs [84]. Null point STM is used to improve the signal-to-noise ratio, and local heat transportation from the sample to tip.

11.7.4 Laser Flash Apparatus

This technique is widely used to measure the TC of bulk materials. Figure 11.5(d) shows the experiment setup. The sample is cut according to the proper dimensions and the sample is heated with the pulse laser. The value of thermal diffusivity is calculated using the thermal diffusion time and thickness of the sample.

11.7.5 Time-Domain Thermoreflectance (TDTR)

It is the pump-probe optical technique used to calculate the value of thermal transport. In this technique, the mode-locked laser is divided into a pump beam and a probe beam as shown in Figure 11.5(e). The pump makes a temperature increase up to 3 K near the metallic surface. On the other side, the probe is in the middle of the pump, which can observe the surface temperature with the variation of optical reflectivity. The amplitude and phase of the probe are calculated as a function of the time delay between the probe and pump. In 1986, this technique was firstly reported [86]. This technique is most useful in calculating the thermal transport at the interface [87].

11.7.6 Other Methods

There are many other novel techniques that are used for analyzing the thermal properties of the samples i.e., Joule heating, the dependence of resistance on temperature (DRT), pulsed photothermal reflectance (PPR), transient plane source (TPS) and infrared microscopy, etc. These techniques should be selected according to the intrinsic properties and geometry of the materials. Carbon nanofillers reinforcement polymer composites are promising materials for TIMs, which are widely used in batteries, super capacitors, IC, and solar cells.

11.8 APPLICATIONS OF THERMAL MANAGEMENT

In the past few decades, with the quick development of microelectronic methods, electronic devices have become more rapid, miniaturization, excellent performance, and a high degree of integration. In the energy storage system, TM is an important requirement for their performance over a suitable range of temperatures. In the low-temperature range, the performance of the devices decays due to the low ionic conductivity. At the high temperature range, the constituent elements tend to age because of the series of side reactions, reliability, and safety issues. Heat dissipation and TC are important issues of modern electronic devices because of the increase in operational frequency and transistor density. CNTs and graphene have drawn significant research interest worldwide due to their excellent heat dissipation and TC. Due to these excellent properties, carbon nanofillers are used in various applications in TM systems as shown in Figure 11.6.

11.8.1 BATTERIES

The fast progress of electric vehicles (EV) and hybrid electric vehicles (HEV), excess heat generation issues become the main challenges, which affects the life span, performance, and temperature distribution of the batteries. Temperature plays a significant role in the efficiency, safety, and reliability of secondary batteries. At higher temperatures, the performance of the secondary batteries

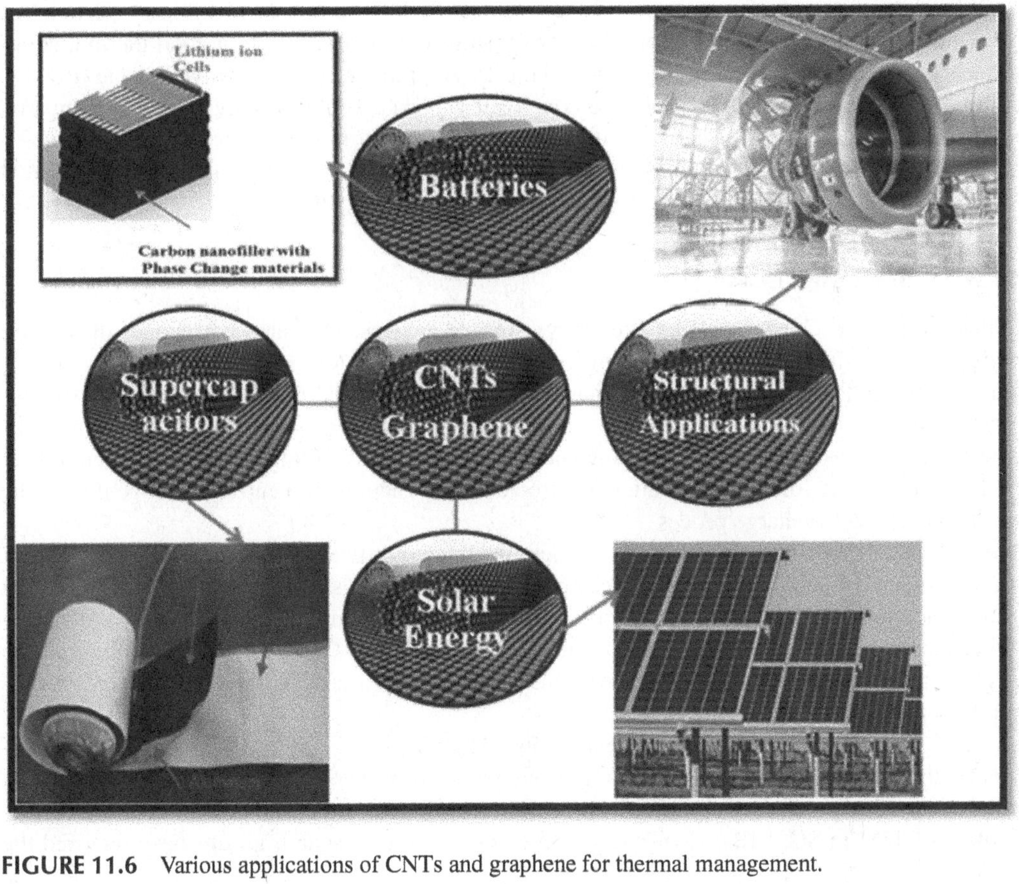

FIGURE 11.6 Various applications of CNTs and graphene for thermal management.

has been limited during the charge/discharge processes. Secondary batteries generate excess heat, which causes an increase in the temperature of the battery pack and affects the high operating current. Efficient lithium-ion batteries (LIBs) are mostly used in modern electric vehicles (EV). The performance of LIBs is mainly dependent upon their operation and storage temperatures. The internal and external factors determine the heat generation in secondary batteries. The operating temperature in the LIBs must be allowable at less than 60°C. Temperature not only affects the life and performance of the LIBs but also can cause safety issues. Therefore, there is a significant demand to design suitable TM systems which control the temperature of the batteries. TM materials have been used for controlling the degradation performance of the batteries at the high temperature. In recent years, the TM using the PCM have a significant research interest in LIBs [88, 89] and these have been extensively used in TM systems. During melting/solidification activity, the low TC of PCM leads to a low value of heat transfer. Carbon-based nanofillers have been used in PCM due to their own high value of TC [90, 91]. Wang et al. [92] reported the composites of paraffin wax and CNTs for the high value of the TM system. The analyzed value of TC increment ratio was achieved 35% in solid and 40% in liquid. Zou et al. [93] reported the comparative performance of CNTs, graphene and CNTs/graphene PCM composites in TM of LIBs pack. The carbon nanofiller PCM showed excellent TC and improved the temperature regulation of the battery pack. Hussain et al. [88] reported the graphene-coated nickel foam with paraffin wax to improve the thermal performance of the PCM-battery. Graphene-coated nickel foam improved the TC by 23 times as compared to paraffin wax. Goli et al. [94] reported the TC of graphene reinforcement paraffin composites. The hybrid material showed improved TC by more than 2 orders of magnitude as compared to traditional PCM. The improvement in properties was due to the easy binding of graphene flakes with paraffin wax resulting in a good thermal coupling between them. Zhang et al. [95] reported the TC of graphene/polydimethylsiloxane (PDMS) composites. The graphene loaded PDMS composites showed 10.8 times higher TC as compared to pure PDMS. Additions of carbon nanofillers reinforcement in composites have lower coefficient of thermal expansion. The carbon nanofiller PCM hybrid composites can be used as talented thermal conductive materials in the TM systems for secondary batteries.

11.8.2 Supercapacitors

Supercapacitors (SCs) are well-known energy storage devices. The main advantages of SCs are high power density, energy density, long cycle life, fast charge/discharge, low cost, easy maintenance, instantaneous high current discharge and environmentally friendly. In the supercapacitor series of exothermic reactions are involved in charge/discharge processes. High operating temperature usually causes accelerated aging and recycling capacity fading as the production of heat is caused by ohmic losses, polarization electrochemical reaction, and entropy change during the reversible charge/discharge process.

The TM system, thermal regulation and temperature have significant research attention because of the significant demand in energy storage applications. TM and systems have become an important issue concerning the life span and performance of SCs and cooling systems with TM is an essential requirement for the development of high-performance SC. Gualous et al. [96] reported the effect of thermal shock and temperature on SC aging and reported that the lifetime of SC significantly depends upon these two factors.

Quick elimination of unnecessary heat produced by integrated circuits (IC) has become necessary for the further improvement of electronic devices. Carbon nanofillers such as CNTs, graphene, graphene oxides and their composites have attracted significant attention for the design and development of TIM in SC, LIBs, solar energy, LED, etc. Various research groups have reported the ultra-compressible electric double-layer capacitor (EDLC) that sustains the performance of capacitive up to 50% compression. Conducting polymer composites, metal oxide, graphene reinforced polymer composites, CNTs reinforced polymer composites enhance the value of energy storage to 300 F/g and 28 F/cm^3 [97–99]. Sun et al. [100] reported the polyaniline/ CNTs functionalized phase change-microcapsules for TM applications of SCs. PANI/CNTs hybrid showed an excellent self-regulation ability of temperature and high heat storage capacity. S Kumar et al. [101] reported the synergetic effect of PANI/CNTs/rGO hybrid composites used in SCs. The hybrid materials improved the specific capacitance and life span of the SCs. Xie et al. [102] reported NiMn layered double hydroxide (Ni/Mn/LDH) supported the Ni/graphene film used in SCs. The Ni/Mn/LDH/Ni/graphene exhibits the improved electrochemical performance of 220 mF/cm^2. The graphene film and Ni layer avoid the problem of overheating. The graphene film could effectively dissipate the heat generated in energy storage devices. Zhao et al. [103] reported the graphene-MnO_2 film used as electrode materials. The graphene-MnO_2 film demonstrated a high value of TC of 614 W/mK as compared to the traditional MnO_2 slurry electrodes. The graphene-MnO_2 hybrid film is beneficial for TM of SCs devices but also has good cycling performance and outstanding rate capacity. The CNTs and graphene and their reinforced polymer material could provide the effective TM approach to promote high-energy storage devices in industrial applications. CNTs and graphene could be attained as a combination of high-performance EDLC and pseudo capacitance to effectively satisfy the efficient global demand for energy storage systems. Understanding the heat influences in the supercapacitor behavior with consideration to capacitance and cycle life is an important factor for use in numerous applications.

11.8.3 Solar Energy

TIMs are significantly used in solar energy, light-emitting diodes, aerospace, microelectronics, electrical and electrical engineering, and various other fields [104]. In electronic devices, demands for enhanced TM are not limited. The power generation technologies for the photovoltaic solar cells are also required for suitable TMMs. The efficiency of modern solar cells has approximately close to 15% in the conversion of light to electricity. Generally, 72% of solar energy is lost as heat. To avoid the degradation performance of the cell, CNTs and graphene have been used as TIMs [105]. From the past few decades, CNTs and graphene have been found suitable materials for electronic devices employing organic semiconductors and used in flexible optoelectronic devices. CNTs and graphene-based devices are currently being followed with anticipation as energy-efficient solid-state lighting and a scalable harvesting of electrical energy from sunlight [106]. Renteria et al. [107] reported magnetically functionalized self-aligning graphene fillers for efficient TM applications. The self-aligning graphene fillers improve heat conduction properties in both curing and non-curing matrix materials. The experiment was performed on the computer chips and reported the decrease in the rise in temperature of 10°C with the use of non-curing TIM material with ~1 wt% of graphene flakes.

Zhang et al [108] reported power conversion efficiency (PEC) of oxygen plasma-treated CNTs/TiO_2-based photoanode and showed 6.34% as compared with 3.63% and 4.66% of pure TiO_2 and CNTs respectively. Graphene has been mixed with other materials to improve the efficiency of the solar cells but consequently trouble the reproducibility of solar cell devices. GO is employed as a hole conductor in inverted planar heterojunction perovskite solar cells [109]. Liu et al. [110] reported the quenching of P3OT to photoluminescence properties after the addition of rGO and indicating some electron/energy transfer. Photovoltaic devices based on rGO-P3OT exhibited 1.4% efficiency. The unique structure and outstanding electronic properties of CNTs and graphene show the electron-accepting material in photovoltaic applications.

11.8.4 Structural Applications

In the modern space and aircraft thermal design, TM is an extremely difficult task due to the frequently rising heat load from complex electrical architectures, microprocessors in electronic systems. In the military aircraft, designers also require the TM system for advanced weapons. CNTs and graphene are excellent TM materials for aircraft and military equipment's applications. Efficient TM and structural design for aircraft application also demand mechanical robust material and lightweight TC which stands for high impact velocity. CNTs and graphene reinforced polymer composites with outstanding mechanical, thermal and electrical properties solve the issue of thermal management in the manufacture of advanced military weapons [111].

11.9 FUTURE PROSPECTIVE

TMMs have been a hot research topic with the fast progress of electronic and optoelectronic devices. The main challenges for the heat transportation properties within the polymer composites are:

1. To obtain the higher value of TC, a significant amount of the filler reinforcement is required in the polymer matrix. The higher concentration of filler reinforcement in the polymer matrix not only causes the brittle property of the composites but also suffer the challenging task for the preparation of the composites.
2. Various complicating factors such as defects, aspect ratios, and aggregations of the TC particles are still difficult tasks.
3. Minimize the interfacial thermal resistance between the polymer matrix and conducting fillers.
4. Thermal measurements are also challenging due to the unreliability of the results.
5. The development of suitable, accurate and rapid measurement methods to evaluate the thermal properties in nanoscale and bulk materials is immediately required.
6. Low cost and large-scale productions are also the requirements for electronic industrial applications.

In future applications of electronic devices, TIM materials will be more focused.

REFERENCES

1. Jia, L.-C., et al., *Highly thermally conductive liquid metal-based composites with superior thermostability for thermal management*. Journal of Materials Chemistry C, 2021. **9**(8): p. 2904–2911.
2. Jyoti, J., B.P. Singh, and S.K. Tripathi, *Recent advancements in development of different cathode materials for rechargeable lithium ion batteries*. Journal of Energy Storage, 2021. **43**: p. 103112.
3. Yao, Y., et al., *Construction of 3D skeleton for polymer composites achieving a high thermal conductivity*. Small, 2018. **14**(13): p. 1704044.
4. Ali, S., et al., *A review of graphene reinforced Cu matrix composites for thermal management of smart electronics*. Composites Part A: Applied Science and Manufacturing, 2021: p. 106357.
5. Moore, A.L. and L. Shi, *Emerging challenges and materials for thermal management of electronics*. Materials today, 2014. **17**(4): p. 163–174.
6. Vadivelu, M., C.R. Kumar, and G.M. Joshi, *Polymer composites for thermal management: a review*. Composite Interfaces, 2016. **23**(9): p. 847–872.

7. Zhang, Y., J.R. Choi, and S.-J. Park, *Enhancing the heat and load transfer efficiency by optimizing the interface of hexagonal boron nitride/elastomer nanocomposites for thermal management applications*. Polymer, 2018. **143**: p. 1–9.
8. Barani, Z., et al., *Thermal properties of the binary-filler hybrid composites with graphene and copper nanoparticles*. Advanced Functional Materials, 2020. **30**(8): p. 1904008.
9. Chen, Y., et al., *Constructing a "pea-pod-like" alumina-graphene binary architecture for enhancing thermal conductivity of epoxy composite*. Chemical Engineering Journal, 2020. **381**: p. 122690.
10. Zhu, Z., et al., *Enhanced thermal conductivity of polyurethane composites via engineering small/large sizes interconnected boron nitride nanosheets*. Composites science and technology, 2019. **170**: p. 93–100.
11. Zhang, J., et al., *Large improvement of thermal transport and mechanical performance of polyvinyl alcohol composites based on interface enhanced by SiO2 nanoparticle-modified-hexagonal boron nitride*. Composites science and technology, 2019. **169**: p. 167–175.
12. Guo, Y., et al., *Constructing fully carbon-based fillers with a hierarchical structure to fabricate highly thermally conductive polyimide nanocomposites*. Journal of Materials Chemistry C, 2019. **7**(23): p. 7035–7044.
13. Guo, Y., et al., *Significantly enhanced and precisely modeled thermal conductivity in polyimide nanocomposites with chemically modified graphene via in situ polymerization and electrospinning-hot press technology*. Journal of Materials Chemistry C, 2018. **6**(12): p. 3004–3015.
14. Wang, Z.-G., et al., *Achieving excellent thermally conductive and electromagnetic shielding performance by nondestructive functionalization and oriented arrangement of carbon nanotubes in composite films*. Composites Science and Technology, 2020. **194**: p. 108190.
15. Lin, M., et al., *Thermally conductive nanostructured, aramid dielectric composite films with boron nitride nanosheets*. Composites Science and Technology, 2019. **175**: p. 85–91.
16. Jyoti, J. and B.P. Singh, *A review on 3D graphene–carbon nanotube hybrid polymer nanocomposites*. Journal of Materials Science, 2021: p. 1–46.
17. Geim, A.K., *Graphene: status and prospects*. Science, 2009. **324**(5934): p. 1530–1534.
18. Sarangdevot, K. and B. Sonigara, *The wondrous world of carbon nanotubes: Structure, synthesis, properties and applications*. Journal of Chemical and Pharmaceutical Research, 2015. **7**(6): p. 916–933.
19. Pop, E., et al., *Thermal conductance of an individual single-wall carbon nanotube above room temperature*. Nano Letters, 2006. **6**(1): p. 96–100.
20. Blackburn, J.L., et al., *Carbon-nanotube-based thermoelectric materials and devices*. Advanced Materials, 2018. **30**(11): p. 1704386.
21. Nika, D.L. and A.A. Balandin, *Phonons and thermal transport in graphene and graphene-based materials*. Reports on Progress in Physics, 2017. **80**(3): p. 036502.
22. Benedict, L.X., S.G. Louie, and M.L. Cohen, *Heat capacity of carbon nanotubes*. Solid State Communications, 1996. **100**(3): p. 177–180.
23. Kunova, O., et al., *Non-equilibrium kinetics, diffusion and heat transfer in shock heated flows of N2/N and O2/O mixtures*. Chemical Physics, 2015. **463**: p. 70–81.
24. Fiamegkou, E., N. Athanasopoulos, and V. Kostopoulos, *Prediction of the effective thermal conductivity of carbon nanotube-reinforced polymer systems*. Polymer Composites, 2014. **35**(10): p. 1997–2009.
25. Zhang, Y., K.Y. Rhee, and S.-J. Park, *Nanodiamond nanocluster-decorated graphene oxide/epoxy nanocomposites with enhanced mechanical behavior and thermal stability*. Composites Part B: Engineering, 2017. **114**: p. 111–120.
26. Balandin, A.A., *Phononics of graphene and related materials*. ACS Nano, 2020. **14**(5): p. 5170–5178.
27. Jyoti, J., et al., *Mechanical, electrical and thermal properties of graphene oxide-carbon nanotube/ABS hybrid polymer nanocomposites*. Journal of Polymer Research, 2020. **27**(9): p. 1–16.
28. Toberer, E.S., L.L. Baranowski, and C. Dames, *Advances in thermal conductivity*. Annual Review of Materials Research, 2012. **42**: p. 179–209.

29. Burger, N., et al., *Review of thermal conductivity in composites: Mechanisms, parameters and theory*. Progress in Polymer Science, 2016. **61**: p. 1–28.
30. Tritt, T.M., *Thermal Conductivity: Theory, Properties, and Applications*, 2005: Springer Science & Business Media.
31. Yao, W.-J. and B.-Y. Cao, *Thermal wave propagation in graphene studied by molecular dynamics simulations*. Chinese Science Bulletin, 2014. **59**(27): p. 3495–3503.
32. Wang, P., R. Xiang, and S. Maruyama, *Thermal Conductivity of carbon nanotubes and assemblies*. Advances in Heat Transfer, 2018. **50**: p. 43–122.
33. Narula, R., et al., *Dominant phonon wave vectors and strain-induced splitting of the 2 D Raman mode of graphene*. Physical Review B, 2012. **85**(11): p. 115451.
34. Lindsay, L., *First principles peierls-boltzmann phonon thermal transport: a topical review*. Nanoscale and Microscale Thermophysical Engineering, 2016. **20**(2): p. 67–84.
35. Liao, B., *Nanoscale Energy Transport; Emerging Phenomena, Methods and Applications*, 2020.
36. Dugdale, J. and D. MacDonald, *Lattice thermal conductivity*. Physical Review, 1955. **98**(6): p. 1751.
37. Allen, P.B. and J.L. Feldman, *Thermal conductivity of glasses: Theory and application to amorphous Si*. Physical Review Letters, 1989. **62**(6): p. 645.
38. Allen, P.B., et al., *Diffusons, locons and propagons: Character of atomie yibrations in amorphous Si*. Philosophical Magazine B, 1999. **79**(11–12): p. 1715–1731.
39. Larkin, J.M. and A.J. McGaughey, *Thermal conductivity accumulation in amorphous silica and amorphous silicon*. Physical Review B, 2014. **89**(14): p. 144303.
40. Hansson, J., et al., *Novel nanostructured thermal interface materials: a review*. International Materials Reviews, 2018. **63**(1): p. 22–45.
41. Georgakilas, V., et al., *Broad family of carbon nanoallotropes: classification, chemistry, and applications of fullerenes, carbon dots, nanotubes, graphene, nanodiamonds, and combined superstructures*. Chemical reviews, 2015. **115**(11): p. 4744–4822.
42. Jyoti, J., et al., *Superior mechanical and electrical properties of multiwall carbon nanotube reinforced acrylonitrile butadiene styrene high performance composites*. Composites Part B: Engineering, 2015. **83**: p. 58–65.
43. Jyoti, J., et al., *Synergetic effect of graphene oxide-carbon nanotube on nanomechanical properties of acrylonitrile butadiene styrene nanocomposites*. Materials Research Express, 2018. **5**(4): p. 045608.
44. Selamneni, V., et al., *Carbon Nanomaterials for Emerging Electronic Devices and Sensors*, in *Carbon Nanomaterial Electronics: Devices and Applications*, 2021, Springer. p. 215–258.
45. Fujii, M., et al., *Measuring the thermal conductivity of a single carbon nanotube*. Physical Review Letters, 2005. **95**(6): p. 065502.
46. Balandin, A.A., et al., *Superior thermal conductivity of single-layer graphene*. Nano Letters, 2008. **8**(3): p. 902–907.
47. Xie, H., A. Cai, and X. Wang, *Thermal diffusivity and conductivity of multiwalled carbon nanotube arrays*. Physics Letters A, 2007. **369**(1–2): p. 120–123.
48. Grujicic, M., G. Cao, and B. Gersten, *Atomic-scale computations of the lattice contribution to thermal conductivity of single-walled carbon nanotubes*. Materials Science and Engineering: B, 2004. **107**(2): p. 204–216.
49. Han, Z. and A. Fina, *Thermal conductivity of carbon nanotubes and their polymer nanocomposites: A review*. Progress in Polymer Science, 2011. **36**(7): p. 914–944.
50. Maultzsch, J., et al., *Phonon dispersion of carbon nanotubes*. Solid State Communications, 2002. **121**(9–10): p. 471–474.
51. Maeda, T. and C. Horie, *Phonon modes in single-wall nanotubes with a small diameter*. Physica B: Condensed Matter, 1999. **263**: p. 479–481.
52. Volkov, A.N. and L.V. Zhigilei, *Heat conduction in carbon nanotube materials: Strong effect of intrinsic thermal conductivity of carbon nanotubes*. Applied Physics Letters, 2012. **101**(4): p. 043113.
53. Neto, A.C., et al., *The electronic properties of graphene*. Reviews of Modern Physics, 2009. **81**(1): p. 109.

54. Mohr, M., et al., *Phonon dispersion of graphite by inelastic x-ray scattering*. Physical Review B, 2007. **76**(3): p. 035439.
55. Al Taleb, A. and D. Farías, *Phonon dynamics of graphene on metals*. Journal of Physics: Condensed Matter, 2016. **28**(10): p. 103005.
56. Maruzhenko, O., et al., *Improving the thermal and electrical properties of polymer composites by ordered distribution of carbon micro-and nanofillers*. International Journal of Heat and Mass Transfer, 2019. **138**: p. 75–84.
57. Huang, C., X. Qian, and R. Yang, *Thermal conductivity of polymers and polymer nanocomposites*. Materials Science and Engineering: R: Reports, 2018. **132**: p. 1–22.
58. Guo, Y., et al., *Factors affecting thermal conductivities of the polymers and polymer composites: A review*. Composites Science and Technology, 2020. **193**: p. 108134.
59. Zhu, Y., K. Chen, and F. Kang, *Percolation transition in thermal conductivity of β-Si3N4 filledepoxy*. Solid State Communications, 2013. **158**: p. 46–50.
60. Wong, C. and R.S. Bollampally, *Thermal conductivity, elastic modulus, and coefficient of thermal expansion of polymer composites filled with ceramic particles for electronic packaging*. Journal of Applied Polymer Science, 1999. **74**(14): p. 3396–3403.
61. Turner, P.S., *The Problem of Thermal-Expansion Stresses in Reinforced Plastics*. 1942.
62. Schapery, R.A., *Thermal expansion coefficients of composite materials based on energy principles*. Journal of Composite Materials, 1968. **2**(3): p. 380–404.
63. Hashin, Z. and S. Shtrikman, *A variational approach to the theory of the elastic behaviour of multiphase materials*. Journal of the Mechanics and Physics of Solids, 1963. **11**(2): p. 127–140.
64. Progelhof, R., J. Throne, and R. Ruetsch, *Methods for predicting the thermal conductivity of composite systems: a review*. Polymer Engineering & Science, 1976. **16**(9): p. 615–625.
65. Tavman, I., *Effective thermal conductivity of isotropic polymer composites*. International Communications in Heat and Mass Transfer, 1998. **25**(5): p. 723–732.
66. Kochetov, R., et al., *Modelling of the thermal conductivity in polymer nanocomposites and the impact of the interface between filler and matrix*. Journal of Physics D: Applied Physics, 2011. **44**(39): p. 395401.
67. Renaud, T., et al., *Thermal properties of model foods in the frozen state*. Journal of Food Engineering, 1992.
68. Nielsen, L.E., *Thermal conductivity of particulate-filled polymers*. Journal of Applied Polymer Science, 1973. **17**(12): p. 3819–3820.
69. Tessema, A., et al., *Effect of filler loading, geometry, dispersion and temperature on thermal conductivity of polymer nanocomposites*. Polymer Testing, 2017. **57**: p. 101–106.
70. Shtein, M., et al., *Graphene-based hybrid composites for efficient thermal management of electronic devices*. ACS Applied Materials & Interfaces, 2015. **7**(42): p. 23725–23730.
71. Shahil, K.M. and A.A. Balandin, *Thermal properties of graphene and multilayer graphene: Applications in thermal interface materials*. Solid State Communications, 2012. **152**(15): p. 1331–1340.
72. Fu, Y., et al., *Graphene related materials for thermal management*. 2D Materials, 2019. **7**(1): p. 012001.
73. Cai, W., et al., *Thermal transport in suspended and supported monolayer graphene grown by chemical vapor deposition*. Nano Letters, 2010. **10**(5): p. 1645–1651.
74. Zhou, H., et al., *High thermal conductivity of suspended few-layer hexagonal boron nitride sheets*. Nano Research, 2014. **7**(8): p. 1232–1240.
75. Sahoo, S., et al., *Temperature-dependent Raman studies and thermal conductivity of few-layer MoS2*. The Journal of Physical Chemistry C, 2013. **117**(17): p. 9042–9047.
76. Peimyoo, N., et al., *Thermal conductivity determination of suspended mono-and bilayer WS 2 by Raman spectroscopy*. Nano Research, 2015. **8**(4): p. 1210–1221.
77. Song, H., et al., *Two-dimensional materials for thermal management applications*. Joule, 2018. **2**(3): p. 442–463.
78. Kim, P., et al., *Thermal transport measurements of individual multiwalled nanotubes*. Physical Review Letters, 2001. **87**(21): p. 215502.

79. Jo, I., et al., *Thermal conductivity and phonon transport in suspended few-layer hexagonal boron nitride*. Nano Letters, 2013. **13**(2): p. 550–554.
80. Sadeghi, M.M., M.T. Pettes, and L. Shi, *Thermal transport in graphene*. Solid State Communications, 2012. **152**(15): p. 1321–1330.
81. Choi, D., et al., *Large reduction of hot spot temperature in graphene electronic devices with heat-spreading hexagonal boron nitride*. ACS Applied Materials & Interfaces, 2018. **10**(13): p. 11101–11107.
82. Liu, D., et al., *Conformal hexagonal-boron nitride dielectric interface for tungsten diselenide devices with improved mobility and thermal dissipation*. Nature Communications, 2019. **10**(1): p. 1–11.
83. Grosse, K.L., et al., *Direct observation of resistive heating at graphene wrinkles and grain boundaries*. Applied Physics Letters, 2014. **105**(14): p. 143109.
84. Shi, L., et al., *Scanning thermal microscopy of carbon nanotubes using batch-fabricated probes*. Applied Physics Letters, 2000. **77**(26): p. 4295–4297.
85. Guo, X., et al., *A review of carbon-based thermal interface materials: Mechanism, thermal measurements and thermal properties*. Materials & Design, 2021. **209**: p. 109936.
86. Paddock, C.A. and G.L. Eesley, *Transient thermoreflectance from thin metal films*. Journal of Applied Physics, 1986. **60** (1): p. 285–290.
87. Costescu, R.M., M.A. Wall, and D.G. Cahill, *Thermal conductance of epitaxial interfaces*. Physical Review B, 2003. **67**(5): p. 054302.
88. Hussain, A., C.Y. Tso, and C.Y. Chao, *Experimental investigation of a passive thermal management system for high-powered lithium ion batteries using nickel foam-paraffin composite*. Energy, 2016. 115 : p. 209–218.
89. Jiang, G., et al., *Experiment and simulation of thermal management for a tube-shell Li-ion battery pack with composite phase change material*. Applied Thermal Engineering, 2017. **120**: p. 1–9.
90. Karthik, M., et al., *Preparation of erythritol–graphite foam phase change composite with enhanced thermal conductivity for thermal energy storage applications*. Carbon, 2015. **94**: p. 266–276.
91. An, F., et al., *Vertically aligned high-quality graphene foams for anisotropically conductive polymer composites with ultrahigh through-plane thermal conductivities*. ACS Applied Materials & Interfaces, 2018. **10**(20): p. 17383–17392.
92. Wang, J., H. Xie, and Z. Xin, *Thermal properties of paraffin based composites containing multi-walled carbon nanotubes*. Thermochimica Acta, 2009. **488**(1–2): p. 39–42.
93. Zou, D., et al., *Thermal performance enhancement of composite phase change materials (PCM) using graphene and carbon nanotubes as additives for the potential application in lithium-ion power battery*. International Journal of Heat and Mass Transfer, 2018. **120**: p. 33–41.
94. Goli, P. and A.A. Balandin. *Graphene-enhanced phase change materials for thermal management of battery packs*. in *Fourteenth Intersociety Conference on Thermal and Thermomechanical Phenomena in Electronic Systems (ITherm)*. 2014. IEEE.
95. Zhang, Y.-F., et al., *Enhanced thermal properties of PDMS composites containing vertically aligned graphene tubes*. Applied Thermal Engineering, 2019. **150**: p. 840–48.
96. Gualous, H., et al., *Supercapacitor ageing at constant temperature and constant voltage and thermal shock*. Microelectronics Reliability, 2010. **50**(9–11): p. 1783–1788.
97. Li, P., et al., *Carbon nanotube-polypyrrole core-shell sponge and its application as highly compressible supercapacitor electrode*. Nano Research, 2014. **7**(2): p. 209–218.
98. Li, P., et al., *Core-double-shell, carbon nanotube@ polypyrrole@ MnO2 sponge as freestanding, compressible supercapacitor electrode*. ACS Applied Materials & Interfaces, 2014. **6**(7): p. 5228–5234.
99. Kim, K.H., Y. Oh, and M.F. Islam, *Mechanical and thermal management characteristics of ultrahigh surface area single-walled carbon nanotube aerogels*. Advanced Functional Materials, 2013. **23**(3): p. 377–383.

100. Sun, Z., et al., *Construction of polyaniline/carbon nanotubes-functionalized phase-change microcapsules for thermal management application of supercapacitors*. Chemical Engineering Journal, 2020. **396**: p. 125317.
101. Kumar, M.S., et al., *Carbon-polyaniline nanocomposites as supercapacitor materials*. Materials Research Express, 2018. **5**(4): p. 045505.
102. Xie, J.-Q., et al., *NiMn hydroxides supported on porous Ni/graphene films as electrically and thermally conductive electrodes for supercapacitors*. Chemical Engineering Journal, 2020. **393**: p. 124598.
103. Zhao, B., et al., *Highly thermally conductive graphene-based electrodes for supercapacitors with excellent heat dissipation ability*. Sustainable Energy & Fuels, 2017. **1**(10): p. 2145–2154.
104. Yu, W., et al., *Advanced thermal interface materials for thermal management*. Engineered Science, 2018. **2**(9): p. 1–3.
105. Tong, X.C., *Advanced materials for thermal management of electronic packaging*. Vol. 30. 2011: Springer Science & Business Media.
106. Hatton, R.A., A.J. Miller, and S. Silva, *Carbon nanotubes: a multi-functional material for organic optoelectronics*. Journal of Materials Chemistry, 2008. **18**(11): p. 1183–1192.
107. Renteria, J., et al., *Magnetically-functionalized self-aligning graphene fillers for high-efficiency thermal management applications*. Materials & Design, 2015. **88**: p. 214–221.
108. Zhang, S., et al., *Synthesis of TiO2 nanoparticles on plasma-treated carbon nanotubes and its application in photoanodes of dye-sensitized solar cells*. The Journal of Physical Chemistry C, 2011. **115**(44): p. 22025–22034.
109. Xu, X., et al., *Surface functionalization of a graphene cathode to facilitate ALD growth of an electron transport layer and realize high-performance flexible perovskite solar cells*. ACS Applied Energy Materials, 2020. **3**(5): p. 4208–4216.
110. Liu, Z., et al., *Organic photovoltaic devices based on a novel acceptor material: graphene*. Advanced Materials, 2008. **20**(20): p. 3924–3930.
111. Mohd Nurazzi, N., et al., *Fabrication, functionalization, and application of carbon nanotube-reinforced polymer composite: an overview*. Polymers, 2021. **13**(7): p. 1047.

12 Functionalized Graphene and Carbon Nanotubes Materials Towards Environmental Applications

Swati Verma, Navneet Kumar, and Jinsub Park

CONTENTS

12.1 INTRODUCTION

Environmental concerns such as water quality management (WQM) and air quality management (AQM) are two global issues which need to be addressed properly for a healthy and prosperous living. Conversely, a desire to accomplish economic development by promoting industrialization and urbanization is the lead cause responsible for these environmental problems [1]. These economic development strategies cause significant damages to the environment by triggering global warming [2]. Additionally, unregulated release of industrial effluents containing metallic/organic pollutants and emission of smoke containing hazardous gases have adverse impacts on environment [3, 4]. These toxic chemicals enter the human system through the food chain or via respiration/inhalation and affect the functioning of the main organs. As an example, discharge of toxic and non-biodegradable chemical dyes from textile, paints, plastics, and leather industries into mainstream waterbodies are known to obstruct aquatic photosynthesis processes because of the decreased light penetration [5]. Likewise, disposal of undesirable by-product(s) produced during the synthesis of valuable chemicals are also identified as major water pollutants. The U.S. Environmental Protection Agency (USEPA) has categorized certain dyes, coloring pigments, drugs, and cosmetic colorants as hazardous waste. USEPA regulates hazardous waste under the Resource Conservation and Recovery Act (RCRA) to ensure that these wastes are managed in ways that protect human health and the environment [6].

The existence of dissolved or suspended contaminants in wastewater and suspended particulate material (PM) in air is responsible for the deterioration of environmental quality. To date, adsorption is the most widely used technique for the removal of toxic pollutants from

DOI: 10.1201/9781003231943-12

wastewater and contaminated air, though several other techniques are also viable [7, 8]. The adsorption technique gained popularity in environment remediation processes owing to its simplicity, cost-effectiveness, and easy operational conditions [9]. The basic mechanism observed in adsorption processes involve accumulation of target pollutant molecules (called adsorbate) on the surface and into the pores of adsorbing material (called adsorbent). To date, activated carbon (AC) is the most effective material for the removal of dissolve impurities from wastewater and contaminated air based on surface adsorption phenomenon [10, 11]. AC is a highly porous and the most used material for home purification applications [12]. It is also considered as reference material to compare performances of other materials. Apart from AC, silica, zeolites, and alumina are some conventional absorbents used for the treatment of air and water via adsorption technique Industrial emission is as major source of water and air pollution as shown in Figure 12.1 [13].

Unregulated discharge of aqueous and gaseous pollutants from industries needs attention to ensure safety of humans and the environment. Advanced carbon materials are suitable to realize remediation of such pollutants.

During the last decade, demand for advanced and smart adsorbent materials for environmental remediations has been geared up due to unexpected environmental and climatic changes. In this context, newly discovered nanostructured and porous adsorbent materials such as metal organic frameworks (MOFs), metal oxide (MO_x), graphene and carbon nanotubes (CNTs) have shown excellent performance towards environmental remediation processes. However, metal corrosion or leaching is the major drawback associated with metal-based adsorbent materials (MOF and MO_x). In contrast, graphene and CNTs are carbon-based materials which have been simulating enormous interest in environmental applications because of exceptional physicochemical properties. These two materials are composed of sp^2-hybridized carbon atom array arrange in a way to yield 1-dimensional CNTs and 2-dimensional graphene sheets. The pristine forms of both these materials are hydrophobic in nature owing to the absence of any polar functional moieties. Therefore, these two materials need to be modified/functionalized to enable their use in

FIGURE 12.1 Industrial emissions as major source of water and air pollution.

environmental remediation applications. The most used approach for the functionalization of pristine graphene and CNTs is the oxidation process. Oxidations establish introduction of diverse oxygen functionalities such as hydroxy (–OH), carboxy (–COOH), carbonyl (–C=O), epoxy (–C–O–C–), etc. in the plane and at the edges of graphene sheets and tubular CNTs. The addition of these moieties facilitates interactions of these two materials with target pollutant molecules and enables their facile removal. Several reports are available in literature on the application of graphene oxide (GO) and oxidized CNTs towards wastewater treatment [14–17]. But exceptionally good hydrophilic nature of GO and oxidized CNTs is the major drawback, which limits their application in water treatment applications by generating post separation difficulties.

To overcome this drawback, strategies were developed to modify/functionalize embedded oxygen moieties of GO and oxidized CNTs. One such strategy to achieve desirable modification is the treatment of these materials with chemical agents to make functionalized materials or in combination with other significant materials to develop composite materials. The significant contribution of organic chemistry reactions for attaining functionalization of these material via covalent and non-covalent approaches cannot be overlooked [18]. The role of nucleophilic reactions in modifying oxygen moieties via amination, esterification, phosphorylation, etc. enables straightforward functionalization of GO and oxidized CNTs structure and improves its performance towards environmental applications [19]. In this chapter, we have assessed selected recent literature reported on the usage of functionalized graphene and CNT materials toward dye and heavy metal removal applications. Detailed discussions on adsorption mechanisms based on experimental parameters and material properties are furnished. At last, the effectiveness of various listed materials has been assessed in terms of FoM values using following formulation.

$$FoM = \frac{Adsroption\,capacity\left(\frac{mg}{g}\right)}{Initial\,concentration\,of\;pollutant\left(\frac{mg}{L}\right)}$$

12.2 MECHANISTIC APPROACH

Adsorption performances of graphene and CNTs-based adsorbents depend majorly on the amount and kind of active sites present on the adsorbent surface. Functional moieties present at the edges and in the planar network structure of these two materials enables removal of pollutants molecules via diverse interactive forces. Adsorption technology with reference to WQM and AQM is a mass transfer process in which molecules of a target pollutants got adhered to a solid surface via chemical and physical interactive forces [20]. The corresponding adsorption processes are refereed as chemisorption and physisorption, respectively. Basic characteristics of chemisorption and physisorption processes with reference to wastewater and bad air treatment protocols are shown in Table 12.1. All these properties have significant impact on the adsorption kinetic and adsorption isotherms profiles and therefore plays an important role in deducing adsorption capacity values of an adsorbent material. In general, chemisorption processes are faster as compared to physisorption processes. Consequently, a chemisorption process is appeared to occur at a relatively faster kinetic rate and attains equilibrium faster as compared to a physisorption process. Constructive interaction between adsorbate molecules prompts multilayers adsorption and thus helps in recording high efficiency. Conversely, adsorption processes governed by chemical forces are selective but might reveal lower adsorption capacity values owing to monolayer adsorption approach. Linear and non-linear expressions of some most used models are given in Table 12.2.

TABLE 12.1
Properties of Chemisorption and Physisorption Processes

Order	Property	Chemisorption	Physisorption
1.	Binding forces	Similar to covalent bonds	Vander Waals, electrostatic, hydrogen bonding,
2.	Enthalpy of adsorption	High, $\Delta H = 40–200$ kJ/mol	Low $\Delta H = 10–40$ kJ/mol
3.	Energy of activation	High like a chemical process	Very low (nearly zero)
4.	Temperature effect	Positive	Negative
5.	Chemical change in adsorbent	Formation of surface compounds	Ideally not
6.	Selectivity	High	Poor
7.	Multilayer process	No	Yes

TABLE 12.2
Non-Linear and Linear Expressions of Various Adsorption Models

Order	Model	Non-linear	Linear	Assumptions	Ref.
1.	Pseudo first order (PFO)	$Q_t = Q_e\left(1-e^{-k_1 t}\right)$	$\ln\left(Q_e - Q_t\right) = lnQ_e - k_1 t$	The rate at which adsorbate molecules are getting adsorbed on the surface is directly proportional to the difference in adsorbate concentration in the two phases i.e., solid adsorbent and aqueous/gaseous phase	[22, 23]
2.	Pseudo second order (PSO)	$Q_t = \dfrac{k_2 Q_e^2 t}{1 + k_2 Q_e t}$	$\dfrac{1}{Q_t} = \dfrac{1}{k_2 Q_e^2} + \dfrac{1}{Q_e}$	The rate limiting step may be chemisorption if the valency forces are involved in the adsorption of adsorbate molecules on the surface of adsorbent	[24, 25]
3.	Intraparticle diffusion (IPD)	$Q_t = K_i * t^{1/2}$	-----	The uptake of adsorbate molecules from liquid solution varies proportionally to $t^{1/2}$ rather than t. It is anticipated that if plot of qt vs. $t^{1/2}$ is linear, then the adsorption process is governed by intraparticle diffusion mechanism. If such is a case, the plot should pass through origin and intraparticle diffusion is the rate determining step. However, if the plot exhibits multilinear characteristic, then the adsorption process is supposed to be governed by two or more forces.	[26, 27]

TABLE 12.2 (Continued)
Non-Linear and Linear Expressions of Various Adsorption Models

Order	Model	Non-linear	Linear	Assumptions	Ref.
4.	Langmuir isotherm	$Q_e = \frac{Q_m K_L C_e}{1+K_L C_e}$	$\frac{C_e}{Q_e} = \frac{C_e}{Q_m} + \frac{1}{K_L Q_e}$	It assumes that the surface of an adsorbent contains finite number of energetically equivalent binding sites which can be singly occupied by the adsorbate molecule to cause monolayer adsorption.	[28]
5.	Freundlich isotherm	$Q_e = K_F C_e^{1/n}$	$lnQ_e = lnK_F + \frac{1}{n} lnC_e$	This model is applicable for multilayer adsorption processes on heterogenous surfaces.	[29]
6.	Temkin isotherm	$Q_e = \frac{RT}{b} \ln(K_T C_{e)}$	$Q_e = \frac{RT}{b} \ln K_T + \frac{RT}{b} lnC_{e)}$	The model assumes that the indirect interaction between adsorbate-adsorbent interactions is responsible for causing decreases in the heat of adsorption of molecule forming an adsorbate layer.	[30]

Many researchers have believed that the linearization of a non-linear pseudo first order (PFO) and pseudo second order (PSO) models leads to bias analysis and thus produce data which violates the variance and normality assumption of the standard least square values [21]. The same is also true for other non-linear models. Therefore, it is recommended to use non-linear expressions of the models for proper evaluation of a material performance. Basic assumptions of these models are also provided alongside for the better understanding of the removal mechanism of graphene and CNT-based adsorbents.

Q_e = equilibrium adsorption capacity; Q_t = adsorption capacity at any time "*t*"; Q_m = maximum adsorption capacity; C_e = adsorbate concentration at equilibrium; *t* = time; *T* = temperature; *R* = Universal gas constant; *b* = Temkin heat constant; *1/n* = heterogeneity parameter; k_1 = PFO rate constant; k_2 = PSO rate constant; K_i = IPD rate constant; K_L = Langmuir constant; K_F = Freundlich constant; and K_T = Temkin constant.

12.3 FUNCTIONALIZED GRAPHENE AND CNT MATERIALS IN WQM

Application of graphene and CNT-based materials in the treatment of textile effluent containing toxic dyes compounds is significant owing to their high surface area, good thermo-mechanical stability, large functional active sites, and low mass transfer resistance [31]. Among these, the number of active sites present on the surface of ametrias is the most important aspects that effects surface properties of these materials and thus their adsorption capacity. Wastewater remediation by graphene or CNT-based adsorbent materials involved different types of interaction as represented in Figure 12.2. Furthermore, the chemical nature of such active sites is the deciding factor in controlling the adsorption process. In this article, we have selected some recently developed functionalized graphene and CNT materials and examined their performances against dye and heavy metal removal. The utility of conventional kinetic and isotherm models in examining adsorption behavior of graphene and CNT based materials has been explained. A consolidated

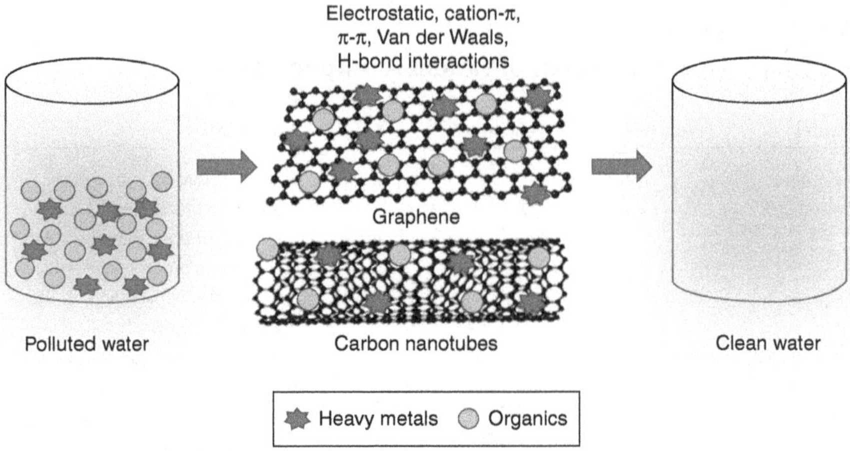

FIGURE 12.2 Schematic for mechanism of wastewater treatment by graphene and carbon nanotubes materials.

list of functionalized graphene and CNT adsorbents used for the removal of dyes and heavy metal from aqueous medium is provided in Tables 12.3 and 12.4. As can be seen, almost all kind of functionalized graphene and CNT materials obey PSO kinetics models suggesting that the removal of target pollutants is majorly governed by chemical interactions by these materials is a chemisorption process. This means that the interactions occurring between these materials and dyes/metal pollutants are covalent in nature and involves sharing or exchange of valency electrons to facilitate removal of target pollutants. Furthermore, Langmuir isotherm is the most suitable model for maximum cases indicating that these adsorbents provide homogeneous surface for the uptake of adsorbent molecules or ions.

12.3.1 Dyes as Pollutants

Among the listed graphene adsorbents in Table 12.3, the highest adsorption capacity values of 3059.20 and 2043.70 mg/g were shown against methyl orange (MO) and amaranth (AM) dyes, respectively by polyethyleneimine-functionalized graphene aerogel (PFGA) [32]. The corresponding FoM values for MO and AM dyes were evaluated as 10.20 and 6.81, respectively. Large magnitude of FoM and adsorption capacity values clearly indicates the effectiveness of PFGA material in removing these dyes. Such high adsorption capacity values were attributed to the electrostatic and π-π interactions prevailing between PFGA and dye molecules. the adsorptive mechanism if PFGA was judges at three pH condition-basic (pH =11), neutral (pH =7), and acidic (pH = 3). At basic pH, all amine groups of grafted polyethyleneimine molecules are fully deprotonated and thus PFGA surface is negative. Consequently, the existing electrostatic repulsion between PFGA and dyes molecules hinders adsorption and produced nearly zero adsorption capacity values. In contrast, protonation of amine groups of polyethyleneimine molecules under acidic and neutral conditions enable removal of dye molecules because of favorable electrostatic interactions. Furthermore, removal of MO dye by PFGA satisfies Freundlich model while that of AM dye follows Langmuir model. This could be attributed to the fact that two molecules of MO could bound to one amino group of polyethyleneimine under in acidic/neutral conditions and causes multilayer removal. In contrast, steric hinderance causes adsorption of one AM molecule to each amino group of polyethyleneimine and thus represents monolayer adsorption process.

TABLE 12.3
Applications of Functionalized Graphene and CNT Materials in Dye Removal

Order	Materials	Target	pH	Kinetic model	Isotherm model	Concentration (mg/L)	Q_m (mg/g)	FoM	Ref.
Functionalized G materials									
1.	GO/silk fibroin	MB	7	PSO	Langmuir	64	1322.71	20.67	[33]
2.	Bi_2O_3@GO	RhB	4	PFO	Langmuir	10	320.00	32.00	[41]
3.	rGO/CTAB	AR265	2	PSO	Langmuir	250	510.67	2.04	[42]
		AO7	2	PSO	Langmuir	250	355.89	1.42	
4.	TRGO/PVA	NR + IC mix	2	PSO	Langmuir	30	900.00	30.00	[43]
5.	PFGA	MO	3	---	Freundlich	300	3059.20	10.20	[32]
		AM	3	---	Langmuir	300	2043.70	6.81	
6.	Inu-GO	MB	5.25	PSO	Langmuir	500	789.00	1.58	[44]
7.	Fe_3O_4-Cooh@ (PAH/GO-COOH)n	MB	---	PSO	---	10	35.96	3.60	[45]
		RhB	---	PSO	---	5	22.12	4.42	
8.	AGO	MB	6	PSO	Langmuir	200	578.00	2.89	[46]
9.	MCGO	MO	Natural	PSO	Langmuir	50	398.08	7.96	[47]
10.	rGO-PIL	MeB	---		Langmuir	350	1910.00	5.46	[48]
11.	3D rGO	MB	7	PSO	Langmuir	300	302.11	1.01	[49]
		AR1	7	PSO	Freundlich	300	277.01	0.92	
12.	Konjac glucomannan/GO	MO	7	PSO	Freundlich	50	51.60	1.03	[50]
		MB	7	PSO	Freundlich	120	92.30	0.77	
13.	Magnetic graphene oxide/PVA-50%	MB	7	PSO	Langmuir	---	270.94	---	[51]
		MV	7	PSO	Langmuir	---	221.23	---	
14.	PAA/MGO	MB	7	PSO	Langmuir	20	291.00	14.55	[52]
15.	rGO-CNT-PPD	MV	7	PSO	Langmuir	30	298.00	9.93	[53]
		MO	3	PSO	Langmuir	30	294.00	9.80	
16.	3D Graphene	MO	7.7		Langmuir	50	27.93	0.56	[54]
17.	Graphene aerogel	MB	5	PSO	Langmuir	20	76.00	3.80	[34]
		MG	5	PSO	Langmuir	20	352.00	17.60	
		RhB	5	PSO	Langmuir	20	111.00	5.55	
		MO	5	PSO	Langmuir	20	16.00	0.80	

(continued)

TABLE 12.3 (Continued)
Applications of Functionalized Graphene and CNT Materials in Dye Removal

Order	Materials	Target	pH	Kinetic model	Isotherm model	Concentration (mg/L)	Q_m (mg/g)	FoM	Ref.
18.	Hollow carbon spheres/graphene hybrid aerogels	RhB	---	PSO	---	200	441.50	2.21	[35]
		MO	---	PSO	---	200	344.10	1.72	
Functionalized CNTs materials									
1.	MWCNT	MB	6	PSO	Langmuir, Freundlich	20	59.70	2.99	[55]
2.	Fe_3O_4—MWCNTs (HNO3)	MB	7	PSO	Langmuir	30	48.06	1.60	[56]
3.	MWCNTs	MB	6	PSO	Langmuir	120	176.02	1.47	[57]
4.	MWCNTs/Gly/β-CD)	MB	8	PSO	Langmuir	---	90.90	---	[58]
		AB 113	7	PSO	Langmuir	---	172.41	---	
		MO	6	PSO	Langmuir	---	96.15	---	
		DR 1	7	PSO	Langmuir	---	500.00	---	
5.	NH_2-MWCNTs@Fe_3O_4	MB	8	PSO	Langmuir, Freundlich	100	178.57	1.79	[40]
6.	MWCNT-COOH-Cysteamine	AB 10B	6	IPD	---	50	131.00	2.62	[59]
7.	CNT/Chitosan	RhB	7	PSO	---	100	67.20	0.67	[60]
		MO	7	PSO	---	100	45.80	0.46	
8.	SWCNT	BR 46	9	PSO	Langmuir	150	38.35	0.26	[61]
9.	SWCNT-COOH	BR 46	9	PSO	Langmuir	150	49.45	0.33	
10.	MWCNTs-COOH	BTB	1	PSO	Langmuir	30	55.30	1.84	[62]
11.	Functionalized CNT/Mg(Al)O	CR	7		Langmuir	800	1250.00	1.56	[38]
12.	MWCNTs-Fe_3C	DR23	3.5	PSO	Freundlich	54	172.40	3.19	[63]
13.	SWCNT	MG	7	PFO	Langmuir, Freundlich	10	28.40	2.84	[37]
14.	SWCNT-NH_2	MG	7	PFO	Langmuir, Freundlich	10	35.16	3.52	
15.	SWCNT-COOH	MG	7	PFO	Langmuir, Freundlich	10	37.06	3.71	
16.	SWCNT	MO	7	PFO	Langmuir, Freundlich	10	41.35	4.14	
17.	SWCNT-NH_2	MO	7	PFO	Langmuir, Freundlich	10	31.77	3.18	
18.	SWCNT-COOH	MO	7	PFO	Langmuir, Freundlich	10	26.50	2.65	
19.	Oxidized MWCNTs	MO	7	PSO	---	20	54.84	2.74	[64]
20.	Functionalized CNTs loaded TiO_2	MO	6.5	PSO	---	5	42.85	8.57	[39]
21.	Thiol-functionalized MWCNT (MWCNT-SH)	MB	6	PSO	Langmuir	40	400.00	10.00	[65]

A similar adsorption mechanism was also observed in the case of graphene oxide-silk fibroin (GO-SF) hybrid material which causes removal of cationic methylene blue (MB) dye through interaction of dyes molecules with N-atoms of amide groups of silk fibroin [33]. The adsorption capacity and FoM values for GO-SF material were estimated as 1322.71 mg/g and 20.67, respectively. Aerogel kind graphene materials were also developed and used for dye removal applications [34, 35]. Graphene aerogel (GA) prepared using glutaraldehyde as crosslinking agents exhibited excellent performance against a variety of cationic dyes. The removal mechanism of all target dyes satisfies Langmuir isotherm model and obey pseudo second order kinetics. The pH_{ZPC} of GA was found to be ~3.3. This means that the surface of GA bears positive and negative surface charge below and above a pH of 3.3, respectively. This significant difference in the surface properties of GA caused adsorption of cationic dyes (methylene blue, malachite green, and rhodamine B) at nearly neutral conditions (pH = ~6). On the other hand, removal of anionic MO dye by GA is poor in the studied pH range (3–8) due to appearance of electrostatic repulsion. As the valence electron are involved in causing electrostatic interaction at the adsorbate-adsorbent interface, the removal mechanism of all target dyes follows PSO and therefor revealed a chemisorption process.

Removal of heavy metals and organics from aqueous system by advanced carbon materials like graphene and carbon nanotubes is facilitated by electrostatic, cation-π, π-π, Van der Waals, and H bond interactions.

Like graphene materials, functionalized CNTs materials also supply degenerate active sites for the adsorption of dye molecules through chemical interactions. It can also be inferred that the removal of most of the dyes by functional side graphene materials occur in monolayer fashion as implied by the suitability of Langmuir isotherm. As can be seen from Table 12.3, thiol functional multi-walled CNTs (MWCNT-SH) showed best performance for MB dye among listed adsorbent and furnish highest FoM value [36]. The Langmuir adsorption capacity of MWCNT-SH was determined as 400 mg/g in the concentration range of 10–40 mg/L. Adsorption isotherm model revealed that the removal phenomenon could be best explained by Langmuir model and obeys PSO kinetics. The suitability of the two models reflects existence of uniform active sites with equal affinity for dye molecule. Another study on functionalized CNTs adsorbents examined the performance of heteroatom functionalized CNTs towards dye removal [37]. The reported study inferred that the functionalization of single-walled CNTs with N (SWCNT-NH_2) and O (SWCNT-COOH) hetero atoms improves its effectiveness in removing cation MG dye but lowered its efficiency against anionic MO dye. At an initial dye concentration of 10 mg/L, non-functionalized SWCNTs exhibited adsorption capacity values of 28.40 mg/g for MG and 41.35 mg/g for MO dyes. The performance of SWCNT was solely ascribed to the hydrophobic and π–π interactions. The adsorption capacity values of SWCNT-NH_2 and SWCNT-COOH for MG dye were observed as 35.16 and 37.06 mg/g, respectively. Consequently, the observed FoM values improved from 2.84 for SWCNTs to 3.52 and 3.71 for SWCNT-NH_2 and SWCNT-COOH, respectively. In contrast, the adsorption capacity values of MO dye on SWCNT, SWCNT-NH_2, and SWCNT-COOH were estimated as 41.35, 31.77, and 26.50 mg/g, respectively. The decreasing trend in adsorption capacity values caused reduction in FoM values from 4.14 for SWCNT to 3.18 for SWCNT-NH_2, and 2.65 for SWCNT-COOH. This study thus demonstrates that addition of carboxylate and amide groups on the surface of SWCNT improves electrostatic, hydrogen bonds, and covalent interactions with cationic MG dye. On the other hand, MO dye removal efficiency of SWCNT was reduced on functionalization because of appearance of electrostatic repulsions between negatively charged dye molecules and surface N/O heteroatoms.

In terms of maximum adsorption capacity, CNT/Mg(Al)O material showed highest capacity value of 1250 mg/g towards CR dye at an initial concentration of 800 mg/L [38]. Studies indicates that the electrostatic interactions between negatively charge CR dye molecules and positively charge LDH layer played key role in facilitating dye removal. Moreover, strong interaction of oxygen functional groups of CNTs with CR molecules accelerated the adsorption process. The FoM value for the process of CR adsorption by the material was evaluated as 1.56. Similarly, functionalized CNTs loaded with TiO_2 revealed excellent performance against anionic MO dye (FoM = 8.57) because of binary interfacial mechanisms i.e., dye-CNTs and dye-TiO_2 [39]. An interesting study reported on amino functionalized MWCNT (NH_2-MWCNT) coupled with Fe_3O_4 to form NH_2-MWCNT@Fe_3O_4 adsorbents revealed suitability of Langmuir as well as Freundlich model in explaining the adoption mechanism [40]. The high performance of NH_2-MWCNT@Fe_3O_4 was accredited to the strong electrostatic interactions of cationic MB dye molecules with $-NH_2$ groups as well as with Fe_3O_4 nanoparticles. Further, facile post-separation of adsorbent-adsorbate complex is facilitated by the presence of magnetic Fe_3O_4 nanoparticles in the material. Therefore, it can be concluded that both information materials and organic functionalism influences performance of graphene and CNTs depending upon the nature of target dye molecule.

12.3.2 Heavy Metals as Pollutants

Like dyes, functionalized graphene and CNT materials have been successfully used in the removal of toxic heavy metal from aqueous solutions. Table 12.4 provides an outlook on some recent studies reported on metal removal using functionalized graphene and CNT material. Removal of diverse metallic pollutants such as Pb, Cd, Hg, As, U, Cu, Se, etc., by the listed materials follows PSO and Langmuir model and thus again establish the major of chemical interactions in these processes. Among graphene materials, best performance towards divalent metallic ions was shown by graphene oxide/polyamidoamine (GO-PAMAMs) material as indicated by high values of adsorption capacities of 568.18, 253.81, 68.68, and 18.29 mg/g for Pb (II), Cd (I), Cu (II), and Mn (II) respectively [66]. The outperformance of GO-PAMAMs material was attributed to the reactivity of terminal functional groups of polyamidoamine dendrimers. This study suggests that the functionalization of graphene material through incorporation of heteroatoms is one of the best approaches to achieve desirable results. Also, the porous network structure of GO-PAMAMs was expected to supply higher number of adsorption sites for the removal of metal ions. Likewise, EDTA-MGO, rGO-PVP, rGO-thymine, AMGO, MGO, chitosan, phos-GO, and CS/GO-SH also signified key contributory effect of hetero atoms moieties in facilitating chemical interactions between adsorbent surface and metal ions.

Functionalized CNTs materials also revealed similar phenomenon in the removal of metallic impurities from aqueous systems. The best performer in the list is Cu/CNT as it showed highest FoM value of 67.70 against As (III) removal as arsenite (AsO_2^-) [67]. Against an initial As (II) concentration of 0.1 mg/L, Cu/CNT revealed high adsorption capacity value 67.70 mg/g which indicates efficiency of the adsorption process. In comparison to Cu/CNT, other metal containing CNT materials such as Fe doped MWCNT and Ce-Fe mixed oxide MWCNT yield significantly lower FoM values of 2.50 and 5.75, respectively [68, 69]. In case of Pb (II) ions, highest FoM value of 12.25 was observed for acid oxidized CNTs [70]. Pb (II) removal is mainly facilitated by three kinds of interactions i.e., Van der Waal, electrostatic, and chemical interactions. Furthermore, rapid aggregation of Pb (II) loaded-acid oxidized CNTs complex was subsequently achieved by the addition of cationic surfactant CTAB. The addition of surfactant did not produce any negative impact on the removal efficiency of acid oxidized CNTs over a wide range of pH and ionic strength. The suitability of functionalized CNTs/metal materials was also recognized in Hg (II) removal and follows PSO and Langmuir adsorption models [71, 72].

TABLE 12.4
Functionalized G and CNT Materials for Metal Removal Applications from Aqueous Medium

Order	Materials	Target	pH	Kinetic model	Isotherm model	Concentration (mg/L)	Q_m (mg/g)	FoM	Ref.
Functionalized G materials									
1.	GO-PDA	Pb (II)	4.0–5.4	---	Langmuir	50	53.6	1.07	[73]
		Cu (II)	5.2–6.8	---	Langmuir	50	24.4	0.49	
		Cd (II)	5.2–6.8	---	Langmuir	50	33.3	0.67	
		Hg (II)	3.5–4.0	---	Langmuir	50	15.2	0.30	
2.	MGO	Se (IV)	6.0–9.0	---	Langmuir	0.3	23.81	79.37	[74]
		Se (VI)	6.0–9.0	---	Langmuir	0.3	15.12	50.40	
3.	rGO-Fe_3O_4-nZVI	Cr (VI)	3	PSO	Langmuir	100	101.2	1.01	[75]
4.	rGO-nZVI	As (III)	7	PSO	Langmuir	10	35.83	3.58	[76]
		As (V)	7	PSO	Langmuir	10	29.04	2.90	
5.	rGO-Acetylacetone	Cd (II)	6	---	Langmuir	49.28	49.28	1.00	[77]
		Co (II)	6	---	Langmuir	27.78	27.28	0.98	
6.	MGO	Fe (II)	5.5	PFO, PSO	Langmuir, Freundlich	80	43.2	0.54	[78]
		Mn (II)	5.5	PFO, PSO	Langmuir, Freundlich	80	16.5	0.21	
7.	GO-PAMAM	Pb (II)	4.5	PSO	Langmuir	200	568.18	2.84	[66]
		Cd (II)	5	PSO	Langmuir	200	253.81	1.27	
		Cu (II)	4.5	PSO	Langmuir	200	68.68	0.34	
		Mn (II)	4	PSO	Langmuir	200	18.29	0.09	
8.	NH_3-GO	U (VI)	6	PSO	Langmuir, Freundlich	100	80.13	0.80	[79]
9.	EDTA-MGO	Pb (II)	4.2	PSO	Freundlich, Temkin	100	508.4	5.08	[80]
		Hg (II)	4.1	PSO	Freundlich, Temkin	100	268.4	2.68	
		Cu (II)	5.1	PSO	Freundlich, Temkin	100	301.2	3.01	
10.	GO-CdS (en)	Cu (II)	6	---	Langmuir	50	137.17	2.74	[81]
11.	rGO-Diatom	Hg (II)	6.5	PSO	Langmuir	400	528	1.32	[82]
12.	CS/GO-SH	Cu (II)	5	PSO	Freundlich	80	425	5.31	[83]
		Pb (II)	5	PSO	Freundlich	80	447	5.59	
		Cd (II)	6	PSO	Freundlich	80	177	2.21	

(continued)

TABLE 12.4 (Continued)
Functionalized G and CNT Materials for Metal Removal Applications from Aqueous Medium

Order	Materials	Target	pH	Kinetic model	Isotherm model	Concentration (mg/L)	Q_m (mg/g)	FoM	Ref.
13.	rGO-Thymine	Hg (II)	6	PSO	Langmuir	250	128	0.51	[84]
14.	GO-Fe_3O_4-EDA	Hg (II)	5.3	PSO	Langmuir	8	127.23	15.90	[85]
15.	GO-APTS-poly(AMPS-co-MA)/Fe_3O_4	Pb (II)	6	PSO	Langmuir	40	310.1	7.75	[86]
		Cu (II)	6	PSO	Langmuir	40	282.25	7.06	
		Co (II)	6	PSO	Langmuir	40	238.35	5.96	
16.	GO/Fe-Mn	Hg (II)	7	PSO	Langmuir	1	32.9	32.90	[87]
17.	GO-PAMAM	Se (IV)	6	PSO	Langmuir	40	60.9	1.52	[88]
		Se (VI)	6	PSO	Langmuir	40	77.9	1.95	
18.	AMGO	Cr (VI)	2	PSO	Langmuir	33	123.4	3.74	[89]
19.	MGO-PMAM	Hg (II)	6	PSO	Langmuir	50	113.71	2.27	[90]
20.	rGO-ZnO	Cu (II)	6	PSO	Langmuir	10	67.39	6.74	[91]
		Co (II)	6	PSO	Langmuir	10	36.35	3.64	
21.	CARGO	U (VI)	5	PSO	Langmuir	250	388	1.55	[92]
22.	rGO-Mg-Al LDH	Pb (II)	4.5	PSO	Langmuir	100	116.2	1.16	[93]
23.	GO-DEA-DIBA	RE (VII)	2	PSO	Langmuir	800	140.82	0.18	[94]
24.	MGO-Ppy	Hg (II)	7	PSO	Langmuir	100	400	4.00	[95]
25.	MGO-Chitosan	Cu (II)	7	PSO	Langmuir	100	217.4	2.17	[96]
26.	GO-TEPA-Ppy	Cr (II)	2	PSO	Langmuir	100	408.48	4.08	[97]
27.	rGO-P(TA-TEPA)-PAM	Cr (VI)	2	PSO	Langmuir	200	394.32	1.97	[98]
28.	Phos-GOF	U (VI)	4	---	Langmuir	10	93.5	9.35	[99]
29.	GOCS	Eu (III)	6	PSO	Freundlich	100	150	1.50	[100]
		Th (IV)	3	PSO	Freundlich	100	220	2.20	
30.	MGOH	As (III)	7.7	PSO	Langmuir	1.5	25.1	16.73	[101]
		As (V)	6.2	PSO	Langmuir	4.53	74.2	16.38	
31.	CD@RGO	Hg (II)	7	PSO	Langmuir	10	82.64	8.26	[102]
32.	CDGF	Cr (VI)	3	PSO	Langmuir	200	130.4	0.65	[103]
33.	GA/TiO_2	U (VI)	5	PSO	Langmuir	20	441.3	22.07	[104]
34.	PGA/P-Fe_2O_3	As (III)	5	PSO	Freundlich	5	60.81	12.16	[105]
		As (V)	5	PSO	Freundlich	5	76.72	15.34	

35.	MWCNT-PDA/GO	Cu (II)	6	PSO	Langmuir	300	318.47	1.06	[106]
		Pb (II)	6	PSO	Langmuir	300	350.87	1.17	
36.	GA/EDA	U (VI)	4	PSO	Langmuir	42	238.67	5.68	[107]
Functionalized CNTs materials									
1.	Fe doped MWCNT	As (III)	7	PSO	Langmuir, Freundlich	80	200	2.50	[68]
		As (V)	7	PSO	Langmuir, Freundlich	80	250	3.13	
2.	Ce-Fe mixed oxide MWCNT	As (III)	7.5	PSO	Langmuir	5	28.74	5.75	[69]
		As (V)	7.5	PSO	Freundlich	5	30.96	6.19	
3.	KA-CNTs	Hg (II)	5.5	PSO	Freundlich	70	250	3.57	[108]
4.	CNT-S	Hg (II)	6	PSO	Langmuir	100	151.51	1.52	[109]
5.	Cu/CNT	As (III)	7	NA	Langmuir	0.1	6.77	67.70	[67]
6.	Iron oxide-SWCNTs	As (V)	4	PSO	Freundlich	5	49.65	9.93	[110]
7.	CNTs	Pb (II)	7	PSO	Langmuir	100	63.5	0.64	[111]
		Cd (II)	7	PSO	Langmuir	30	52.4	1.75	
		Hg (II)	7	PSO	Langmuir	2	4.4	2.20	
8.	L-CNTs	Pb (II)	7	PSO	Langmuir	10	23.6	2.36	[112]
9.	MWCNTs	Pb (II)	5	PSO	Langmuir	80	104.2	1.30	[113]
10.	MWCNT-g-VP	Pb (II)	6	NA	Langmuir	10	15.9	1.59	[36]
11.	P-CNTs	Pb (II)	5	NA	Langmuir	100	27.3	0.27	[114]
12.	Acidified functionalized MWCNT	Pb (II)	9	NA	Langmuir, Freundlich	100	166	1.66	[115]
		Cu (II)	9	NA	Langmuir, Freundlich	100	123	1.23	
		Ni (II)	9	NA	Langmuir, Freundlich	100	95	0.95	
		Cd (II)	9	NA	Langmuir, Freundlich	100	101	1.01	
13.	acid-oxidized CNT	Pb (II)	10.5	NA	NA	18.2	223	12.25	[70]
14.	Dispersed CNT	Cu (II)	5.6	NA	Langmuir	10	67.8	6.78	[116]
		Pb (II)	5.6	NA	Langmuir	10	92.3	9.23	
15.	CNT/Chitosan	Cu (II)	7	PSO	NA	100	52	0.52	[60]
		Zn (II)	7	PSO	NA	100	16.2	0.16	
		Fe (III)	7	PSO	NA	100	60.1	0.60	
		Pb (II)	7	PSO	NA	100	50.6	0.51	
		Cr (VI)	2	PSO	NA	100	175.43	1.75	
		Cr (VI)	2	PSO	NA	200	166.66	0.83	

(*continued*)

TABLE 12.4 (Continued)
Functionalized G and CNT Materials for Metal Removal Applications from Aqueous Medium

Order	Materials	Target	pH	Kinetic model	Isotherm model	Concentration (mg/L)	Q_m (mg/g)	FoM	Ref.
16.	$MgFe_2O_4$ doped MWCNTs	Hg (II)	7.94	PSO	Langmuir	60	41.66	0.69	[71]
17.	pTSA-Pani@CNT	Hg (II)	7.6	PSO	Langmuir	60	100	1.67	[117]
18.	MSWCNT-CoS	Hg (II)	5.26	PSO	Langmuir	2000	1666	0.83	[72]
19.	oxFMWCNT3h	Cr (VI)	2.5	Elovich	Freundlich	3.5	5.42	1.55	[118]
		Cu (II)	5	Elovich	Freundlich	1.5	14.09	9.39	
		Ni (II)	5	Elovich	Freundlich	3.5	8.46	2.42	
	oxFMWCNT6h	Cr (VI)	2.5	Elovich	Freundlich	3.5	5.1	1.46	
		Cu (II)	7	Elovich	Freundlich	10	31.25	3.13	

TABLE 12.5
Adsorption Capacities of Diverse VOCs on Functionalized Graphene and CNTs-Based Adsorbents

Order	Materials	Target VOC	Flow rate (mL/min)	Concentration (mg/L)	Q_m (mg/g)	Ref.
1.	GO	Benzene	40	50	216.2	[119]
	GO	Toluene	40	50	240.6	
	rGO	Benzene	40	50	276.4	
	rGO	Toluene	40	50	304.4	
2.	Cu-BTC@GO	Ethanol			635	[121]
		Toluene			838	
3.	ZIF-8/GO	Toluene	15		123	[122]
4.	MIL-101@GO	N-hexane	10		1042	[123]
5.	G-GND/S	Formaldehyde	400	3.6	22	[124]
6.	GP	Toluene		30	1.9	[125]
	rGOMW	Toluene		30	7	
	rGOMWKOH	Toluene		30	14.4	
	GP	Acetaldehyde		30	0.32	
	rGOMW	Acetaldehyde		30	0.8	
	rGOMWKOH	Acetaldehyde		30	1.23	
7.	GO/MIL-101	Acetone			1167	[126]
8.	GO/CNF	Butanone			419	[127]
		Benzene			289	
9.	MIL-101/GO	Chloroform			2368	[128]
10.	CNTs – iron oxide	Benzene	40	1	987.58	[129]
	CNTs	Benzene	40	1	517.27	
11.	CNT/ACF	Formaldehyde	2600		62.49	[120]

12.4 FUNCTIONALIZED GRAPHENE AND CNT MATERIALS IN AQM

Gaseous pollutant emissions from automobiles and industries have worsened quality of outdoor. air we breathe. Similarly, release of toxic compounds from household articles such as furniture, plastics, combustion appliances, and paints have produced negative impact on the quality of house air. The effective removal of such toxic gases (mainly VOCs) is essential as they all are associated with carcinogenic effects. Adsorption is the most successful, simple, portable, and cost-effective technique for the removal of toxic gases. Unlike water remediation, not many studies are reported on the usage of graphene and CNT based materials in the removal of VOCs. An attempt has been in made in Table 12.5 to summarize some studies based on the removal of VOCs by graphene and CNT materials. The chemical structure of VOC has a major effect on adsorption capacity value. As can be seen from table, the adsorption capacity values for GO and rGO towards Benzene (B) was found to be 216.2 and 276.4 mg/g, respectively [119]. However, GO and rGO showed higher toluene adsorption capacity of 240.6 and 304.4 mg/g, respectively. This could be ascribed to the presence of side alkyl chain in toluene which enhances electron cloud density of the aromatic ring by virtue of electron donating effect. Moreover, the side methyl group in toluene can interact with the oxygen functional groups of rGO and GO via hydrogen bonds and exhibited higher adsorption values. Composite material of activated carbon fiber with CNTs (CNT/ACF) also revealed improved formaldehyde (FA) removal as compared to ACF 19.11 to 62.49 mg/g [120]. The higher activity of CNT/ACF materials (62.49 mg/g) in comparison to ACF (19.11 mg/g) was ascribed to the physicochemical interactions occurring between CNT and polar FA molecules. Additionally, high pressure resistant in case of CNT/ACF materials produce more hinderance in the passage of FA molecules and thereby increases the possibilities of FA encounter with the inner surface of CNTs. Gaseous adsorption on surfaces is also influenced by soft parameters such as flow rate of gaseous steam and humidity. This is because low flow rate allows gaseous molecule to reside on the surface of adsorbent for larger time to cause more effective removal. On the other hand, presence of humidity or water vapor along with the target gaseous pollutant increases the chance of competitive interaction sand thereby decrease the overall effectiveness of the process.

12.5 CONCLUSIONS

Graphene and CNT materials have attracted widespread attention in environmental remediation such as dye, heavy metal, and VOCs removal owing to their large surface area and tunable surface characteristic. Introduction of diverse functional moieties comprising of hetero-atoms such as oxygen, nitrogen, phosphorus, and sulfur can be used as a promising approach to produce facile adsorption of aqueous and gaseous pollutants. These functionalities govern pollutants removal through hydrogen bonding, Van der Waals, electrostatic, cation-π, and π-π interactions. The applicability of various isotherm and kinetic models suggest that chemical interactions are majorly responsible for the adsorptive characteristics of graphene and CNT materials. Future research in this area should be aimed at producing effective graphene and CNT material for real world applications. Emphasis should be given on broad spectrum graphene and CNT materials that could simultaneously remove different types of organic and inorganic impurities. Further, adsorbents with short equilibration time and large capacities should be developed for quick remediation. Also, the surface properties of the adsorbent should not change drastically with pH to achieve consistency removal in wide pH range. In the last, greenness of graphene and CNT is the biggest boon that allows their usage in environmental remediation applications.

REFERENCES

1. Wang, Q., M. Su, and R. Li, *Toward to economic growth without emission growth: The role of urbanization and industrialization in China and India*. Journal of Cleaner Production, 2018. **205**: p. 499–511.
2. Opoku, E.E.O. and M.K. Boachie, *The environmental impact of industrialization and foreign direct investment*. Energy Policy, 2020. **137**: p. 111178.
3. Rajaram, T. and A. Das, *Water pollution by industrial effluents in India: Discharge scenarios and case for participatory ecosystem specific local regulation*. Futures, 2008. **40**(1): p. 56–69.
4. Lin, Y., R. Huang, and X. Yao, *Air pollution and environmental information disclosure: An empirical study based on heavy polluting industries*. Journal of Cleaner Production, 2021. **278**: p. 124313.
5. Gita, S., et al., *Toxic Effects of Selected Textile Dyes on Elemental Composition, Photosynthetic Pigments, Protein Content and Growth of a Freshwater Chlorophycean Alga Chlorella vulgaris*. Bulletin of Environmental Contamination and Toxicology, 2019. **102**(6): p. 795–801.
6. www.epa.gov/.
7. Bonilla-Petriciolet, A., D.I. Mendoza-Castillo, and H.E. Reynel-Ávila, *Adsorption Processes for Water Treatment and Purification*, 2017: Springer.
8. Yue, X., et al., *Mitigation of indoor air pollution: A review of recent advances in adsorption materials and catalytic oxidation*. Journal of Hazardous Materials, 2021. **405**: p. 124138.
9. Naushad, M., et al., *Adsorption of textile dye using para-aminobenzoic acid modified activated carbon: Kinetic and equilibrium studies*. Journal of Molecular Liquids, 2019. **296**: p. 112075.

10. Wong, S., et al., *Recent advances in applications of activated carbon from biowaste for wastewater treatment: A short review*. Journal of Cleaner Production, 2018. **175**: p. 361–375.
11. Bhave, P.P. and D. Yeleswarapu, *Removal of indoor air pollutants using activated carbon—A review*. Global Challenges in Energy and Environment, 2020: p. 65–75.
12. Sweetman, M.J., et al., *Activated Carbon, Carbon Nanotubes and Graphene: Materials and Composites for Advanced Water Purification*. C, 2017. **3**(2).
13. Gupta, V.K., et al., *Low-Cost Adsorbents: Growing Approach to Wastewater Treatment—a Review*. Critical Reviews in Environmental Science and Technology, 2009. **39**(10): p. 783–842.
14. Apul, O.G. and T. Karanfil, *Adsorption of synthetic organic contaminants by carbon nanotubes: A critical review*. Water Research, 2015. **68**: p. 34–55.
15. Boukhalfa, N., et al., *Kinetics, thermodynamics, equilibrium isotherms, and reusability studies of cationic dye adsorption by magnetic alginate/oxidized multiwalled carbon nanotubes composites*. International Journal of Biological Macromolecules, 2019. **123**: p. 539–548.
16. Li, Y., et al., *Comparative study of methylene blue dye adsorption onto activated carbon, graphene oxide, and carbon nanotubes*. Chemical Engineering Research and Design, 2013. **91**(2): p. 361–368.
17. Liu, H. and H. Qiu, *Recent advances of 3D graphene-based adsorbents for sample preparation of water pollutants: A review*. Chemical Engineering Journal, 2020. **393**: p. 124691.
18. Georgakilas, V., et al., *Functionalization of Graphene: Covalent and Non-Covalent Approaches, Derivatives and Applications*. Chemical Reviews, 2012. **112**(11): p. 6156–6214.
19. Xu, J., et al., *A review of functionalized carbon nanotubes and graphene for heavy metal adsorption from water: Preparation, application, and mechanism*. Chemosphere, 2018. **195**: p. 351–364.
20. Králik, M., Adsorption, chemisorption, and catalysis. Chemical Papers, 2014. **68**(12): p. 1625–1638.
21. Kajjumba, G.W., et al., Modelling of adsorption kinetic processes—errors, theory and application. *Advanced Sorption Process Applications*, 2018: p. 187–206.
22. Yuh-Shan, H., *Citation review of Lagergren kinetic rate equation on adsorption reactions*. Scientometrics, 2004. **59**(1): p. 171–177.
23. Bonelli, B., et al., *Nanomaterials for the Detection and Removal of Wastewater Pollutants*, 2020: Elsevier.
24. Blanchard, G., M. Maunaye, and G. Martin, *Removal of heavy metals from waters by means of natural zeolites*. Water research, 1984. **18**(12): p. 1501–1507.
25. Azizian, S., *Kinetic models of sorption: a theoretical analysis*. Journal of colloid and Interface Science, 2004. **276**(1): p. 47–52.
26. Wu, F.-C., R.-L. Tseng, and R.-S. Juang, *Initial behavior of intraparticle diffusion model used in the description of adsorption kinetics*. Chemical Engineering Journal, 2009. **153**(1): p. 1–8.
27. Fadi, A., *Kinetics and Diffusion Analysis for the Removal of Cadmium Ion from Aqueous Solutions Using Chitosan-iso-Vanillin Sorbent*. Russian Journal of Physical Chemistry A, 2019. **93**(13): p. 2628–2634.
28. Liu, Y., *Some consideration on the Langmuir isotherm equation*. Colloids and Surfaces A: Physicochemical and Engineering Aspects, 2006. **274**(1): p. 34–36.
29. Singh, A.K., *Engineered Nanoparticles: Structure, Properties and Mechanisms of Toxicity*, 2015: Academic Press.
30. Ayawei, N., A.N. Ebelegi, and D. Wankasi, *Modelling and Interpretation of Adsorption Isotherms*. Journal of Chemistry, 2017. **2017**: p. 3039817.
31. Gopinath, K.P., et al., *Environmental applications of carbon-based materials: a review*. Environmental Chemistry Letters, 2021. **19**(1): p. 557–582.
32. Shu, D., et al., *Prominent adsorption performance of amino-functionalized ultra-light graphene aerogel for methyl orange and amaranth*. Chemical Engineering Journal, 2017. **324**: p. 1–9.
33. Wang, S., et al., *Preparation and characterization of graphene oxide/silk fibroin hybrid aerogel for dye and heavy metal adsorption*. Composites Part B: Engineering, 2019. **163**: p. 716–722.
34. Tang, S., et al., *Dye adsorption by self-recoverable, adjustable amphiphilic graphene aerogel*. Journal of Colloid and Interface Science, 2019. **554**: p. 682–691.
35. Hou, P., et al., *Hollow carbon spheres/graphene hybrid aerogels as high-performance adsorbents for organic pollution*. Separation and Purification Technology, 2019. **213**: p. 524–532.

36. Ren, X., et al., *Plasma Induced Multiwalled Carbon Nanotube Grafted with 2-Vinylpyridine for Preconcentration of Pb (II) from Aqueous Solutions*. Plasma Processes and Polymers, 2011. **8**(7): p. 589–598.
37. Derakhshan, M.S. and O. Moradi, *The study of thermodynamics and kinetics methyl orange and malachite green by SWCNTs, SWCNT-COOH and SWCNT-NH2 as adsorbents from aqueous solution*. Journal of Industrial and Engineering Chemistry, 2014. **20**(5): p. 3186–3194.
38. Yang, S., et al., *Enhanced adsorption of Congo red dye by functionalized carbon nanotube/mixed metal oxides nanocomposites derived from layered double hydroxide precursor*. Chemical Engineering Journal, 2015. **275**: p. 315–321.
39. Ahmad, A., et al., *Adsorption of methyl orange by synthesized and functionalized-CNTs with 3-aminopropyltriethoxysilane loaded TiO2 nanocomposites*. Chemosphere, 2017. **168**: p. 474–482.
40. Ahamad, T., et al., *Effective and fast adsorptive removal of toxic cationic dye (MB) from aqueous medium using amino-functionalized magnetic multiwall carbon nanotubes*. Journal of Molecular Liquids, 2019. **282**: p. 154–161.
41. Das, T.R., et al., *Bismuth oxide decorated graphene oxide nanocomposites synthesized via sonochemical assisted hydrothermal method for adsorption of cationic organic dyes*. Journal of Colloid and Interface Science, 2018. **509**: p. 82–93.
42. Yusuf, M., et al., *Synthesis of CTAB intercalated graphene and its application for the adsorption of AR265 and AO7 dyes from water*. Journal of Colloid and Interface Science, 2017. **493**: p. 51–61.
43. Xiao, J., et al., *Multifunctional graphene/poly(vinyl alcohol) aerogels: In situ hydrothermal preparation and applications in broad-spectrum adsorption for dyes and oils*. Carbon, 2017. **123**: p. 354–363.
44. Qi, Y., et al., *Natural polysaccharides-modified graphene oxide for adsorption of organic dyes from aqueous solutions*. Journal of Colloid and Interface Science, 2017. **486**: p. 84–96.
45. Guo, R., et al., *Sandwiched Fe3O4/carboxylate graphene oxide nanostructures constructed by layer-by-layer assembly for highly efficient and magnetically recyclable dye removal*. ACS Sustainable Chemistry & Engineering, 2018. **6**(1): p. 1279–1288.
46. Chen, L., et al., *High performance agar/graphene oxide composite aerogel for methylene blue removal*. Carbohydrate Polymers, 2017. **155**: p. 345–353.
47. Jiang, Y., et al., *Magnetic chitosan–graphene oxide composite for anti-microbial and dye removal applications*. International Journal of Biological Macromolecules, 2016. **82**: p. 702–710.
48. Zhao, W., et al., *Functionalized graphene sheets with poly (ionic liquid) s and high adsorption capacity of anionic dyes*. Applied Surface Science, 2015. **326**: p. 276–284.
49. Kim, H., et al., *Adsorption isotherms and kinetics of cationic and anionic dyes on three-dimensional reduced graphene oxide macrostructure*. Journal of Industrial and Engineering Chemistry, 2015. **21** : p. 1191–1196.
50. Gan, L., et al., *Konjac glucomannan/graphene oxide hydrogel with enhanced dyes adsorption capability for methyl blue and methyl orange*. Applied Surface Science, 2015. **357**: p. 866–872.
51. Cheng, Z., et al., *One-step fabrication of graphene oxide enhanced magnetic composite gel for highly efficient dye adsorption and catalysis*. ACS Sustainable Chemistry & Engineering, 2015. **3**(7): p. 1677–1685.
52. Zhang, J., et al., *Poly (acrylic acid) functionalized magnetic graphene oxide nanocomposite for removal of methylene blue*. RSC Advances, 2015. **5**(41): p. 32272–32282.
53. Sarkar, C., C. Bora, and S.K. Dolui, *Selective dye adsorption by pH modulation on amine-functionalized reduced graphene oxide–carbon nanotube hybrid*. Industrial & Engineering Chemistry Research, 2014. **53**(42): p. 16148–16155.
54. Labiadh, L. and A.R. Kamali, *3D graphene nanoedges as efficient dye adsorbents with ultra-high thermal regeneration performance*. Applied Surface Science, 2019. **490**: p. 383–394.
55. Wang, S., et al., *Synergistic and competitive adsorption of organic dyes on multiwalled carbon nanotubes*. Chemical Engineering Journal, 2012. **197**: p. 34–40.
56. Ai, L., et al., *Removal of methylene blue from aqueous solution with magnetite loaded multi-wall carbon nanotube: kinetic, isotherm and mechanism analysis*. Journal of Hazardous Materials, 2011. **198**: p. 282–290.

57. Li, Y., et al., *Comparative study of methylene blue dye adsorption onto activated carbon, graphene oxide, and carbon nanotubes*. Chemical Engineering Research and Design, 2013. **91**(2): p. 361–368.
58. Mohammadi, A. and P. Veisi, *High adsorption performance of β-cyclodextrin-functionalized multi-walled carbon nanotubes for the removal of organic dyes from water and industrial wastewater*. Journal of Environmental Chemical Engineering, 2018. **6**(4): p. 4634–4643.
59. Sadegh, H., et al., *Synthesis of MWCNT-COOH-Cysteamine composite and its application for dye removal*. Journal of Molecular Liquids, 2016. **215**: p. 221–228.
60. Jin, J., et al., *Magnetic-responsive CNT/chitosan composite as stabilizer and adsorbent for organic contaminants and heavy metal removal*. Journal of Molecular Liquids, 2021. **334**: p. 116087.
61. Moradi, O., *Adsorption behavior of basic red 46 by single-walled carbon nanotubes surfaces*. Fullerenes, Nanotubes and Carbon Nanostructures, 2013. **21**(4): p. 286–301.
62. Ghaedi, M., et al., *Oxidized multiwalled carbon nanotubes as efficient adsorbent for bromothymol blue*. Toxicological & Environmental Chemistry, 2012. **94**(5): p. 873–883.
63. Konicki, W., et al., *Adsorption of anionic dye Direct Red 23 onto magnetic multi-walled carbon nanotubes-Fe3C nanocomposite: Kinetics, equilibrium and thermodynamics*. Chemical Engineering Journal, 2012. **210**: p. 87–95.
64. Zhao, D., et al., *Adsorption of methyl orange dye onto multiwalled carbon nanotubes*. Procedia Environmental Sciences, 2013. **18**: p. 890–895.
65. Robati, D., et al., *Adsorption behavior of methylene blue dye on nanocomposite multi-walled carbon nanotube functionalized thiol (MWCNT-SH) as new adsorbent*. Journal of Molecular Liquids, 2016. **216**: p. 830–835.
66. Zhang, F., et al., *Preparation of graphene-oxide/polyamidoamine dendrimers and their adsorption properties toward some heavy metal ions*. Journal of Chemical & Engineering Data, 2014. **59**(5): p. 1719–1726.
67. Luan, H., B. Teychene, and H. Huang, *Efficient removal of As (III) by Cu nanoparticles intercalated in carbon nanotube membranes for drinking water treatment*. Chemical Engineering Journal, 2019. **355**: p. 341–350.
68. Alijani, H. and Z. Shariatinia, *Effective aqueous arsenic removal using zero valent iron doped MWCNT synthesized by in situ CVD method using natural α-Fe2O3 as a precursor*. Chemosphere, 2017. **171**: p. 502–511.
69. Chen, B., et al., *Surfactant assisted Ce–Fe mixed oxide decorated multiwalled carbon nanotubes and their arsenic adsorption performance*. Journal of Materials Chemistry A, 2013. **1**(37): p. 11355–11367.
70. Gao, L., et al., *Directing carbon nanotubes from aqueous phase to o/w interface for heavy metal uptaking*. Environmental Science and Pollution Research, 2015. **22**(18): p. 14201–14208.
71. Verma, B. and C. Balomajumder, Magnetic magnesium ferrite–doped multi-walled carbon nanotubes: an advanced treatment of chromium-containing wastewater. Environmental Science and Pollution Research, 2020: p. 1–11.
72. Alijani, H. and Z. Shariatinia, *Synthesis of high growth rate SWCNTs and their magnetite cobalt sulfide nanohybrid as super-adsorbent for mercury removal*. Chemical Engineering Research and Design, 2018. **129**: p. 132–149.
73. Dong, Z., et al., *Bio-inspired surface-functionalization of graphene oxide for the adsorption of organic dyes and heavy metal ions with a superhigh capacity*. Journal of Materials Chemistry A, 2014. **2**(14): p. 5034–5040.
74. Fu, Y., et al., *Water-dispersible magnetic nanoparticle–graphene oxide composites for selenium removal*. Carbon, 2014. **77**: p. 710–721.
75. Lv, X., et al., *Nanoscale zero-valent iron (nZVI) assembled on magnetic Fe3O4/graphene for chromium (VI) removal from aqueous solution*. Journal of Colloid and Interface Science, 2014. **417**: p. 51–59.
76. Wang, C., et al., *Removal of As (III) and As (V) from aqueous solutions using nanoscale zero valent iron-reduced graphite oxide modified composites*. Journal of Hazardous Materials, 2014. **268**: p. 124–131.

77. Xu, M., et al., *Facile synthesis of soluble functional graphene by reduction of graphene oxide via acetylacetone and its adsorption of heavy metal ions.* Nanotechnology, 2014. **25**(39): p. 395602.
78. Yan, H., et al., Rapid removal and separation of iron (II) and manganese (II) from micropolluted water using magnetic graphene oxide. ACS Applied Materials & Interfaces, 2014. **6**(12): p. 9871–9880.
79. Verma, S. and R.K. Dutta, *A facile method of synthesizing ammonia modified graphene oxide for efficient removal of uranyl ions from aqueous medium.* RSC Advances, 2015. **5**(94): p. 77192–77203.
80. Cui, L., et al., *EDTA functionalized magnetic graphene oxide for removal of Pb (II), Hg (II) and Cu (II) in water treatment: adsorption mechanism and separation property.* Chemical Engineering Journal, 2015. **281**: p. 1–10.
81. Jiang, T., et al., *Adsorption behavior of copper ions from aqueous solution onto graphene oxide–CdS composite.* Chemical Engineering Journal, 2015. **259**: p. 603–610.
82. Kabiri, S., et al., *Graphene-diatom silica aerogels for efficient removal of mercury ions from water.* ACS Applied Materials & Interfaces, 2015. **7**(22): p. 11815–11823.
83. Li, X., et al., *Studies of heavy metal ion adsorption on Chitosan/Sulfydryl-functionalized graphene oxide composites.* Journal of Colloid and Interface Science, 2015. **448**: p. 389–397.
84. Liu, L., et al., *Enhancing the Hg (II) removal efficiency from real wastewater by novel thymine-grafted reduced graphene oxide complexes.* Industrial & Engineering Chemistry Research, 2016. **55**(24): p. 6845–6853.
85. Liu, M., et al., *Study on the adsorption of Hg (II) by one-pot synthesis of amino-functionalized graphene oxide decorated with a Fe 3 O 4 microsphere nanocomposite.* RSC Advances, 2016. **6**(88): p. 84573–84586.
86. Sahraei, R., K. Hemmati, and M. Ghaemy, *Adsorptive removal of toxic metals and cationic dyes by magnetic adsorbent based on functionalized graphene oxide from water.* RSC Advances, 2016. **6**(76): p. 72487–72499.
87. Tang, J., et al., *Preparation of a novel graphene oxide/Fe-Mn composite and its application for aqueous Hg (II) removal.* Journal of Hazardous Materials, 2016. **316**: p. 151–158.
88. Xiao, W., et al., *Dendrimer functionalized graphene oxide for selenium removal.* Carbon, 2016. **105**: p. 655–664.
89. Zhao, D., et al., *Facile preparation of amino functionalized graphene oxide decorated with Fe3O4 nanoparticles for the adsorption of Cr (VI).* Applied Surface Science, 2016. **384**: p. 1–9.
90. Ma, Y.-X., et al., *Preparation of polyamidoamine dendrimers functionalized magnetic graphene oxide for the adsorption of Hg (II) in aqueous solution.* Journal of Colloid and Interface Science, 2017. **505**: p. 352–363.
91. Ranjith, K.S., et al., *Multifunctional ZnO nanorod-reduced graphene oxide hybrids nanocomposites for effective water remediation: Effective sunlight driven degradation of organic dyes and rapid heavy metal adsorption.* Chemical Engineering Journal, 2017. **325**: p. 588–600.
92. Verma, S. and R.K. Dutta, *Development of cysteine amide reduced graphene oxide (CARGO) nano-adsorbents for enhanced uranyl ions removal from aqueous medium.* Journal of Environmental Chemical Engineering, 2017. **5**(5): p. 4547–4558.
93. Varadwaj, G.B.B., et al., *Facile synthesis of three-dimensional Mg–Al layered double hydroxide/partially reduced graphene oxide nanocomposites for the effective removal of Pb2+ from aqueous solution.* ACS Applied Materials & Interfaces, 2017. **9**(20): p. 17290–17305.
94. Xiong, Y., et al., *Improving Re (VII) adsorption on diisobutylamine-functionalized graphene oxide.* ACS Sustainable Chemistry & Engineering, 2017. **5**(1): p. 1010–1018.
95. Zhou, C., et al., *Adsorption of mercury (II) with an Fe 3 O 4 magnetic polypyrrole–graphene oxide nanocomposite.* RSC Advances, 2017. **7**(30): p. 18466–18479.
96. Hosseinzadeh, H. and S. Ramin, *Effective removal of copper from aqueous solutions by modified magnetic chitosan/graphene oxide nanocomposites.* International Journal of Biological Macromolecules, 2018. **113**: p. 859–868.
97. Liang, Q., et al., *Facile one-pot preparation of nitrogen-doped ultra-light graphene oxide aerogel and its prominent adsorption performance of Cr (VI).* Chemical Engineering Journal, 2018. **338**: p. 62–71.

98. Zhang, Z., et al., *One-pot preparation of P (TA-TEPA)-PAM-RGO ternary composite for high efficient Cr (VI) removal from aqueous solution.* Chemical Engineering Journal, 2018. **343**: p. 207–216.
99. Cai, Y., et al., *Fully phosphorylated 3D graphene oxide foam for the significantly enhanced U (VI) sequestration.* Environmental Pollution, 2019. **249**: p. 434–442.
100. Huang, Z.-W., et al., *Adsorption of Eu (III) and Th (IV) on three-dimensional graphene-based macrostructure studied by spectroscopic investigation.* Environmental Pollution, 2019. **248**: p. 82–89.
101. Liang, J., et al., *Facile construction of 3D magnetic graphene oxide hydrogel via incorporating assembly and chemical bubble and its application in arsenic remediation.* Chemical Engineering Journal, 2019. **358**: p. 552–563.
102. Qiu, P., et al., *Adsorption of low-concentration mercury in water by 3D cyclodextrin/graphene composites: Synergistic effect and enhancement mechanism.* Environmental Pollution, 2019. **252**: p. 1133–1141.
103. Wang, Z., et al., *Cyclodextrin functionalized 3D-graphene for the removal of Cr (VI) with the easy and rapid separation strategy.* Environmental Pollution, 2019. **254**: p. 112854.
104. Yu, S., et al., *Three-dimensional graphene/titanium dioxide composite for enhanced U (VI) capture: insights from batch experiments, XPS spectroscopy and DFT calculation.* Environmental Pollution, 2019. **251**: p. 975–983.
105. Yu, X., et al., *Ultrafast and deep removal of arsenic in high-concentration wastewater: A superior bulk adsorbent of porous Fe2O3 nanocubes-impregnated graphene aerogel.* Chemosphere, 2019. **222**: p. 258–266.
106. Zhan, W., et al., *Green synthesis of amino-functionalized carbon nanotube-graphene hybrid aerogels for high performance heavy metal ions removal.* Applied Surface Science, 2019. **467**: p. 1122–1133.
107. Zhao, D., et al., *A simple method for preparing ultra-light graphene aerogel for rapid removal of U (VI) from aqueous solution.* Environmental Pollution, 2019. **251**: p. 547–554.
108. AlOmar, M.K., et al., *Allyl triphenyl phosphonium bromide based DES-functionalized carbon nanotubes for the removal of mercury from water.* Chemosphere, 2017. **167**: p. 44–52.
109. Gupta, A., S. Vidyarthi, and N. Sankararamakrishnan, *Enhanced sorption of mercury from compact fluorescent bulbs and contaminated water streams using functionalized multiwalled carbon nanotubes.* Journal of Hazardous Materials, 2014. **274**: p. 132–144.
110. Jyoti, J., et al., *Significant improvement in static and dynamic mechanical properties of graphene oxide carbon nanotube acrylonitrile butadiene styrene hybrid composites.* Journal of Materials Science, 2018. **53** (4): p. 2520–2536.
111. Zhang, C., et al., *Adsorption behavior of engineered carbons and carbon nanomaterials for metal endocrine disruptors: experiments and theoretical calculation.* Chemosphere, 2019. **222**: p. 184–194.
112. Li, Z., J. Chen, and Y. Ge, *Removal of lead ion and oil droplet from aqueous solution by lignin-grafted carbon nanotubes.* Chemical Engineering Journal, 2017. **308**: p. 809–817.
113. Mubarak, N.M., et al., *Rapid adsorption of toxic Pb (II) ions from aqueous solution using multiwall carbon nanotubes synthesized by microwave chemical vapor deposition technique.* Journal of Environmental Sciences, 2016. **45**: p. 143–155.
114. Tian, X., et al., *Metal impurities dominate the sorption of a commercially available carbon nanotube for Pb (II) from water.* Environmental Science & Technology, 2010. **44**(21): p. 8144–8149.
115. Farghali, A., et al., *Functionalization of acidified multi-walled carbon nanotubes for removal of heavy metals in aqueous solutions.* Journal of Nanostructure in Chemistry, 2017. **7**(2): p. 101–111.
116. Tian, Y., et al., *Methods of using carbon nanotubes as filter media to remove aqueous heavy metals.* Chemical Engineering Journal, 2012. **210**: p. 557–563.
117. Kumar, R., et al., *Adsorption modeling and mechanistic insight of hazardous chromium on para toluene sulfonic acid immobilized-polyaniline@ CNTs nanocomposites.* Journal of Saudi Chemical Society, 2019. **23**(2): p. 188–197.

118. Šolić, M., et al., *Comparing the adsorption performance of multiwalled carbon nanotubes oxidized by varying degrees for removal of low levels of copper, nickel and chromium (VI) from aqueous solutions*. Water, 2020. **12**(3): p. 723.
119. Yu, L., et al., *Adsorption of VOCs on reduced graphene oxide*. Journal of Environmental Sciences, 2018. **67**: p. 171–178.
120. Yang, S., et al., *Enhancement of formaldehyde removal by activated carbon fiber via in situ growth of carbon nanotubes*. Building and Environment, 2017. **126**: p. 27–33.
121. Yan, J., et al., *Improved ethanol adsorption capacity and coefficient of performance for adsorption chillers of Cu-BTC@ GO composite prepared by rapid room temperature synthesis*. Industrial & Engineering Chemistry Research, 2016. **55**(45): p. 11767–11774.
122. Chu, F., et al., *Adsorption of toluene with water on zeolitic imidazolate framework-8/graphene oxide hybrid nanocomposites in a humid atmosphere*. RSC Advances, 2018. **8**(5): p. 2426–2432.
123. Sun, X., et al., *Synthesis and adsorption performance of MIL-101 (Cr)/graphite oxide composites with high capacities of n-hexane*. Chemical Engineering Journal, 2014. **239**: p. 226–232.
124. Wu, L., et al., *Facile synthesis of 3D amino-functional graphene-sponge composites decorated by graphene nanodots with enhanced removal of indoor formaldehyde*. Aerosol and Air Quality Research, 2015. **15**(3): p. 1028–1034.
125. Kim, J.M., et al., *Toluene and acetaldehyde removal from air on to graphene-based adsorbents with microsized pores*. Journal of Hazardous Materials, 2018. **344**: p. 458–465.
126. Zhou, X., et al., *A novel MOF/graphene oxide composite GrO@ MIL-101 with high adsorption capacity for acetone*. Journal of Materials Chemistry A, 2014. **2**(13): p. 4722–4730.
127. Guo, Z., et al., *Electrospun graphene oxide/carbon composite nanofibers with well-developed mesoporous structure and their adsorption performance for benzene and butanone*. Chemical Engineering Journal, 2016. **306**: p. 99–106.
128. Zheng, Y., et al., *Ultrahigh adsorption capacities of carbon tetrachloride on MIL-101 and MIL-101/graphene oxide composites*. Microporous and Mesoporous Materials, 2018. **263**: p. 71–76.
129. Abbas, A., et al., *Benzene removal by iron oxide nanoparticles decorated carbon nanotubes*. Journal of Nanomaterials, 2016. **2016**.

13 Carbon Nanotubes and Graphene

The Novel Materials for Terahertz Detection

Subhash Nimanpure, Guruvandra Singh, and Mukesh Jewariya

CONTENT

13.1 INTRODUCTION

The last decade has shown an incredible evolution in the applications particularly in far-infrared and terahertz (THz) systems and their various applications in several fields, as novel and superior-power sources are existing, and the perspective of THz photonics for progressive potential physics research and industrial applications have been well established. Numerous recent innovations in the terahertz area have advanced the research into the cutting-edge stage where advanced killer applications of terahertz are identified such as identification figure print common tissue. As an example of landmark accomplishment, we are able to point out meticulous development of three-dimensional (3-D) THz imaging and high-power THz generation using $LiNbO_3$ [1, 2]. Devices used to exploit this particular wavelength region are set to end up with increasing numbers of

DOI: 10.1201/9781003231943-13

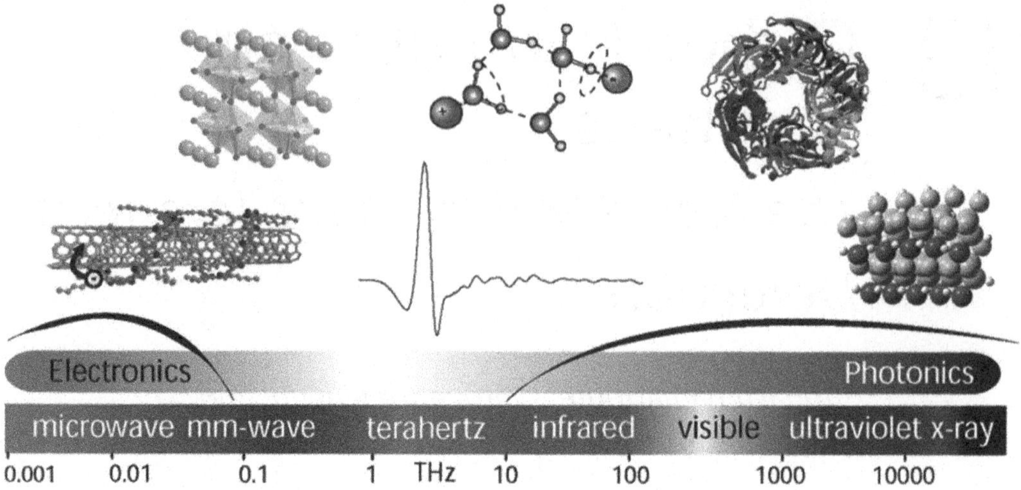

FIGURE 13.1 An electromagnetic spectrum showing terahertz region [6]. Copyright © 2011, American Chemical Society.

vitality in a numerous variety of human pastime applications (e.g., Imaging, security, explosion detection, fingerprints, drugs & chemical identification, biology, and so on). At present, infrared (IR) and THz band are used in the latest scenario along with nanomaterials such as non-destructive and non-contact measurement [3], this is totally based on the reality that THz frequencies lie in excitations in nanoelectronics gadgets and collective dynamics. Although lots of compact, electrically driven, solid-kingdom technology have been developed for generation and detection of THz waves [4, 5], all of them are afflicted by drawbacks which are presently even restricting the significant utilization of THz photonics as shown in Figure 13.1.

Newly, carbon-based nanomaterials known as CNTs and, graphene have developed as unexpected low-dimensional materials with a diversity of excellent properties in term of electronic and photonics [7–10], specially those preferably well-matched for THz devices [11–13]. CNTs have an outstanding capability to fascinate electromagnetic (EM) waves in an ultrawide EM spectral range from microwave to the ultraviolet (UV), with respect to interband (excitonic) and intraband (free carrier) absorbance processes [12, 14–17]. The group of single-wall carbon nanotubes (SWCNTs) and multi-wall carbon nanotubes (MWCNTs) with diverse chirality's are therefore able to absorb EM radiation. Basically, we can also say that CNTS can absorb EM radiation at any frequency in the entire EM spectrum specially from the electronics band to the photonic band, and the same property is also associated with the graphene [18–20]. This wide band property of CNTS, combined with excellent mobility charge carriers, make them fast and ultra-broadband photodetectors [9, 21, 22]. Moreover, the superb opto-electronic characteristics of graphene make it suitable for a brand-new stand for many photonic applications [4], consisting of speedy photodetectors [5], electrodes in photovoltaic modules [23], plasmonic gadgets, optical modulators, and ultrafast lasers [24–26]. Graphene-based terahertz detectors play a crucial part in the improvement of modern technology in lots of fields, including defense, 3-D real-time imaging, military, space, medicine, astronomy, and optical communications. Graphene has established himself an excellent promising candidate for mid and far-infrared (may consider in THz range) plasmonic [27] and paying attention toward outstanding deal as a platform for plasmonic based IR and THz detectors. By means of dipping the operation frequency down to the THz domain, the resonant excitation of plasmons can be accomplished if the momentum rest charge will be underneath the plasmon frequency, calls for very high electron mobility. Therefore, in graphene-based THz detectors, the plasma waves are generally over damped, and the graphene-based devices show evidence of most effective broadband (non-resonant) photo response. As a result, several programs counting on resonant plasmon excitation continue to be proved experimentally but have been unrealized till now.

The graphene flakes had been synthesized and studied by Novoselov et al. [28]. In this study, it is observed that the enhancement in the photo response and improvement in the graphene-based opto-electronic device performance. A very unique type of band shape of graphene [27], in which conduction and valence band contact are different near the Dirac point, gives almost steady photon absorption for visible light to far-infrared pulse [29]. This Dirac point provides extraordinarily excessive mobility [30] and fast carrier relaxation [25], which allows us to understand ultrafast mechanisms (especially which happens at picoseconds time scale) of carrier dynamics in opto-electronic devices for a large spectral bandwidth. Various detectors used for near-infrared radiation (NIR), are used for telecommunication frequencies as much as 16 GHz, had been established [31]. In those detectors a surge of photocurrent is triggered by an asymmetry within the detector, which is furnished through unique metal contacts made to the graphene flake. In recent times detectors primarily based on graphene kind of 2D materials, had been advanced for devices operating on THz frequencies as well. The graphene-based detector works at room temperature with ultrafast (picosecond) response time, consisting of the capability to function an ultra-broadband detector from visible to terahertz region [32–36].

13.2 SYNTHESIS OF CNTS

A carbon nanotube can be thought of as graphene that has been rolled into a cylindrical shape, creating a one-dimensional structure with axial harmony. The six pentagonal graphene sheets have caps on each end that are nanotubes. The caps are positioned ideally to fit the long cylindrical section. Carbon nanotubes are roughly a nanometer in diameter and limited microns long [37]. There are various systems used to fabricate and develop the MWCNTs and SWCNTs like electric arc discharge techniques, laser vaporization and chemical vapor deposition techniques. These are well-known techniques to produce a widespread variety of CNT and are described below.

13.2.1 Arc Discharge Technique

Iijimia used a variety of techniques to generate CNTs, one of which being the arc discharge method. [37]. Direct current delivers advanced yields of CNTs, which are deposited on the cathode. One of the important situations in the arc discharge technique is that the balance of the arc distance maintains constantly about 1 mm through the graphite electrodes [38]. With the help of the helium gas, the synthesis of MWCNTs can be achieved by the arc discharge method [39, 40]. Here, a graphite rod holding a metal catalyst, is used as the anode through pure graphite cathode and it produces the SWCNTs in the form of soot [41, 42]. It was found that the presence of hydrogen gas in the growing zone can provide the best synthesis of MWCNTs with strong crystallinity across a few coexisting carbon nanoparticles [43–48].

13.2.2 Laser Vaporization Technique

The Smalley group [49] established the laser vaporization technique for fullerene and CNTs manufacturing. It is an effective route for the fabrication of SWCNTs within a narrow distribution across the laser evaporation concept. Within the technique, a part of graphite target is heated by laser irradiation with high temperature under inert atmosphere. When we used a pure graphite target a good quality of MWCNTs could be obtained. Moreover, the yield and quality of the CNTs were found to depend on the reaction temperature condition. The best quality of CNTs is attained at 1200°C reaction temperature. In contrast, at lower temperature the structure of CNTs was degraded and showed many defects. The yield of CNTs strongly depended on the category of metal catalyst which was used at the time of CNTs production. It was also found that it is increased with furnace temperature and among other factors [50].

13.2.3 Chemical Vapor Deposition

One of the greatest methods for creating CNTs of excellent quality is chemical vapor deposition (CVD). This method is often used in the semiconductor industry for production of thin-film-based systems. CVD is a system for manufacturing materials in which the chemical component in the vapor phase responds to form a solid film within the surfaces, called substrate [51]. This is discussed in further detail in the next section.

13.3 SYNTHESIS OF GRAPHENE

The numerous approaches to growth or creating graphene, yields distinctive graphene flakes or layer structure, with a reduction in mobility. Thus, while evaluating graphene using various spectroscopic techniques viz UV, FTIR, Raman and THz, etc., the outward appearance and quality of graphene should be taken into account. Some of the manufacturing strategies are excellent, with incredible upgrades in the domain size of graphene recently. For that reason, research on graphene may be of great interest and it is proved to be an exceptional material for device fabrication for modern miniaturized industries.

In past years, numerous attempts had been performed to supply a single (mono) layer of graphite. Fernandez-Moran had exfoliated roughly 15 layers of graphene by the beginning of 1960 [36]. Iijima cut graphite crystals with adhesive tape [23]. Few researchers had tried the oxidized graphite and few others had also attempted to peel off a single layer using an AFM cantilever, but they had failed [25, 29–32, 52].

13.3.1 Mechanical Method

Geim and Novoselov's have used adhesive tape with a force (300 nN/μm^2) [53] to feat the interlayer Van der Waals interaction in graphite. The stripped off portions of graphite and graphene was transported to a silicon dioxide (SiO_2) substrate. Using this approach, mono- and few-layer graphene is obtained, and used to study graphene's electronic behavior. Most researchers believe that mechanical exfoliation yields the high-satisfactory graphene, with extraordinarily nobilities [54]. Moreover, its growth and production take time, and is not possible in mass production. Despite the reduced quality of graphene, researchers switched their focus from mechanical exfoliation to chemical techniques of graphene generation in order to maximize the exploitation of the exceptional electronic properties of graphene in device fabrications.

13.3.2 Epitaxial Growth

The heat processing of silicon carbide (SiC-0001) substrate is required for the epitaxial growth of graphene layers, which results in the sublimation of the Si atoms and the reorganization and graphitization of the carbon layers [55]. The observation of anomalous quantum Hall effect in graphene layers, gives a confidence that mono-layers of graphene can be produced at SiC surface, in reality 10–20 layers were formed [56, 57].

13.3.3 Chemical Vapor Deposition

One of the best methods for producing graphene in large quantities and for its technological utility is chemical vapor deposition (CVD). The chemical vapor deposition of hydrocarbons on transition metallic faces is capable of harvesting very thin graphitic layers [58]. Motivated by these works, graphene enhances the usage of CVD. The usage of hydrocarbon like ethylene, a carbon source, and ruthenium as the substrate people tried to get graphene or graphite [22]. Using these methods, a graphene film exceeding 100 μm have been achieved.

13.3.4 Wet Chemical Process

Although the mechanical exfoliation method uses adhesive to conquer the Van der Waals interplay among graphene layers, moist chemical substances can weaken by placing reactants into the interlayer space [59]. The wet chemical techniques are performed in suspension; they are used for mass production without difficulty. Many methods using wet chemical techniques have proven to yield very good quality graphene.

13.4 THZ-BASED OPTICAL CHARACTERIZATION TECHNIQUES FOR CNTS AND GRAPHENE

13.4.1 Pump-Probe Spectroscopy

The understanding of carrier dynamics of graphene is vital for designing and developing of new graphene-based devices in application of opto-electronics [33]. A probe pulse is utilized to probe the heated carriers with sub-picoseconds (ps) resolution after a pump pulse (usually at 800 nm) photo-excites the graphene. The ultrafast relaxation dynamics of low energy carriers near to the Dirac point may be studied using THz as a probe.

Dawlaty et al. observed the graphene with 6 and 37 layers of epitaxial grown, by with 780 nm pump and terahertz (THz) as probe [60]. They found that by excitating graphene with 780 nm pump beam, we found that the transmission spectrum probed by THz is quickly relaxed (photo-triggered bleaching) because of state blocking. Fast relaxation (70–120 fs) and a slower time (0.4–1.7 ps) were both seen. The photoexcited carrier populations' equilibration by intraband scattering from one carrier to another is attributed to the quick relaxation time scale. Later, a thermalization through carrier-phonon scattering to the larger time constant was shown.

13.4.2 THz-Time Domain Spectroscopy for Graphene and Observation of Conductivity

Terahertz spectroscopy has been employed as a contactless, non-destructive technique for characterizing 2D materials, such as graphene. Terahertz spectroscopy of graphene and its layered structure has become a topic of particular interest. These materials at terahertz frequencies, obeys a Debye model, the charge carrier conductivity is successfully resolute on the order of nanometers. Therefore, the sheet conductivity obtained using terahertz spectroscopic measurements gives averaged nanoscale conductivity and is less impinged by the microscale scattering phenomena. THz spectroscopy offers a way into these measurements with ultra-high time resolution and without complications of making electrical contacts on the sample. As a result, charge transport measurements using THz-TDS can result in a more relevant estimate of such conductive layers' intrinsic carrier transport properties. Utilizing THz-TDS and far-IR fourier transform interferometer (far-IR FTIR) spectroscopy, this approximation has been verified on both CVD and epitaxially produced graphene [60].

13.5 GRAPHENE AS A DETECTOR

Graphene-based detectors are generally categorized into two different types: thermal (bolometer based) or photon detectors (photovoltaic or photocurrent based). There is another detector which works on thermopile where an extra current is applied to generate photo thermoelectric effect (Seebeck impact) to produce a net electric current because of electron diffusion. The separation of electron-hole pairs and the electric fields that result from their generation at the junctions of p-type and n-type doped portions of graphene are the main components of photovoltaic (PV)-based detectors as shown in Figure 13.2. By creating an extra electric field with a source-drain bias voltage, it is possible to achieve the same effect. Since graphene is a semi-metal and generates a significant dark current, this effect is typically avoided. Because of the exchange of one high-energy electron-electron pair for multiple lower-energy electron-electron pairs inside graphene, the photodetection efficiency may be improved [33, 61].

Figure 13.3 shows a typical layout of graphene phototransistor with a short-circuit current where light is incident on its gate part. When there is no bias voltage on supply and the drain, minimum amount of photocurrent is recorded when light is incident on center of the graphene gate. When light is shone at the metal-graphene interface region, a significant quantity of photocurrent is created. This phenomenon is referred to as the PV effect. Since no electric field is created at the gate's center, there is no photocurrent there.

In addition, the photothermal electric effect (PTE; Seebeck effect) plays a crucial role in the production of photocurrent in graphene [33, 61–64]. Since the optical phonon energy (~200 meV) in graphene is very high, so in this case hot electrons created by optical field in graphene are of the order of picoseconds. As shown in Figure 13.4, incident light radiation causes temperature differences among charge carriers. As a result, hot electrons are produced as a result of photons

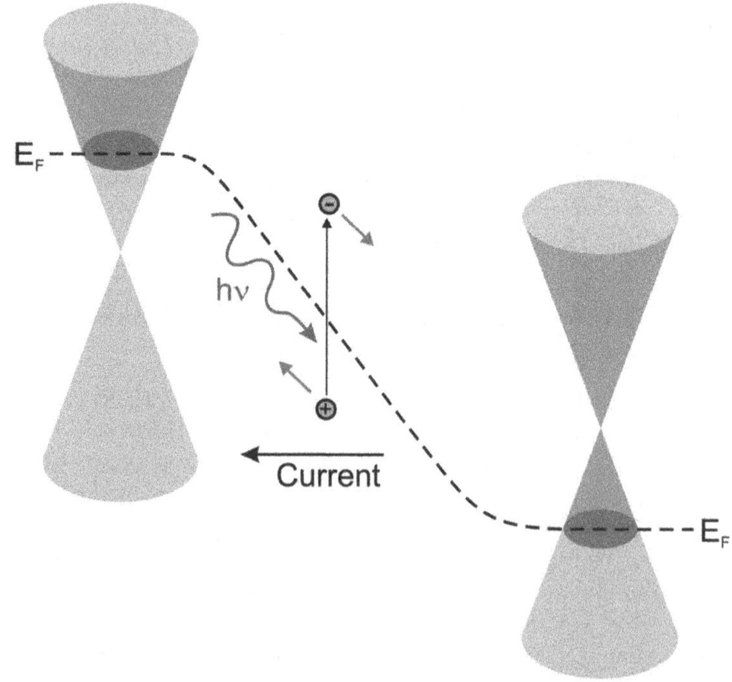

FIGURE 13.2 Separation of electron and hole by an internal electric field [62]. [Reprinted with permission from [ref # Adv. opt. photon, 11, 314-379 (2019)] © The Optical Society.

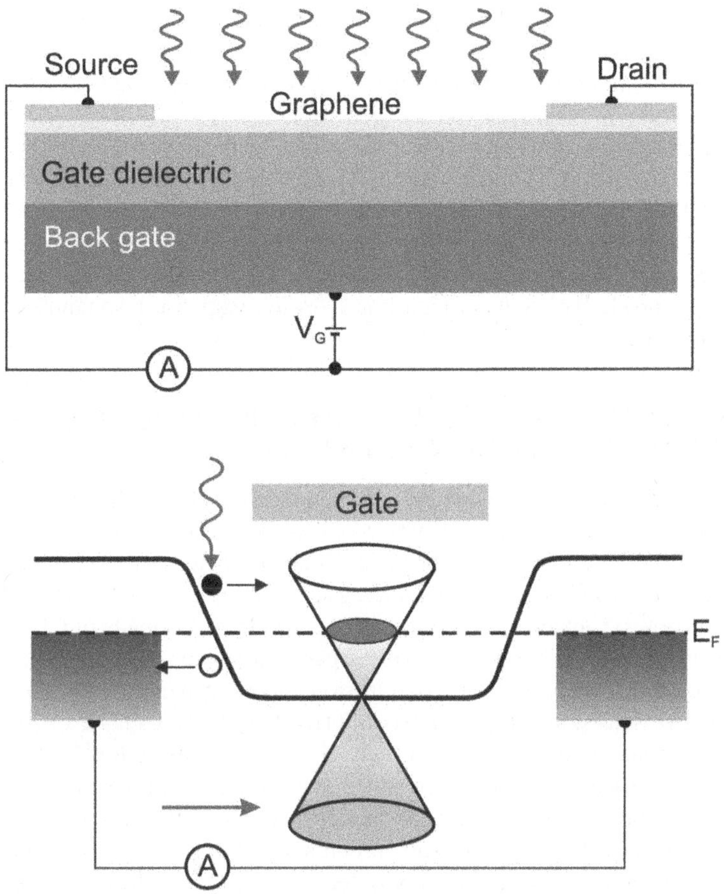

FIGURE 13.3 Graphene phototransistor: (a) transistor construction and (b) A diagram showing how photocurrent is produced [62]. [Reprinted with permission from [ref # Adv. opt. photon, 11, 314-379 (2019)] © The Optical Society.

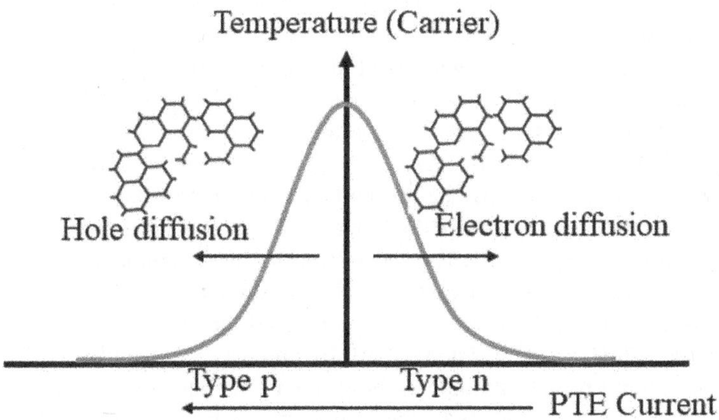

FIGURE 13.4 Photocurrent generation in a graphene p–n junction [62].

diffused as a result of the temperature gradient among carriers, which produces photocurrent. Even though disorder-assisted collisions may result in a significant speedup, the heated electrons and the lattice are brought into equilibrium through the slower scattering between charge carriers and acoustic phonons.

To detect THz radiation, graphene can be used as a field effect transistor (FET). Wojciech Knap et al. reported that the FETs can be used as THz detectors for imaging application [65]. The detection through FETs is because of the nonlinear behavior of the transistor, which results in the rectification of the ac current through the incident light. Consequently, a photo current gives rise to a dc voltage between source and drain. The detection system takes advantage of the plasma waves' interference inside the cavity, which improves the response resonantly [33, 61, 63, 64, 66–69] of detection mechanism.

13.6 THZ-TIME DOMAIN SPECTROSCOPY FOR FLEXIBLE CNTS AND POTENTIAL TO DETECTION OF THZ PULSE

Jewariya and his group develop flexible carbon nanotubes-based terahertz detecting materials [11]. A short summary is presented here. CNTs which are made up of graphene and in the form of SWCNTs and MWCNTs, can be used in the development of nano structures and that can be used in the variety of electrical and photonic engineering applications [8, 70–72]. The visible, infrared, and THz areas can now be detected using innovative photodetectors made of CNTs and graphene [9, 21]. Realistically, graphene has a remarkable capacity for both intraband (free charge carrier) and interband (excitonic) absorption of THz electromagnetic waves [15, 69, 73, 74] makes it a suitable for terahertz detection. The CNT based THz devices are capable of providing lots of room for the development of responsivity and sensitivity of THz detection.

Flexible terahertz devices are sturdily demanded as these are one of the promising technologies for room temperature operation for THz optoelectronics devices. We verified a newly developed flexible paper (MWCNT-FP) with multi-wall carbon nanotubes that is freestanding, allowing for the improvement in response time and sensitivity for THz detection. The CNTs were grown using chemical vapor deposition, and flexible CNTs paper was made using these CNTs. Figure 13.5 depicts an example of MWCNT-FP and how it interacts with THz incident pulses. After the electron-photon interaction, THz pulses incident on MWCNT-FP in the normal direction continue to flow through the specimens due to the temporal delay between the pump and probe and accumulate on the detector, resulting in transmitted statistics being recorded.

Figure 13.6 proves the test setup for THz measurement. A femtosecond laser (1560 nm, 100 MHz, 60 fs) is built and used in the test setup (Teraflash time domain THz source through Toptica Photonics) for terahertz pulses with spectral resolution of roughly 67 GHz. Femtosecond pulse is divided into two components, with the pump traveling toward the THz emitter (Photo Conducting Antenna-PCA) and the probe traveling toward the detector (Photo Conducting Antenna-PCA). Both the emitter and the detector (high bandwidth > 6 THz and 30 W power) use InGaAs/InP photoconductive switches; the former has a 100 m strip-line antenna while the latter has a 25 m dipole antenna. Terahertz-time domain spectroscopy is used to examine the transmission/abortion capabilities of MWCNT-FP in a dry environment at ambient temperature.

Femtosecond pulses are used to create a transient cutting-edge inside the emitter, which results in a broad spectrum in the terahertz range. The generated terahertz pulses passed through the MWCNT-FP sample, which has dimensions of 10, 10, and 0.1 mm (length, width, and thickness, respectively). The "pump-probe" module is how the PCA detector operates. A probe pulse measures the influence of the incident terahertz pulse's modulation as it is being transmitted through the MWCNT-FP. The recorded time component of the terahertz pulse is then transformed

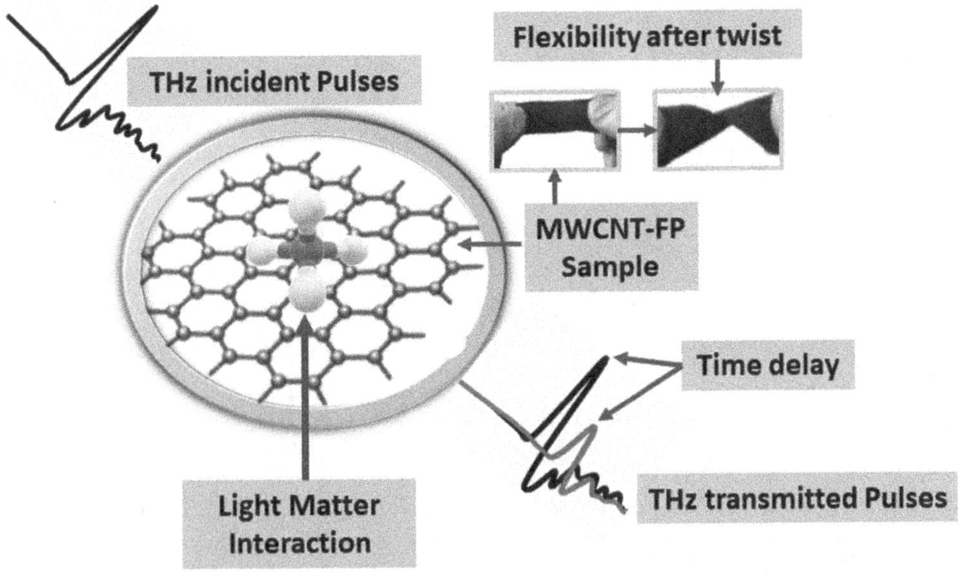

FIGURE 13.5 The representation of MWCNT-FP and interplay with THz incident pulses [11]. [5343170257210 (Springer) [Journal of Electronic Materials] Dynamic Optical Study of Flexible Multiwall Carbon Nanotube Paper Using Terahertz Spectroscopy by Subhash Nimanpure et. al. [copy right], 2021.

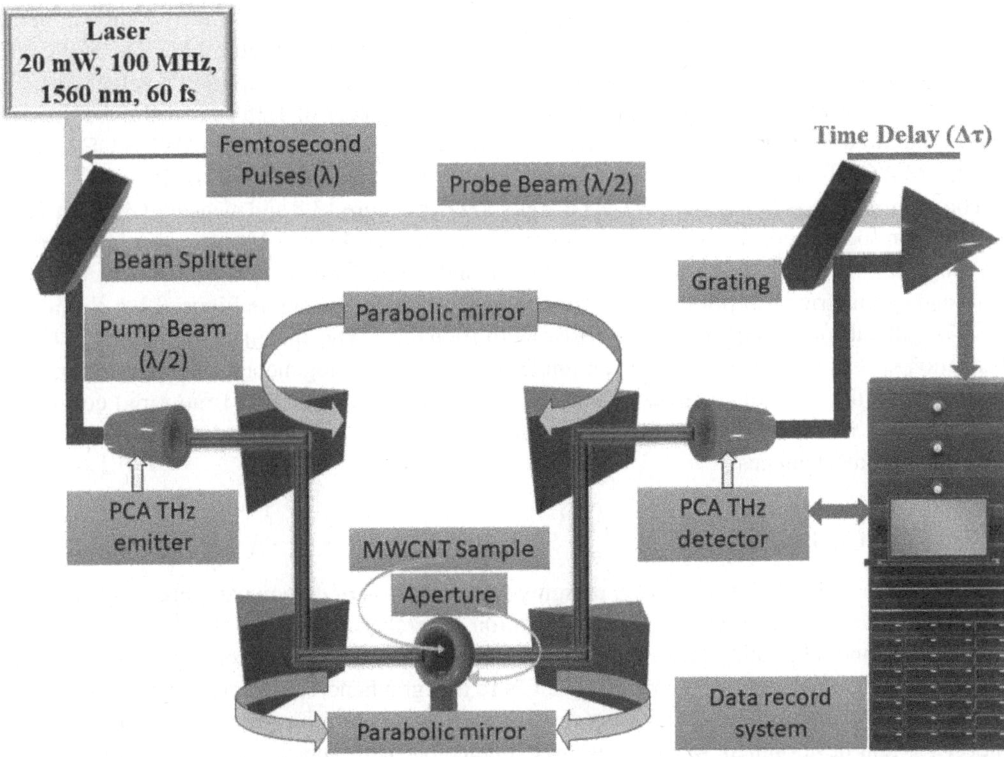

FIGURE 13.6 Test setup for THz-time domain spectroscopy system [11]. [5343170257210 (Springer) [Journal of Electronic Materials] Dynamic Optical Study of Flexible Multiwall Carbon Nanotube Paper Using Terahertz Spectroscopy by Subhash Nimanpure et. al. [copy right], 2021.

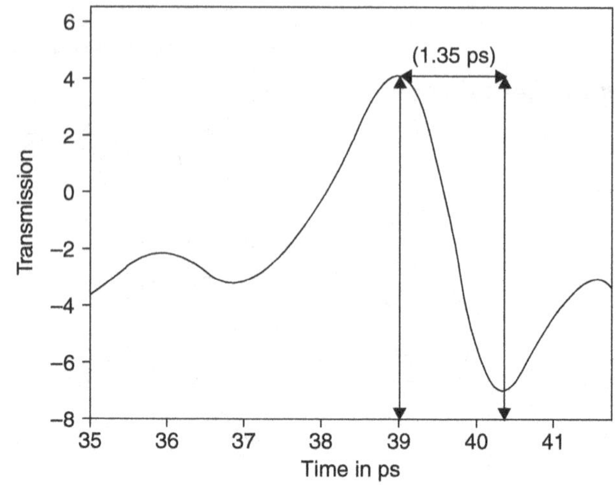

FIGURE 13.7 MWCNT-FP THz-time domain spectroscopy [11]. [5343170257210 (Springer) [Journal of Electronic Materials] Dynamic Optical Study of Flexible Multiwall Carbon Nanotube Paper Using Terahertz Spectroscopy by Subhash Nimanpure et. al. [copy right], 2021.

using the Fast Fourier transform (FFT) to obtain the frequency component. The THz-TDS setup at room temperature is normalized by further characterizing and using a clean silicon substrate of the same size as a reference. The THz pulse emitted by the MWCNT-FP shows a delay in the results. The amplitude of the electrical discipline transmission of the THz pulse is shown on the Y-axis together with the delay time (ps) on the X-axis as shown in the Figure 13.7. The MWCNT-FP specimen is constantly established response time approximately 1.35 ps and it is far proven in Figure 13.7. This response feature suggests rise and fall instances of 1.35 ps constrained through the one hundred MHz bandwidth. The determined quick response time of MWCNT-FP shows that this MWCNT-FP or grapheme-based materials are feasible to used as THz detector.

The absorption coefficient of MWCNT-FP as shown in Figure 13.8 and measured through THz-time domain spectroscopy at room temperature. The frequency domain THz signals are detectable in a range of 0.02 to 4.5 THz, which is quite broadband as compared to many other detectors. The sharp and symmetric absorption peak occurs roughly at 1 THz, as seen in Figure 13.8. From 0.02 to 1 THz, the absorption coefficient will rise with frequency. The specimen embedded in CNTs absorption widths may be affected by a number of factors. Inhomogeneous effects resulting from differences in the immediate environment and homogeneous broadening from rapid collisional dephasing and vibrational energy exchange among carbon atoms or molecules are other possible line widening mechanisms.

13.7 CONCLUSION

The study of these book chapter is very obliging for fabrication of carbon nanotubes and graphene across the different structures and their application in terahertz detection. It is well known that every growth and fabrication techniques have its own advantages and direct impact on the application in different prospect. It is fact that the CNTs and graphene-based semiconducting devices are the most promising type of devices for millimeter and sub-millimeter waves detection. The emergence and development of graphene-based detectors with appropriate fabrication facilities can enable a new detector system used for terahertz opto-electronics. Various properties like

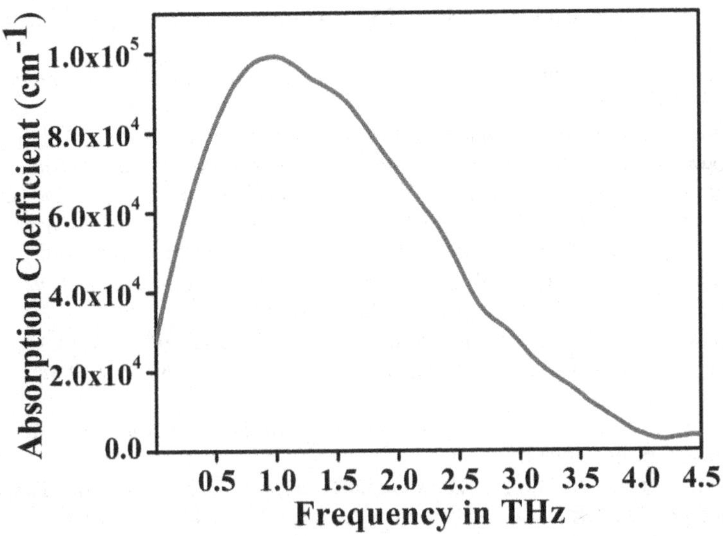

FIGURE 13.8 Absorption range of MWCNT-FP with respect to THz frequency [11]. [5343170257210 (Springer) [Journal of Electronic Materials] Dynamic Optical Study of Flexible Multiwall Carbon Nanotube Paper Using Terahertz Spectroscopy by Subhash Nimanpure et. al. [copy right], 2021.

scalability in CNTs, integration of it with semiconducting and plasmonic-based electronics, can make it attractive materials for future of THz detection system. Our team directly employs flexible carbon nanotube paper as substrate and investigates critical capabilities closer to THz wave detection in the transmission mode. The MWCNT-FP response time of close to 1.35 ps has been repeatedly demonstrated. An amazing property of a freestanding flexible carbon nanotube-based device is excessive absorption in the broad terahertz frequency region, which is measured between 0.02 and 4.5 THz. The current results demonstrate that the MWCNTs, which is a freestanding paper, has enormous potential for THz detector development and is remarkably adaptable to the advancement of THz wearable scanner for real-time non-invasive imaging and other potential applications like biomonitoring, surveillance, and telecommunication.

REFERENCES

1. Jewariya, M., et al., *Fast three-dimensional terahertz computed tomography using real-time line projection of intense terahertz pulse*. Optics Express, 2013. **21**(2): p. 2423–2433.
2. Jewariya, M., M. Nagai, and K. Tanaka, *Ladder Climbing on the Anharmonic Intermolecular Potential in an Amino Acid Microcrystal via an Intense Monocycle Terahertz Pulse*. Physical Review Letters, 2010. **105**(20): p. 203003.
3. Nimanpure, S., et al., *Investigation of cerium oxide thin film thickness using THz spectroscopy for non-destructive measurement*. Journal of Optics, 2021. **50**(1): p. 90–94.
4. Li, X., et al., *Graphene and related two-dimensional materials: Structure-property relationships for electronics and optoelectronics*. Applied Physics Reviews, 2017. **4**(2): p. 021306.
5. Xia, F., et al., *Ultrafast graphene photodetector*. Nature Nanotechnology, 2009. **4**(12): p. 839–843.
6. Baxter, J.B. and G.W. Guglietta, *Terahertz Spectroscopy*. Analytical Chemistry, 2011. **83**(12): p. 4342–4368.
7. Dresselhaus, M.S.D., G.; Avouris, P., *Carbon Nanotubes, Synthesis, Structure, Properties, and Applications*. Vol. 80 Topics in Applied Physics. 2001, Berlin: Springer.

8. Avouris, P., Z. Chen, and V. Perebeinos, *Carbon-based electronics*. Nature Nanotechnology, 2007. **2**(10): p. 605–615.
9. Avouris, P., M. Freitag, and V. Perebeinos, *Carbon-nanotube photonics and optoelectronics*. Nature Photonics, 2008. **2**(6): p. 341–350.
10. Tonouchi, M., *Cutting-edge terahertz technology*. Nature Photonics, 2007. **1**(2): p. 97–105.
11. Nimanpure, S., et al., *Dynamic Optical Study of Flexible Multiwall Carbon Nanotube Paper Using Terahertz Spectroscopy*. Journal of Electronic Materials, 2021. **50**(10): p. 5625–5631.
12. Ren, L., et al., *Terahertz Dynamics of Quantum-Confined Electrons in Carbon Nanomaterials*. Journal of Infrared, Millimeter, and Terahertz Waves, 2012. **33**(8): p. 846–860.
13. Hartmann, R.R., J. Kono, and M.E. Portnoi, *Terahertz science and technology of carbon nanomaterials*. Nanotechnology, 2014. **25**(32): p. 322001.
14. Nimanpure, S., et al., *Investigation of dynamic optical study of Bi2Te3 topological insulators thin film based on MWCNT flexible paper using terahertz spectroscopy*. Optical Materials, 2021. **121**: p. 111490.
15. Nanot, S., et al., *Optoelectronic Properties of Single-Wall Carbon Nanotubes*. Advanced Materials, 2012. **24**(36): p. 4977–4994.
16. Taniguchi, T., et al., *Long-range order and spin-liquid states of polycrystalline Tb${}_{2+x}$Ti${}_{2\ensuremath{-}x}$O${}_{7+y}$*. Physical Review B, 2013. **87**(6): p. 060408.
17. Hartmann, R.R., I.A. Shelykh, and M.E. Portnoi, *Excitons in narrow-gap carbon nanotubes*. Physical Review B, 2011. **84**(3): p. 035437.
18. Ando, T., Y. Zheng, and H. Suzuura, *Dynamical Conductivity and Zero-Mode Anomaly in Honeycomb Lattices*. Journal of the Physical Society of Japan, 2002. **71**(5): p. 1318–1324.
19. Nair, R.R., et al., *Fine Structure Constant Defines Visual Transparency of Graphene*. Science, 2008. **320**(5881): p. 1308–1308.
20. Mak, K.F., et al., *Measurement of the Optical Conductivity of Graphene*. Physical Review Letters, 2008. **101**(19): p. 196405.
21. Léonard, F.O. The physics of carbon nanotube devices [e-book] / François Léonard. 1st edition ed. *Micro & Nano Technologies,* 2009, Norwich, NY: William Andrew.
22. Bonaccorso, F., et al., *Graphene photonics and optoelectronics*. Nature Photonics, 2010. **4**(9): p. 611–622.
23. Bae, S., et al., *Roll-to-roll production of 30-inch graphene films for transparent electrodes*. Nature Nanotechnology, 2010. **5**(8): p. 574–578.
24. Liu, M., et al., *A graphene-based broadband optical modulator*. Nature, 2011. **474**(7349): p. 64–67.
25. Grigorenko, A.N., M. Polini, and K.S. Novoselov, *Graphene plasmonics*. Nature Photonics, 2012. **6**(11): p. 749–758.
26. Sun, Z., et al., *Graphene Mode-Locked Ultrafast Laser*. ACS Nano, 2010. **4**(2): p. 803–810.
27. Ju, L., et al., *Graphene plasmonics for tunable terahertz metamaterials*. Nature Nanotechnology, 2011. **6**(10): p. 630–634.
28. Novoselov, K.S., et al., *Electric Field Effect in Atomically Thin Carbon Films*. Science, 2004. **306**(5696): p. 666–669.
29. Yan, H., et al., *Tunable infrared plasmonic devices using graphene/insulator stacks*. Nature Nanotechnology, 2012. **7**(5): p. 330–334.
30. Fei, Z., et al., *Gate-tuning of graphene plasmons revealed by infrared nano-imaging*. Nature, 2012. **487**(7405): p. 82–85.
31. Woessner, A., et al., *Highly confined low-loss plasmons in graphene–boron nitride heterostructures*. Nature Materials, 2015. **14**(4): p. 421–425.
32. Alonso-González, P., et al., *Acoustic terahertz graphene plasmons revealed by photocurrent nanoscopy*. Nature Nanotechnology, 2017. **12**(1): p. 31–35.
33. Koppens, F.H.L., et al., *Photodetectors based on graphene, other two-dimensional materials and hybrid systems*. Nature Nanotechnology, 2014. **9**(10): p. 780–793.
34. Cai, X., et al., *Sensitive room-temperature terahertz detection via the photothermoelectric effect in graphene*. Nature Nanotechnology, 2014. **9**(10): p. 814–819.

35. Auton, G., et al., *Terahertz Detection and Imaging Using Graphene Ballistic Rectifiers*. Nano Letters, 2017. **17**(11): p. 7015–7020.
36. Otsuji, T., et al., *A grating-bicoupled plasma-wave photomixer with resonant-cavity enhanced structure*. Optics Express, 2006. **14**(11): p. 4815–4825.
37. Iijima, S., *Helical microtubules of graphitic carbon*. Nature, 1991. **354**(6348): p. 56–58.
38. Ebbesen, T.W., *Carbon Nanotubes*. Physics Today, 1996. **49**(6): p. 26–32.
39. Ebbesen, T.W. and P.M. Ajayan, *Large-scale synthesis of carbon nanotubes*. Nature, 1992. **358**(6383): p. 220–222.
40. Colbert, D.T., et al., *Growth and Sintering of Fullerene Nanotubes*. Science, 1994. **266**(5188): p. 1218–1222.
41. Iijima, S. and T. Ichihashi, *Single-shell carbon nanotubes of 1-nm diameter*. Nature, 1993. **363**(6430): p. 603–605.
42. Bethune, D.S., et al., *Cobalt-catalysed growth of carbon nanotubes with single-atomic-layer walls*. Nature, 1993. **363**(6430): p. 605–607.
43. Ando, Y. and S. Iijima, *Preparation of Carbon Nanotubes by Arc-Discharge Evaporation*. Japanese Journal of Applied Physics, 1993. **32**(Part 2, No.1A/B): p. L107–L109.
44. Ando, Y., *The Preparation of Carbon Nanotubes*. Fullerene Science and Technology, 1994. **2**(2): p. 173–180.
45. Wang, M., et al., *Carbon Nanotubes Grown on the Surface of Cathode Deposit by Arc Discharge*. Fullerene Science and Technology, 1996. **4**(5): p. 1027–1039.
46. Ando, Y., X. Zhao, and M. Ohkohchi, *Production of petal-like graphite sheets by hydrogen arc discharge*. Carbon, 1997. **35**(1): p. 153–158.
47. Journet, C., et al., *Large-scale production of single-walled carbon nanotubes by the electric-arc technique*. Nature, 1997. **388**(6644): p. 756–758.
48. Wang, X.K., et al., *Carbon nanotubes synthesized in a hydrogen arc discharge*. Applied Physics Letters, 1995. **66**(18): p. 2430–2432.
49. Guo, T., et al., *Uranium Stabilization of C_{28}: A Tetravalent Fullerene*. Science, 1992. **257**(5077): p. 1661–1664.
50. Guo, T., et al., *Self-Assembly of Tubular Fullerenes*. The Journal of Physical Chemistry, 1995. **99**(27): p. 10694–10697.
51. Öncel, Ç. and Y. Yürüm, *Carbon Nanotube Synthesis via the Catalytic CVD Method: A Review on the Effect of Reaction Parameters*. Fullerenes, Nanotubes and Carbon Nanostructures, 2006. **14**(1): p. 17–37.
52. Chen, J., et al., *Optical nano-imaging of gate-tunable graphene plasmons*. Nature, 2012. **487**(7405): p. 77–81.
53. Geim, A.K. and K.S. Novoselov, *The rise of graphene*, in *Nanoscience and Technology*. p. 11–19.
54. Ni, G.X., et al., *Fundamental limits to graphene plasmonics*. Nature, 2018. **557**(7706): p. 530–533.
55. Berger, C., et al., *Electronic Confinement and Coherence in Patterned Epitaxial Graphene*. Science, 2006. **312**(5777): p. 1191–1196.
56. Reina, A., et al., *Layer Area, Few-Layer Graphene Films on Arbitrary Substrates by Chemical Vapor Deposition*. Nano Letters, 2009. **9**(8): p. 3087–3087.
57. Li, D., W. Windl, and N.P. Padture, *Toward Site-Specific Stamping of Graphene*. Advanced Materials, 2009. **21**(12): p. 1243–1246.
58. Kim, K.S., et al., *Large-scale pattern growth of graphene films for stretchable transparent electrodes*. Nature, 2009. **457**(7230): p. 706–710.
59. Blake, P., et al., *Graphene-Based Liquid Crystal Device*. Nano Letters, 2008. **8**(6): p. 1704–1708.
60. Docherty, C.J. and M.B. Johnston, *Terahertz Properties of Graphene*. Journal of Infrared, Millimeter, and Terahertz Waves, 2012. **33**(8): p. 797–815.
61. Low, T. and P. Avouris, *Graphene Plasmonics for Terahertz to Mid-Infrared Applications*. ACS Nano, 2014. **8**(2): p. 1086–1101.
62. Rogalski, A., *Graphene-based materials in the infrared and terahertz detector families: a tutorial*. Advances in Optics and Photonics, 2019. **11**(2): p. 314–379.

63. Ryzhii, V., et al., *Terahertz and infrared photodetection using p-i-n multiple-graphene-layer structures*. Journal of Applied Physics, 2010. **107**(5): p. 054512.
64. Ryzhii, V., et al., *Terahertz and infrared photodetectors based on multiple graphene layer and nanoribbon structures*. Opto-Electronics Review, 2012. **20**(1): p. 15–25.
65. Knap, W., et al., *Field Effect Transistors for Terahertz Detection: Physics and First Imaging Applications*. Journal of Infrared, Millimeter, and Terahertz Waves, 2009. **30**(12): p. 1319–1337.
66. Kim, J.Y., et al., *Far-infrared study of substrate-effect on large scale graphene*. Applied Physics Letters, 2011. **98**(20): p. 201907.
67. Tomadin, A., et al., *Photocurrent-based detection of terahertz radiation in graphene*. Applied Physics Letters, 2013. **103**(21): p. 211120.
68. Ryzhii, V., et al., *Double graphene-layer plasma resonances terahertz detector*. Journal of Physics D: Applied Physics, 2012. **45**(30): p. 302001.
69. Fateev, D.V., K.V. Mashinsky, and V.V. Popov, *Terahertz plasmonic rectification in a spatially periodic graphene*. Applied Physics Letters, 2017. **110**(6): p. 061106.
70. Balandin, A.A., *Thermal properties of graphene and nanostructured carbon materials*. Nature Materials, 2011. **10**(8): p. 569–581.
71. Tarasov, M., et al., *Carbon nanotube-based bolometer*. JETP Letters, 2006. **84**(5): p. 267–270.
72. Segawa, Y., H. Ito, and K. Itami, *Structurally uniform and atomically precise carbon nanostructures*. Nature Reviews Materials, 2016. **1**(1): p. 15002.
73. Chen, S.-L., et al., *Efficient real-time detection of terahertz pulse radiation based on photoacoustic conversion by carbon nanotube nanocomposite*. Nature Photonics, 2014. **8**(7): p. 537–542.
74. Bagsican, F.R.G., et al., *Terahertz Excitonics in Carbon Nanotubes: Exciton Autoionization and Multiplication*. Nano Letters, 2020. **20**(5): p. 3098–3105.

Index